Chemoinformatics: Directions Toward Combating Neglected Diseases

Editor

Teodorico C. Ramalho

Universidade Federal de Lavras
Campus Universitário - UFLA
Dept. de Química
37200-000
Lavras-MG - Brazil

Co-Editors

Matheus P. Freitas

Federal University of Lavras - UFLA
Brazil

Elaine F.F. da Cunha

Federal University of Lavras - UFLA
Brazil

eBooks End User License Agreement

Please read this license agreement carefully before using this eBook. Your use of this eBook/chapter constitutes your agreement to the terms and conditions set forth in this License Agreement. Bentham Science Publishers agrees to grant the user of this eBook/chapter, a non-exclusive, nontransferable license to download and use this eBook/chapter under the following terms and conditions:

CONTENTS

CHAPTERS

FOREWORD

Chemoinformatics: Directions Towards Combating Neglected Diseases is the fruit of an original project from a group of Brazilian young scientists very concerned on the threat represented by neglected diseases today. These illnesses are so called because they mostly affect poor people from the third world being usually forgotten by the big pharmaceutical companies, that seems more interested in P&D against more profitable diseases like obesity, Alzheimer disease, Parkinson illness or sexual dysfunctions, than in tuberculosis or protozoan caused diseases, such as malaria, leishmaniosis, toxoplamosis and Chagas disease. As a consequence basically only the governmental agencies invest today in P&D of chemotherapy against neglected diseases. Such a behavior from the pharmaceutical companies could soon prove to be a terrible mistake because, as an outcome of globalization, these diseases are more frequently knocking the doors of the first world nations. MDR tuberculosis, for instance is quickly becoming a worldwide public healthy emergency while malaria is back, thanks to the resistance developed by *Plasmodium falciparum* against the chemotherapy available at present. Also the facilities of traveling today besides the huge amount of tourists attracted by the exotic rain forests worldwide (the main sources of all kind of neglected disease) have contributed to the spreading out of neglected diseases into EUA and Europe. If the developed nations take to long to wake up to this problem their populations will soon become victims again of diseases they have long left behind.

This project brings light back to this issue showing to the scientific community worldwide how the chemoinformatic techniques could be successfully employed to the design of new and more promising chemotherapy against such illnesses. Techniques like QSAR, SAR, homology modeling, molecular dynamics and docking are essential tools in the modern medicinal chemistry and have proved to be very efficient in the drug design against neglected diseases. This e-book, besides making a revision of the main aspects of these diseases, also describes several examples available in literature of successful applications of these techniques on studies of molecular targets on the parasites responsible for causing neglected diseases. The editors and authors are mostly theoretical chemists and hope, to highlight this emerging problem also to show the power of chemoinformatic techniques as cheap and efficient tools to the drug design, motivating young scientists, like them, to face the challenge represented by the fight against these terrible illnesses affecting more than a third of the world's population today.

Tanos Celmar Costa França
Military Institute of Engineering, Pça General Tibúrcio
Rio de Janeiro
Brazil

PREFACE

Low-income populations in developing regions of Africa, Asia and the Americas have been particularly injured by a group of tropical infections denominated neglected diseases. Comparative to important illnesses, neglected diseases do not enjoy significant research funding and are not considerably important targets for the "Big Pharma" in terms of development of new drugs, though these infections can make widespread diseases like AIDS more deadly. Although the known preventive measures or acute medical treatments for some of these diseases, the fact of being especially endemic in poorer areas claims for attention. This eBook is devoted to inspire those who are dealing with drug development to make an effort towards combating neglected diseases.

In this line, this eBook is directed towards anyone who is interested in or deals with medicinal chemistry focused on neglected diseases, from students to advanced researchers. Despite affecting millions of people around the word, causing many deaths and having a great and limiting influence on the life quality, the selection of new molecular targets and the development of more efficient drugs against those diseases are scarce. Furthermore, surprisingly little detailed computational work on this subject has appeared. However, it should be kept in mind that computers are an essential tool in modern medicinal chemistry. Nowadays, computational approaches and 3D visualization are not used simply to depict pretty pictures of molecules in biological systems; these powerful computational tools allow one to obtain insights on the interaction between enzyme-substrate, reaction mechanisms, statistical behavior of molecules and much more, at the molecular level, contributing significantly to solve problems in biological systems. In this context, the study of chemoinformatics is crucial. This new field of science was officially introduced at the end of the last century and it can be defined as "the application of computer to a range of problems in the field of chemistry".

This eBook hereby attempts to explore an open problem in the literature with a stimulating discussion on the state of the art knowledge in this important research field – neglected diseases – pointing out perspectives on using molecular modeling and theoretical approaches. Therefore, 9 chapters were selected containing the participation of different international experts on the subject. We hope that this eBook can play an attractive role for chemists, physicists and biologists in promoting the diffusion of knowledge in the field of chemoinformatics applied to neglected diseases.

Teodorico C. Ramalho
Universidade Federal de Lavras
Campus Universitário - UFLA
Dept. de Química
37200-000
Lavras-MG - Brazil

Matheus P. Freitas
Federal University of Lavras - UFLA
Brazil

Elaine F.F. da Cunha
Federal University of Lavras - UFLA
Brazil

List of Contributors

Teodorico C. Ramalho, Matheus P. Freitas, Elaine F. F. da Cunha and Daiana Teixeira Mancini

Department of Chemistry, Federal University of Lavras, P.O. Box 3037, 37200-000, Lavras, MG, Brazil
E-mail: teo@ufla.br

Rodrigo A. Cormanich

Chemistry Institute, State University of Campinas, P.O. Box 6154, 13083-970, Campinas, SP, Brazil
E-mail: rodrigolsilveira@gmail.com

Alicia Ponte-Sucre, Emilia Díaz and Maritza Padrón-Nieves

Laboratory of Molecular Physiology, Institute of Experimental Medicine, Faculty of Medicine, Universidad Central de Venezuela, Caracas, Venezuela. Phone: +58 212 605 3665. Telefax: +58 212 693 4351
E-mail: aiponte@gmail.com

Tanos C. C. França

Chemical Engineering Department, Military Institute of Engineering, Pça General Tibúrcio, 80 – Urca, 22290-270, Rio de Janeiro/RJ, Brazil
E-mail: tanos@ime.eb.br

Magdalena Nascimento Rennó

Universidade Federal do Rio de Janeiro - Campus Macaé, Rua Aloísio da Silva Gomes, nº 50, Granja dos Cavaleiros, Macaé – RJ. CEP: 27930-560.
E-mail: mnrenno@uol.com.br

Mushtaque S. Shaikh, Vijay M. Khedkar and Evans C. Coutinho

Department of Pharmaceutical Chemistry, Bombay College of Pharmacy, Kalina, Santacruz (E), Mumbai, 400 098. India; Tel: +91 022 26670905, Tele-fax: +91 022 26670816;
E-mail: evans@bcpindia.org

Alan Wilter Sousa da Silva

Department of Biochemistry, University of Cambridge, 80 Tennis Court Road, CB2 1GA, United Kingdom.
E-mails: awd28@mole.bio.cam.ac.uk and alanwilter@gmail.com

Adriana de Oliveira Gomes, Alessandra Mendonça Teles de Souza, Alice Maria Rolim Bernardino, Helena Carla Castro and Carlos Rangel Rodrigues

Laboratório de Modelagem Molecular e QSAR (ModMolQSAR), Faculdade de Farmácia, Universidade Federal do Rio de Janeiro, Rio de Janeiro, RJ. Telefax: +55 21 2560 9897.
E-mail: rangelfarmacia@gmail.com

Carlton Anthony Taft

Centro Brasileiro de Pesquisas Físicas, Rua Dr. Xavier Sigaud, 150, Urca, 22290-180, Rio de Janeiro, Brazil.
E-mail: catff@terra.com.br

Carlos Henrique Tomich de Paula da Silva

School of Pharmaceutical Sciences of Ribeirão Preto, University of São Paulo, 14040-903, Ribeirão Preto, Brazil
E-mail: tomich@fcfrp.usp.br

New Approaches to the Development of Anti-Protozoan Vaccine and Drug Candidates: A Review of Patents

Elaine F.F. da Cunha, Teodorico C. Ramalho[*], Daiana T. Mancini and Matheus P. Freitas

Department of Chemistry, Federal University of Lavras, P.O. Box 3037, 37200-000, Lavras, MG, Brazil

Abstract: Protozoan infections are parasitic diseases that affect hundreds and millions of people worldwide, but have been largely neglected for drug development because they affect poor people in poor regions of the world. Most of the current drugs used to treat these diseases are decades old and have many limitations, including the emergence of drug resistance. This review will focus on the most recent developments, from 2001 to 2008, published in the field of patents and publications, paying particular attention to promising compounds acting against trypanosomiasis, leishmaniasis, malaria, toxoplasmosis, amebiasis, giardiasis, balantidiasis and pneumocystosis, their chemistry and biological evaluation, and to new chemical and pharmaceutical processes.

Keywords: Drug candidates, neglected diseases, review of patents.

1. INTRODUCTION

Neglected parasitic diseases are a group of tropical infections which are especially endemic in low-income populations in developing regions of many countries. Despite these diseases affecting millions of people around the world, causing many deaths and having a great and limiting influence on the quality of life, the selection of new molecular targets and the development of more efficient drugs against such diseases are scarce (Table **1**).

Currently, among them, zoonotic diseases attract much attention. Those are defined as diseases shared by animals and humans. In fact, wildlife serves as a reservoir for many diseases common to domestic animals and humans. Generally, disease is more easily prevented than treated. This discussion reviews common zoonotic diseases, including those ailments that are often erroneoulsy cited as being closely linked to wildlife.

Table 1: The main neglected parasitic diseases. (Adapted from ref [1]).

Diseases	Organisms	Scope	Therapy Needs
Malaria	*Plasmodium* spp.	500 million infections annually	Circumventing drug resistance
Leishmaniasis	*Leishmania* spp.	2 million infections annually	Safe, orally bioavailable drugs, especially for the visceral form of the disease
Trypanosomiasis(sleeping sickness, Chagas´disease)	*T. brucei* (sleeping sickness)*T. cruzi* (Chagas´disease)	HAT: 300,000 cases annually Chagas: 16 million existing infections	Safe, orally bioavailable drugs, especially for the chronic phases of disease
Schistosomiasis	*Schistosoma* spp	>200 million existing infections	Backup drug should resistance arise to praziquantel
Giardiasis	*Giardia lamblia*;	Millions of cases of diarrhea annually	Well-tolerated drugs
Amebiasis	*Entamoeba histolytica*	Millions of cases of diarrhea annually	Well-tolerated drugs
Pneumocystosis	*Pneumocystis carinii or Pneumocystis jiroveci*	-	Trimethoprim-sulfamethoxazole often in conjunction with corticosteroids

*Address correspondence to Teodorico C. Ramalho: Department of Chemistry, Federal University of Lavras, P.O. Box 3037, 37200-000, Lavras, MG, Brazil. E-mail: teo@ufla.br

Several species of protozoans infect humans and inhabit the body as commensals or parasites. Protozoa have traditionally been divided based on their means of locomotion, although this characterictic is no longer believed to represent genuine relationships [2, 3]: a) Flagellates (Zoomastigophora): Flagellates are characterized by having one or more flagella. Parasitic species generally have more flagella than those that are free living. Pathogens: *Giardia intestinalis, Trichomonas vaginalis, Trypanosoma cruzi, Leishmania donovani* http://home.austarnet.com.au/wormman/wlimages.htm - Ginte; b) Amoebae (actinopods, rhizopoda): Amoebae may be divided into several morphological categories based on the form and structure of the pseudopods. It can live in a number of places within the human body, but most are found in the intestine. Those in which the pseudopods are supported by regular arrays of microtubules are called actinopods, and forms in which they are not, are called rhizopods, further divided into lobose, filose, and reticulose amoebae. Pathogens: *Entamoeba histolytica*; c) Sporozoans (Apicomplexa, myxozoa, Microsporidia): the members of this group share an "apical complex" of microtubules at one end of the cell (hence the name that many prefer to the old name of sporozoans). All the members of the phylum are parasites. They do not have a set body plan like the other parasitic protozoans, although they are characterized by having complex life-cycles with an alternation of sexual and asexual generations. The most well known of all the sporozoans are the organisms which cause the disease malaria - *Plasmodium falciparum, Plasmodium vivax, Plasmodium malariae* and *Plasmodium ovale* – and pneumocystosis; d) Ciliates (ciliophora): These are larger protozoans, growing to over 100μm. It has hundreds of tiny cilia which beat in unison to propel it through the water. Often cilia are fused together in rows or tufts (called cirri) and are used for special functions such as food gathering. In addition to locomotion, the Paramecium and other ciliates like the Stentor use cilia to sweep food down into their central channel or gullet. There is only one species of pathogenic ciliate known to parasitise humans: *Balantidium coli*; e) Microsporidia: they are quite difficult to diagnose, very little work has been done into their importance in human disease, although they are known to be a major cause of productivity loss in aquaculture facilities such as prawn farms. Microsporidia have also been implicated in causing disease in immunocompromised hosts.

In this paper, we focused on some diseases, such as trypanosomiasis, leishmaniasis, toxoplasmosis, giardiasis, amoebiasis, balantidiasis and malaria. In this context, we believe that a review of patents on the highlighted inventions and innovative ideas in this field could be of potential importance for scientists, managers and decision-makers in the pharmaceutical industry.

1.1. Treatment of African Trypanosomiasis

By way of review, African trypanosomiasis is a parasitic disease of people and animals, caused by protozoa of species *Trypanosoma brucei gambiense* or *T. brucei rhodesiense* and transmitted by the tsetse fly bite. The fly is infected when it bites during the parasitemic phases and the trypanosome develops in the vector, culminating in infection of its saliva.

The chemotherapy of Human African Trypanosomiasis (HAT) is restricted to our clinically approved drugs: suramin (**1**), pentamidine (**2**), melarsoprol (**3**) and eflornithine (**4**). Suramin and pentamidine are restricted to the treatment of early-stage HAT prior to Central Nervous System (CNS) infection, while melarsoprol is used once the parasites have penetrated the CNS. Eflornitine is not effective against rhondesiense infections. These drug regimens can cause toxicity and poor efficacy. Thus, the development of novel compounds to provide novel treatment againt HAT is necessary.

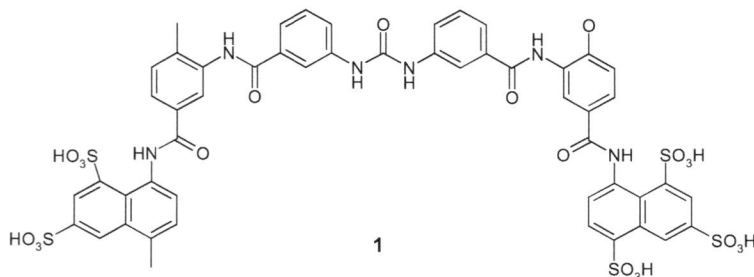

1

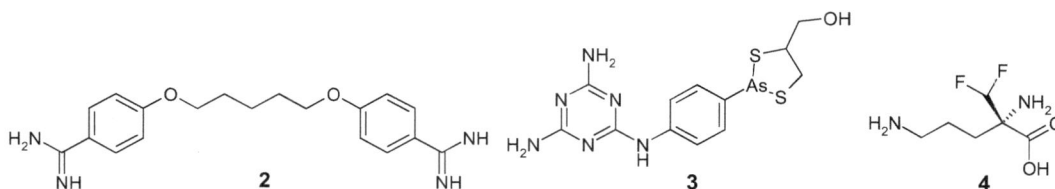

The major lysosomal cysteine proteinase of HAT is a target candidate for novel chemotherapy for sleeping sickness. Lysosomal cysteine proteases, generally known as the cathepsins, were discovered in the first half of the 20th century. Quibell *et al.*, at the Amura Therapeutic Ltd, described a series of tetrahydrofuro [3, 2-b]pyrrol-3-one analogues as cathepsin K inhibitor [4-10]. Cathepsin K exhibits the highest capability to degrade components of the extracellular matrix [11]. Compounds of general formula **5** may be conveniently considered as a combination of three building blocks (P1, P2, and P3) that respectively occupy the S1, S2 and S3 binding sites of the protease. The compounds **6** [9], **7** [10] and **8** [8] exhibited more than 80% inhibition at a concentration of 1 μM.

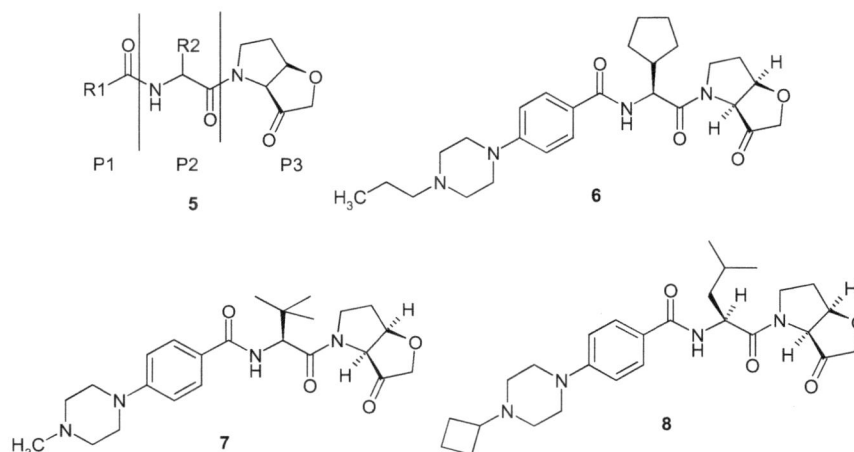

Cysteine protease inhibitors were also described by Merck Frosst Canada Ltd. for the treatment of trypanosomiasis [12-16]. Compounds such as **9** and **10** were claimed in the patent, however no biological data were presented. In addition, SmithKline Beecham Corporation described a novel substituted 3,7-dioxo-azepan-4-ylamide [17], diazepino[1,2-b]phthalazine **11** [18], oxo-amino-sulfonyl-azapanes **12** [19], 5-substituted-6-oxo-[1,2]diazepane, **13 e 14** [20], their pharmaceutical compositions, processes for their preparation and methods of their use for inhibiting a protease, particularly a serine protease and a cysteine protease such as cathepsin K and falcipain, and for the treatment of parasitic infections including trypanosomiasis. For instance, the compound **12** is obtained according to **Scheme 1**.

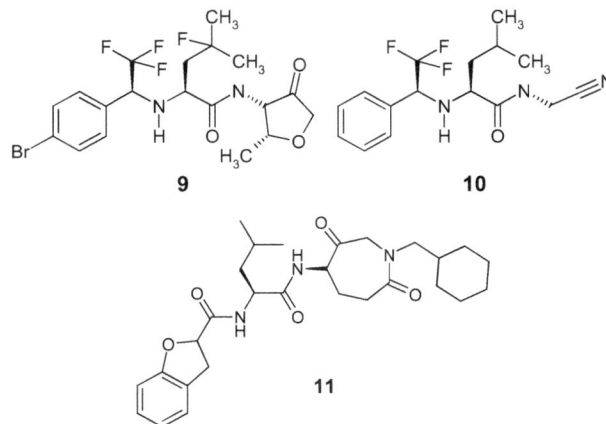

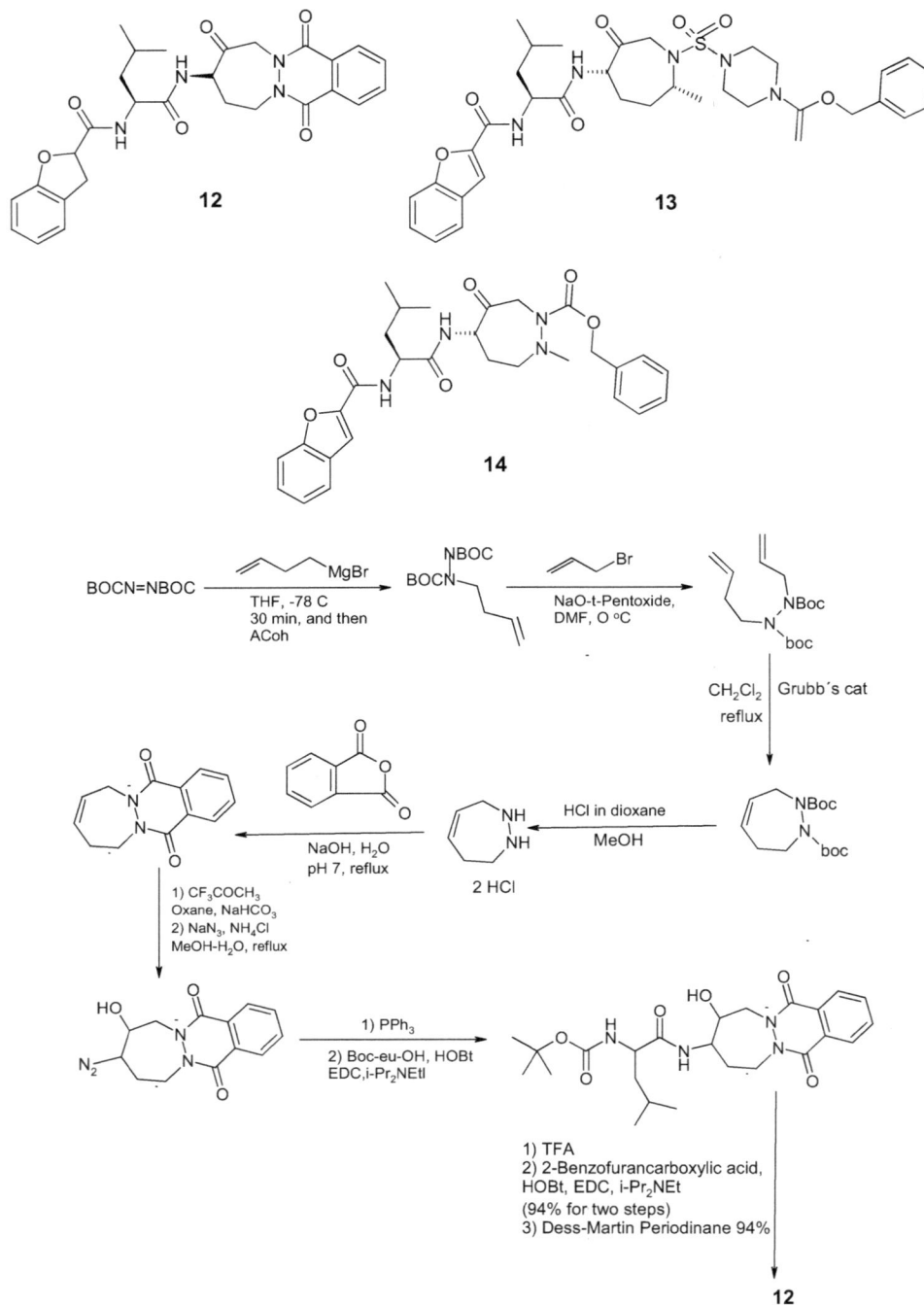

12

13

14

Scheme 1

Based on the **2** structure, Xavier University of Louisiana disclosed bisbenzamidines and bisbenzamidoxime analogs as compounds effective against infection by parasitic hemoflagellates, in particular *Trypanosoma brucei gambiense* and *Trypanosoma brucei rhodesiense* [21]. It has been postulated that aromatic bisbenzamidines are selectively taken up by transporters into trypanosomes, where they bind strongly to DNA in the nucleus and mitochondria inhibiting the topoisomerases II, thus the synthesis of DNA, RNA, and proteins by the parasites will be inhibited [22]. In mice infected with *T. b. brucei*, treatment with the compound **15** at a dose of 5 mg/kg/day resulted in a mean survival in days, of 18 (excluding cured animals) and 5 of the 6 treated animals were cured.

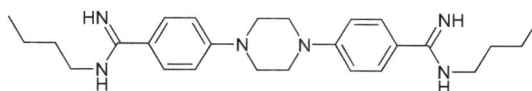

15

1.2. Treatment of American Trypanosomiasis

American tripanosomiasis is a tropical parasitic disease caused by the flagellate protozoan *Trypanosoma cruzi*. The disease may also be spread through an insect vector, blood transfusion and organ transplantation, ingestion of food contaminated with parasites, and from a mother to her fetus. Nifurtimox (**16**) or benznidazole (**17**) is used in the acute phase of Chagas disease and in reactivation in immunosupressed patients. Another irrefutable sign refers to the treatment of victims of laboratory accidents, which demands a very early start for the administration. The action mechanism of **17** may depend on the formation of nitro anion radicals, but precise details are not known. Little is also known regarding the mode of action of benznidazole. Because of the adverse events associated with these medications, patients being treated require careful monitoring. In a patent application, Reed *et al.*, at the Infectious Disease Research Institute, claimed a fusion protein useful for the diagnosis of *Trypanosoma cruzi* infection within a biological sample [23, 24]. More specifically, the invention relates to the use of *T. cruzi* antigenic polypeptides and fusion polypeptides (TcF and ITC-6) in methods for screening individual and blood supplies for *T. cruzi* infection. Also included are details of a kit which uses the invention to detect *T. cruzi* involving the fusion protein. It is a protein created through the joining of two or more genes which originally coded for separate proteins. The reactivity of sera from *T. cruzi*-infected individuals and control sera from non-infected individuals against the fusion polypeptides TcF and ITC-6 was determined by ELISA. The data shows that the invention is capable of recognizing sera that are negative or low with TcF [23, 24].

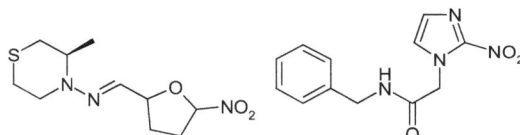

16   **17**

Quibell and Watts, at Amura Therapeutics Limited, described novel tetrahydrofuro[3,2-b]pyrrol-3-one compounds (**18**), their salts, hydrates, complexes or prodrugs. A process for their preparation, compositions comprising them and their use as inhibitors of CAC1 cysteine proteases, particularly cathepsin K, in the treatment of Chagas disease were also described [4-10]. Compound **18**, exhibited good primary DMPK (dystrophia myotonica-protein kinase) properties along with promising activity in an *in vitro* cell-based human osteoclast assay of bone resorption. Compound **19** exhibited >80% inhibition at a concentration of 1 μM. In addition, hexahydrofuro[3,2-b]pyridine-3-one, hexahydro-2-oxa-1,4-diazapentalene and hexahydropyrrolo[3,2-c]pyrazole were also synthetized and made avaliable pharmacologicaly [4-10]. *In vitro* activity was observed against a range of CAC1 cysteinyl proteinases, however tetrahydrofuro[3,2-b]pyrrol analogs showed a better *in vitro* activity value [25].

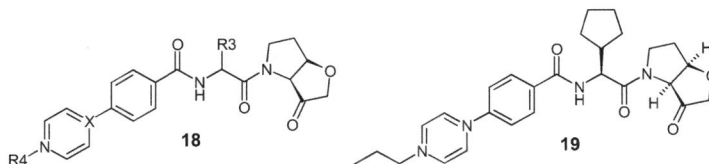

18   **19**

In a further patent application with cathepsin K inhibitors against Chagas disease or American trypanosomiasis, SmithKline Beecham Corporation reported new promising trioxo-[1,2]thiazepanylamide derivatives [26]. These compounds are substituted asymmetrical imidodicarbonimide diamides derived from hydroxylamines. They include the structure **20** as one of several compounds specifically claimed, but biological data are not presented in this patent.

20

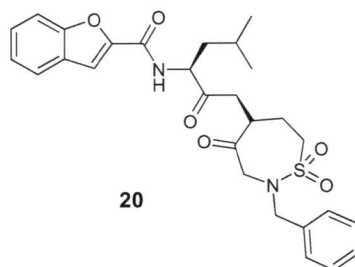

1.3. Treatment of Giardiasis

Giardiasis is a disease caused by *Giardia lamblia* and only the cyst form is infectious by the oral route. People catch *Giardia* by eating food or drinking water which has been contaminated by the organism - usually from feces. When there is a lot of *Giardia* present, this generates inflammation, which causes nausea, stomachache and diarrhea.

Giardiasis has been recently included in the WHO Neglected Diseases Initiative [27]. *Giardia lamblia* infection in humans is frequently misdiagnosed. Multiple stool examinations are recommended, since the cysts and trophozoites are not shed consistently. Given the difficult nature of testing to find the infection, including many false negatives, some patients should be treated on the basis of empirical evidence; treatment based on symptoms. Human infection is conventionally treated with metronidazole (**21**), tinidazole (**22**) or nitazoxanide (**23**) [28]. Nitroimidazoles are the most effective drugs available and no drug was reported to be unsafe, causing only mild to moderate and transient side effects. Suk *et al.* reported the design, synthesis, and activity of a potent and non-toxic second generation anti-giardial agent that was designed as an inhibitor of cyst wall synthase [29]. In the formation process of the *Giardia* cyst wall, the UDP-GalNAc is polymerized by cyst wall synthase into a polysaccharide, which, in conjunction with polypeptides, forms the filamentous outer cyst wall of *Giardia* [30]. The compound **24**, a phosphonoxin, was the more potent inhibitor of *Giardia* cyst formation and should have clinical potential as a new anti-giardia drug.

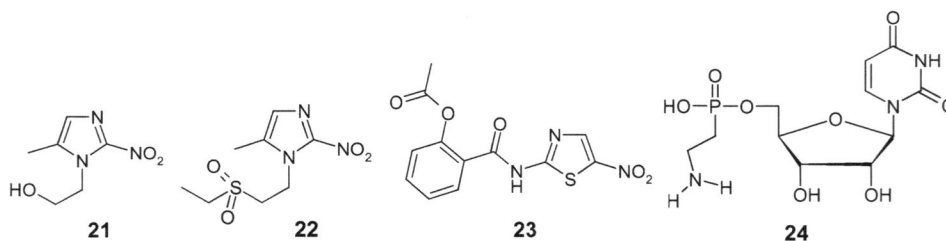

21 **22** **23** **24**

Currently (2001-2008), there are no patents of novel anti-giardiasis componds. However, oligonucleotide molecules for detection of *G. lamblia* were disclosed by Macquarie Research Limited [31]. Preferably, the oligonucleotide molecule hybridises specifically unique to 18S rDNA/rRNA sequences of *G. lamblia* under medium to high stringency conditions. The sample can be environmental, water sources, waste materials, medical and body fluids.

1.4. Treatment of Balantidiasis

Balantidiasis is a disease caused by the large ciliated protozoan, *Balantidium coli* and tetracycline or iodoquinol is usually used in the treatment. Human infection occurs when human contact with pigs is particularly close. Ingestion of cysts or trophozoites from the infected animal or human feces is the infectious form.

Balantidiasis disease is treated with metronidasole (**21**) or tetraciclin (**25**) for 5-10 days. It is usually effective, but without antibiotic mortality levels. In the treatment of acute balantidiasis, tetracycline is the drug of choice to eliminate *B. coli* trophozoites in humans. The disease is especially dangerous in immunocompromised patients as pulmonary parenchyma involvement is possible. *B. coli* is rare in Western

Europe and North America and alternative drugs for the treatment of balantidiasis may be 5-nitroimidazole drugs such as tinidazole (**22**) and ornidazole (**26**) [32]. As well as in the case of *G. lamblia*, currently (2001-2008), there are no new patents for anti-balantidiasis drugs.

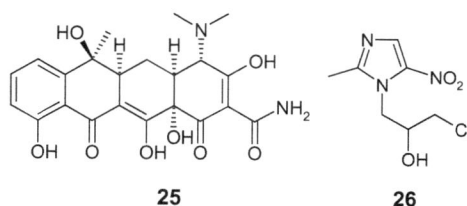

25   **26**

1.5. Treatment of Amebiasis

Amebiasis is an infection of the intestine (gut) caused by an amoeba called *Entamoeba histolytica*, that is found in contaminated food or drink. Virulent strains such as *Entamoeba coli* and *Entamoeba hartmanni*, can produce mild diarrhea and dysentery. Only a few cysts are needed to cause infection. Amoebic cysts resist iodine and chlorine if concentration of these chemicals is too low.

Different drugs are available to treat amebiasis. Oral antiparasitic medication is the standard treatment. For asymptomatic infections, iodoquinol or paromomycin are drugs of choice. For mild, moderate, or severe intestinal disease, and for extraintestinal infections (*e.g.*, hepatic abscess) the drugs of choice are metronidazole (**21**) or tinidazole (**22**), immediately followed by treatment with iodoquinol (**27**), paromomycin (**28**), or diloxanide furoate (**29**). Cen and Lv, Jiangsu Hansoh Pharmaceutical Co Limited, described novel optically pure α-substituted 2-methyl-5-nitroimidazole-1-ethanol derivatives and their use for the treatment of amoebiasis [33]. In an assay testing resistance activity in rats, compound **30** showed ED_{50} value 7 mg/kg against *Amoeba dysenteriae*, whereas metronidazole showed 25 mg/kg, and ornidazole showed 10 mg/kg.

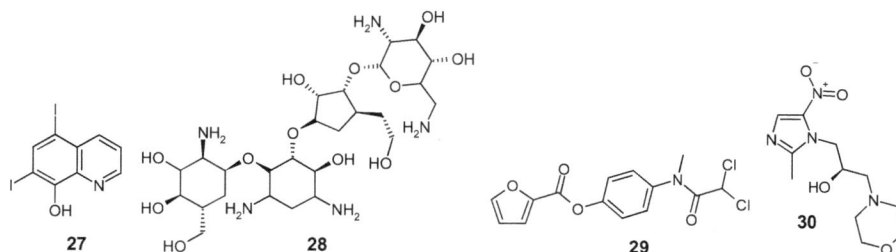

27   **28**   **29**   **30**

Thiosemicarbazones and their metal complexes have been extensively studied during recent years due to their wide variety of biological activities [34]. Certain drugs even show enhanced activity when administered as their metal chelates. Copper (II) is a biologically active essential metal ion; its chelating ability and positive redox potential allow participation in biological transport reactions. Sharma *et al.* reported the synthesis of 5-nitrofuran 2-carboxaldehyde thiosemicarbazones (**31**) and their subsequent bidentate Cu (II) complexes (**32**). These compounds were screened for their anti-amoebic activities against *HK-9* strain of *Entamoeba histolytica* in *in vitro* experiments and found that the chelation induced significant changes in the biological activity of the ligands and some copper showed better IC_{50} value than metronidazole *in vitro* [34].

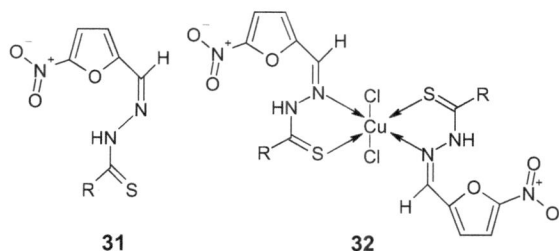

31   **32**

1.6. Treatment of *Plasmodium falciparum* Infection

Malaria is a mosquito-borne disease caused by small, one-celled parasites called Plasmodium that infect and destroy red blood cells. Four different Plasmodium can cause disease in humans: *Plasmodium falciparum, Plasmodium malariae, Plasmodium ovale, and Plasmodium vivax*. People contract malaria from the bite of an infected mosquito. A mosquito can become infected when it bites a person who has malaria organisms in the blood.

The effectiveness of antimalarial drugs differs with different species of the parasite and with different stages of the parasite's life cycle. Drugs include chloroquine, mefloquine, primaquine, quinine, pyrimethamine-sulfadoxine (Fansidar), and doxycycline. Two aspects have stimulated new efforts regarding the development of chemotherapy, vaccines and sanitary studies on malaria: the resistance to currently available antimalarial drugs and the inefficacy of antimalarial vaccines [35]. Clark *et al.*, at the SmithKline Beecham Corporation, described a process able to prepare commercial quantities of benzofuran-2-carboxylic acid {(S)-3-methyl-1-[(4S,7R)-7-methyl-3-oxo-1-(pyridine-2-sulfonyl)-azepan-4-yl-carbamoyl]-butyl}-amide, **33**, and analogs [36]. The method starting from the cycloaddition of 3-chloro-1-butene (**34**) with potassium phthalimide (**35**) in the presence of an alkali metal carbonate to form compound **36** N-(alfa-methylallyl) phthalimide. The compound **36** reacts with alkyldiamine followed by azeotropic distillation with ethanol and gaseous HCl treatment to give (R)-3-amino-1-butene hydrochloride, **37**. 2-Chlorosulfonyl pyridine is coupled with **37** in the presence of a thialkylamine base to form the pyridine sulfonamide **38**. Mitsunobu reactions of (2S,3R)-1,2-epoxy-4-penten-3-ol to form the nitrogen-protected epoxide fragment **40**. Reaction of sulfonamide fragment **39** and epoxide fragment **40** to provide **33**. There are twelve oxidation reaction steps to yield the final product (**Scheme 2**). The compound **33** is the Relacatib used for treatment of malaria.

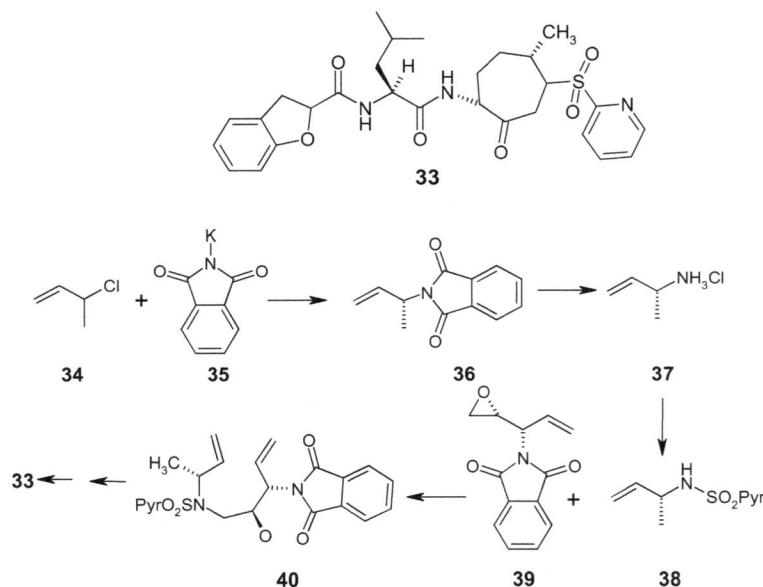

33

34   **35**   **36**   **37**

33 ←   **40**   **39**   **38**

Scheme 2

Smith Kline Beecham Corporation has also patented new promising dihydrotriazine derivatives [37]. These compounds derived from oxo-amino-sulfonyl-azapanes **41** also are Cathepsin k inhibitors and useful for the treatment of *P. falciparum* infection. Each R2 is H, C_{1-8}alkenyl, C_{2-8}alkynyl, or substituted C_{1-4}alkyl. R3 is independently H, C_{1-8}alkyl, C_{2-8}alkenyl, C_{2-8}alkynyl or substituted C_{1-6}alkyl. R4 is H, optionally substituted C_{1-7}alkyl, C_{2-7}alkenyl, C_{2-7}alkynyl, C_{3-7}cycloalkyl, or C_{3-7}cycloalkenyl. R5 is optionally substituted C_{3-7}cycloalkyl, C_{3-7}cycloalkenyl, aryl, heterocycloalkyl. Additionally, SmithKline Beecham Corp claimed novel dioxo-azepan (**42**) [37] and amino-azepan (**43**) [38-43] derivatives and their salts, a process for their preparation, compositions containing them and their use in methods for inhibiting proteases, particularly cathepsin K and falcipain [37,44]. For instance, the scheme for preparation of the compound **42** with substituent CH$_2$-cyclohexyl is shown in **Scheme 3**. Those compounds were tested for treating diseases characterized by bone loss such as osteoporosis, periodontitis and gingivitis, and diseases characterized by parasitic infections particularly malaria, trypanosomiasis and leishmaniasis. In the patent [45], the Corporation related a method of treatment of parasitic diseases, especially malaria. The compound **44** exhibited a *Ki* value of 1.9 nM against the falcipain cysteine protease.

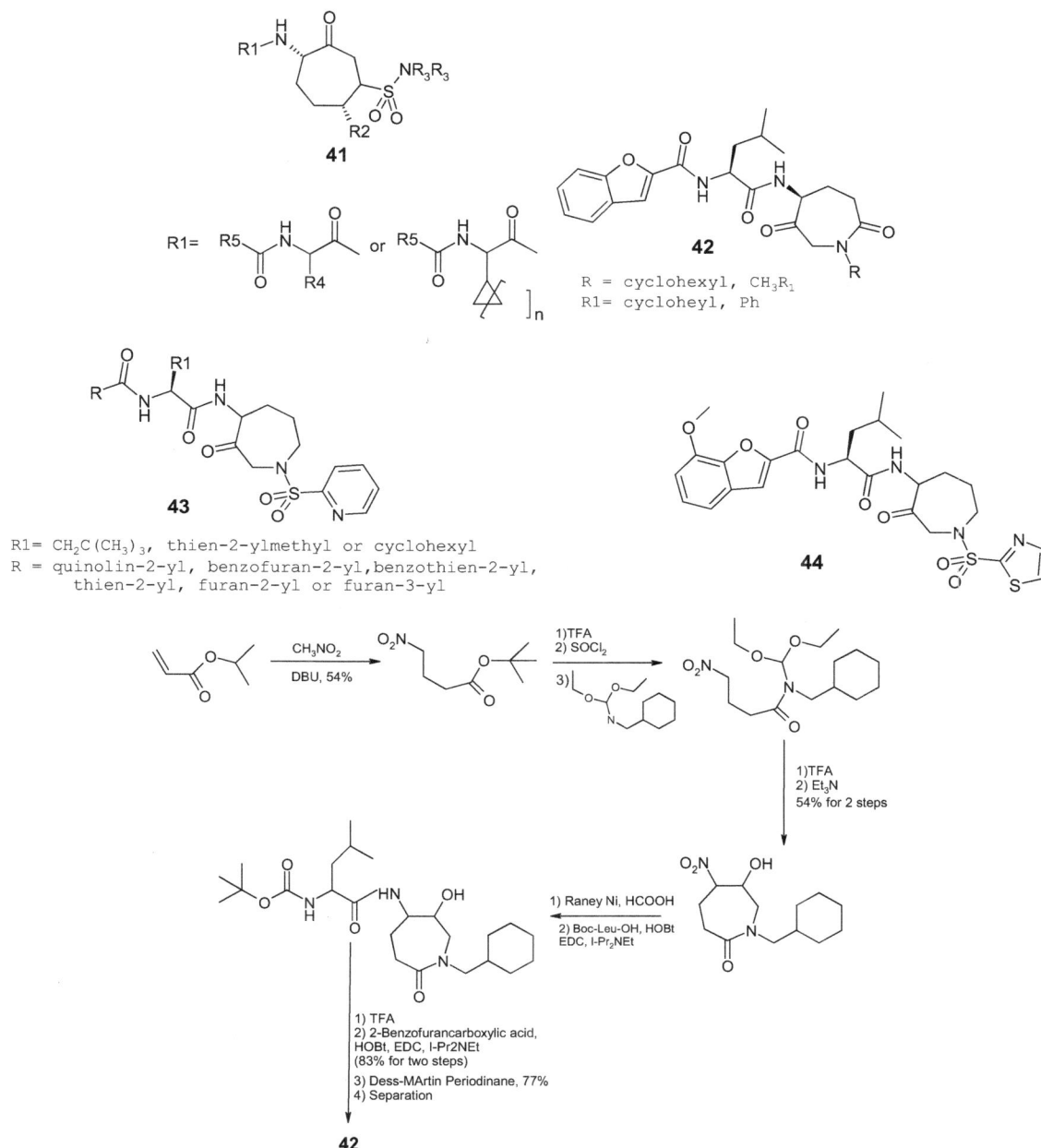

Scheme 3

US Army Medical Research Institute of Infectious Diseases patented novel tricyclic compounds that reduced side effects and restore the clinical efficacy of antimalarial drugs, including mefloquine (**45**) and chloroquine (**46**) [46]. Their salts and prodrugs, and their use as chemosensitizing agents and for modulating resistance to a drug, particularly an antimalarial drug, is also claimed. IC_{50s} were determined for each candidate compound alone and in combination with chloroquine. The fractional inhibitory concentration indices (FIC, actual IC_{50} of one compound in the presence of a second compound but is expressed as a fraction of its IC_{50} when used alone) for 1:1 combinations were determined using a chloroquine-resistant *P. falciparum* W2 clone (Table **2**). The combination of the compounds **47b** and **47e** and chloroquine showed the best MDR-reversing activity. Chloroquine with compounds containing saturared (**47i** and **47m**) and unsaturated seven-membered central rings (**47q** and **47r**) possessed similar activity. The compounds **47s** and **47t** were not as potent as their tricyclic ring counterparts. An FIC index of less than 1.0 represents synergy or potentiation and an FIC index greater than 1.0 represents antagonism (Table **2**).

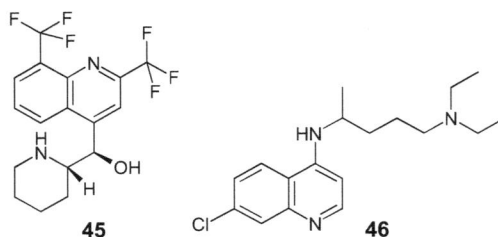

Table 2: FIC (fractional inhibitory concentration – 1:1 combination of drug and chloroquine) indices of new modulators in *P. falciparum* W2 clone.

Compound	X	n	Y	R1	R2	FIC
47a	S	4	-	CH_3	CH_3	0.23
47b	S	4	-	C_2H_5	C_2H_5	0.23
47c	S	4	Pyrrolidinyl	-	-	0.21
47d	S	4	Piperidinyl	-	-	0.49
47e	S	4	Morpholinyl	-	-	0.39
47f	S	4	4-methylpiperazinyl	-	-	0.40
47g	C_2H_4	4	-	CH_3	CH_3	0.23
47h	C_2H_4	4	-	C_2H_5	C_2H_5	0.39
47i	C_2H_4	4	Pyrrolidinyl	-	-	0.32
47j	C_2H_4	4	Piperidinyl	-	-	0.45
47k	C_2H_4	4	Morpholinyl	-	-	0.31
47l	C_2H_4	4	4-methylpiperazinyl	-	-	0.52
47m	C_2H_4	4	-	C_2H_5	C_2H_5	0.53
47n	C_2H_4	4	Pyrrolidinyl	-	-	0.33
47o		4	-	C_2H_5	C_2H_5	0.48
47p		4	Pyrrolidinyl	-	-	0.45
47q	C_2H_4	5	-	C_2H_5	C_2H_5	0.44
47r	C_2H_4	5	Pyrrolidinyl	-	-	0.48
47s	C_2H_4	6	-	C_2H_5	C_2H_5	0.48
47t	C_2H_4	6	Pyrrolidinyl	-	-	0.57

Mepha Pharma AG studied the administration of the combination of artesunate (**48**) and mefloquine (**30**) on humans with acute, uncomplicated *P. falciparum* malaria. Artemisinin and its derivatives are the most rapidly acting antimalarial drugs, however the duration of antimalarial activity is short [47]. Thus, in those studies, patients were randomized to accordingly receive in: Group A- artesunate 200 mg/day and mefloquine 250 mg/day simultaneously once daily for 3 days; Group B - artesunate 200 mg/day and mefloquine 750 mg (no dose on the 1st day, 250 mg on the 2nd day and 500 mg on the 3nd day) once daily for 3 days. The cure rate after 14 days and 28 days was monitored. The results show that group A had a 100% cure rate and group B had a cure rate of 98% after 14 days. Also in group A, the low incidence rate of early vomiting and overall vomiting was observed.

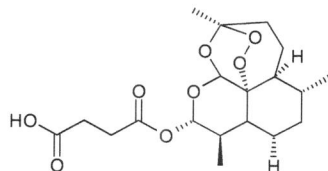

48

Three companies, Universidade de Lisboa, Cenix BioScience GmbH and Instituto De Medicina Molecular, reported the use of an inhibitor of a scavenger receptor class protein (ScarB1) for the therapy and/or prophylaxis of a malaria infection [48]. ScarB1 appears to mediate HDL-transfer and uptake of cholesterol. It was observed that inhibitors of ScarB1 function inhibit the growth of protozoa in liver cells, thus, inhibitors of ScarB1 can be used to treat infectious diseases involving liver cells and because the ScarB1 is expressed in erythrocytes, in hematopoietic cells. Human hepatoma cells were treated and the influence of the compounds on proliferation and infection with plasmodium sporozoites was calculated as % of the plate mean for all samples, with the mean set to 100%. In each compound a score was assigned between 0 and 4 for inhibition of infection. The compounds **49**, **50** and **51** showed score 4 (corresponding to an IC$_{50}$ of 1 μM or lower), 1 (IC$_{50}$ between 3 and 4 μM) and 0 (IC$_{50}$ between 4 μM or large), respectively.

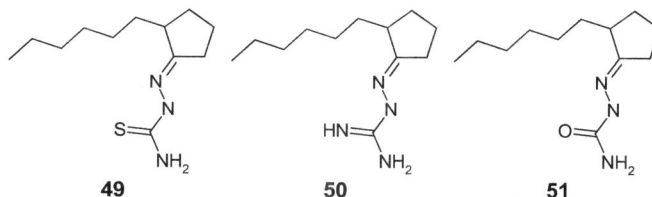

49   **50**   **51**

Fansidar is a combination of sulfadoxin (**52**) and pyrimethamine (**53**) used for the treatment of chloroquine (**47**) resistant *P. falciparum* [49]. Council of Scientific and Industrial Research patented a kit of α,ß-arteether (an artemisinin derivative), for increasing the sites of action on the parasites and thus will be more effective in their controll. The result shows that intramuscular α,ß-arteether (7.5 mg/kg) and the combination of sulfadoxin and pyrimethamine (5 mg/kg) produced 100% curative efficacy compared to the individual compounds alone [49].

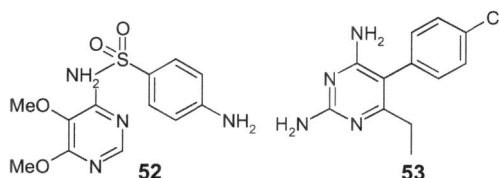

52   **53**

The Holding Company For Biological Products And Vaccines developed a scorpion venom (*Pandinus imperator)* drug that has the ability to stop the development of asexual life cycle from ring to schizont stage [50]. Doses of the venom stimulated the immune system producing specific immune response against malaria parasite. The venom is free of side effects and does not cause histamine release. Curiously, the scorpion venom is used as a carrier to deliver radioactive iodine into tumour cells left behind after surgery has removed the bulk of the tumour.

The *P. falciparum* erythrocytic stage causes several million deaths yearly, primarily in Africa. Thus, National Institutes of Health developed a novel vaccine formulation comprising a *P. falciparum* erythrocyte binding protein, BAEBL [51]. BAEBL polynucleotide is used to induce an immune response to a Plasmodium parasite, because it has homology to EBA-175, a *P. falciparum* receptor that binds specifically to glycophorin A on erythrocytes. Like EBA-175, the erythrocyte receptor for BAEBL is destroyed by neuraminidase and trypsin, however BAEBL can bind to erythocytes that lack glycophorin A and Gerbich erythrocytes (predent glycophorin C and absent glycophorin D) bind BAEBL much more weakly than normal erythrocytes. The compound of this invention can be employed in admixture with conventional excipients and can be combined with other active agents, *e.g.*, vitamins. In addition, Jensen *et al.* isolated VAR4, VAR5 or VAR6 polypeptide (from *var* genes) and its nucleic acid for use as vaccines for the diagnosis and treatment of malaria [52]. The proteins are responsible for inducing the immunoglobulin G (IgG) antibody with specificity to *P. falciparum* erythocyte membrase protein 1 (PfEMP1) on infected red blood cells. Thus, severe malaria is caused by parasites expressing a subset of PfEMP1 molecules and antibodies directed against these will be responsible for the protection against severe disease acquired early in life.

Yet in vaccine preparation, Claudianos *et al.*, at the Imperial College Innovations Limited, claimed novel secreted scavenger receptor (PfSR) proteins from the *P. falciparum* genome database, and their use in the preparation of vaccines against malaria and in diagnosis [53]. PfSR protein was shown to have 64% homology to proteins found in the mouse malarial parasites *P. berghei* (PbSR) and *P. yoelii* (PySR), whilst nucleic acid comparisons showed 74% homology. It contains Limulus clotting factor C, lipid binding, cysteine rich scavenger receptor and pentraxin domains. The protein causes disruption of the normal expression of the proteins, so the malaria parasite dies at an early stage in its life cycle. Parasite disruption *via* drugs or vaccines is therefore more likely to be effective when parasite numbers are low, *i.e.* at the early infective stages: sporozoites in human and ookinete in mosquito.

Walter and Eliza Hall Institute of Medical Research developed a vaccine for inducing an immune response to *P. falciparum* using a glycosylphosphatidylinositol (GPI) inositolglycan domain or its derivative or equivalent [54]. GPI has been identified as a candidate toxin of parasite origin. Experiments using GPI in mice with cerebral malaria showed relatively low parasitemia levels between days 5 to 12 post-infection.

1.7. Treatment of Leishmaniasis

Cutaneous leishmaniasis are *Leishmania mexicana* and *L. brasiliensis* in the Americas, and *L. tropica* in the Old World; and of visceral leishmaniasis, *L. donovani*, *L. infantum*, and *L. chagasi*. Treatment remains inadequate because of drug toxicity (sodium antimony gluconate), long courses required, and frequent need for hospitalization. Sandfly vectors transmit cutaneous leishmaniasis.

For the treatment of *Leishmaniasis* the currently used drugs are limited to four. The first line compounds are the two pentavalent antimonials, sodium stibogluconate and meglumine antimoniate. They were used for the first time in 1947 and 1950, respectively. Failures and relapses occur in all forms of *leishmaniasis* and constitute approximately 10-25% of cases. If these drugs are not effective, the second line compounds of pentamidine (**2**) and amphotericin B (**54**) are used, which were introduced in 1940 and 1959, respectively.

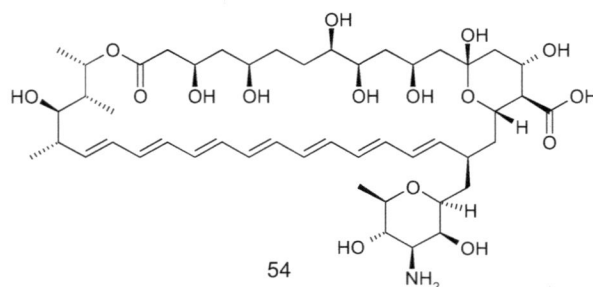

54

Fuertes *et al.*, at the Mologen Holding AG Dana-Farber Cancer Institute, claimed to have discovered new DNA-expression construct for treatment of infections with leishmaniasis and a corresponding vaccine [55]. The DNA expression constructs may code for one or more Leishmania antigens, including the p36 LACK antigen. p36 Antibodies were determined in mice following immunization.

Torrado Duran *et al.* (Universidad Complutense de Madrid) described a novel method of obtaining water-dispersible albumin microspheres containing amphotericin B [56]. The invention is based on the atomisation of the albumin with the drug that is microencapsulated for treatment of leishmaniosis and not requiring the pre-formation of an emulsion, avoiding the use of oils and organic solvents. This composition is less toxic, has fewer side effects than current amphotericin B formulations, such as Fungizone®, and is easy to produce on a industrial scale. The product is suitable for parenteral administration or intravenously.

Preparation containing diaminazen-diaceturate (**55**) and/or pentamidine and chloroquine (**47**) and/or artemisinin (**56**) or a derivative for treating leishmaniasis is related by Tropmed GmbH [57-59]. The inclusion of procaine or lidocaine in the formulation reduces the pain of injection and the fever associated with malaria. A representative tablet composition comprises the following compounds: 350 mg of diminazene diaceturate (**55**), 250 mg of chloroquine (**47**) and 400 mg of antipyrine (**57**).

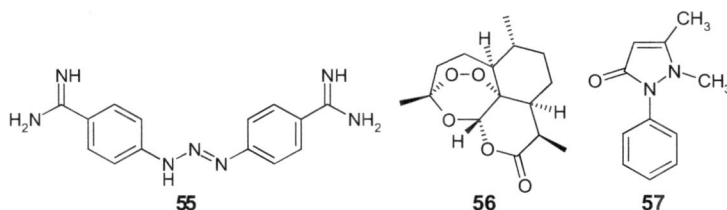

The Institut de Recherche pour le Developpement reported the use of niacin (vitamin B_3, **58a**, R = OH) or nicotinamide (**58b**, R = NH_2) for treating parasitic diseases. Nicotinamide is currently in trials as therapy to prevent cancer recurrence and insulin-dependent (type I) diabetes. In particular, they can be used to treat protozoan parasitic diseases such as leishmaniosis in immunodepressed patients [60]. It has been recently demonstrated that nicotinamide is a substrate of SIR2-like enzymes *in vitro* [61]. The data presented that nicotinamide strongly inhibited the proliferation of both promastigotes and amastigotes with promastigote forms showing less sensitivity to nicotinamide than amstigotes. Also the patent reported the composition and administration in combination with another active agent selected from *e.g.*, amphotericin B, paromomycin, melarsopol and antimonials, using oral, intravenous, topical or intralesional routes. In addition, the Institut described in a patent, the nucleic acid constructs, which comprise nucleic acid encoding an immunogenic protein from a promastigote or amastigote of Leishmania [62, 63].

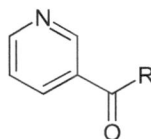

58a R=OH
58b R= NH2

Amine-borane compounds exhibit high similarity to organic compounds mainly due to the atomic radii and the characteristics of the B-N bond, which resemble those of carbon-carbon bonds. Thus, novel families of amine-borane compounds, including fluorinated aminoboranes, novel bis-aminoboranes and aminoboranes having saturated and unsaturated ling alkyl chains are provided by Soreq Nuclear Research Center Israel Atomic Energy Commission and Yissum Research Development Co. [64]. Processes for preparing, pharmaceutical compositions and methods utilizing these novel compounds are also provided. Radiolabeled aminoboranes and uses thereof in radioimaging (*e.g.*, PET) and radiotherapy are further provided. They are used for the treatment of leishmaniasis and demonstrated excellent antimicrobial activity with reduced adverse side effects and improved pharmacokinetic profile. The *in vitro* anti-lishmenial effect of

Coumpound **59** (dimethyl-undecyl-amine cyanoborane) was determined on promastigotes and amastigotes. According to Table **3**, the measured radioactivity, which is indicative of the parasite, diminishes rapidly with the increased dosage of amine-borane, indicating a strong anti-leishmenial effect.

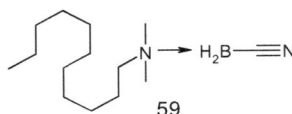

59

Table 3: Concentration of the amine-borane is given in µg/ml *versus* the counts per minute recorded for the parasite sample.

mg/ml	cpm
0	8699
3.6	8003
11	7898
33	537
100	138

In a further patent, Soreq Nuclear Research Center Israel Atomic Energy Commission and Yissum Research Development Co. disclose modified polymer conjugates. The conjugate of a polymer (*e.g.* polysaccharide) and a drug (*e.g.* amphotericin B, **54**) reduced the toxicity relative to the unmodified parent compound, increased the solubility and retained substantially the same degree of therapeutic activity of the unmodified parent compound [65]. It is an antiparasitic composition against *L. donovani* and formulated in the form of a nanoparticle, micellar dispersion and a liposome. The test was performed in dextran-AmB imine (**60**) conjugate with or without ethanolamine (Table **4**). The result shows that the conjugate had an IC$_{50}$ value of 0.25 µg/ml compared to the Dextran-AmB imine alone (1.2 µg/ml).

Table 4: *In vitro* activity against *Leishmania donovani*, cytotoxicily and hemolysis of conjugates.

Compound	Antiparasitic Activity[a] IC$_{50}$ (µgAmB/ml)	Toxicity[b] IC$_{50}$ (µgAmB/ml)	Hemolysis[c] (µgAmB/ml)
Free AmB (**54**)	0.05	9	16
Dextran-AmB amine (**60**)	1.2	1400	>500
Dextran-AmB imine (**61**)	0.3	200	250
Dextran-AmB– ethanolamine imine	0.25	400	>500

[a]IC$_{50}$ values were derived from the activity test of AmB and different dextran-AmB conjugates against *L. donovani*.
[b]IC$_{50}$ values were derived from cytoxicity test AmB and different dextran-AmB canjugates against the murine RAW 264.7 cell line. [c]Hemolysis was evaluated visually after 1h incubation at 37° C with Sheep erythocytes [65].

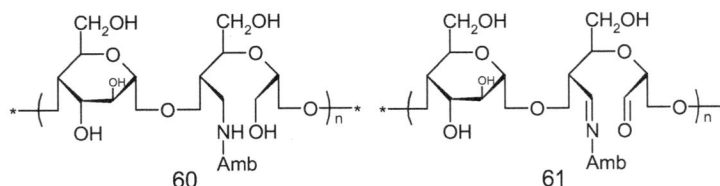

Valtion Teknillinen Tutkimuskeskus reported the use of betulin (**62**) derivatives or their salts against leishmaniasis in pharmaceutical industry applications [66]. Betulin is the most abundant triterpenoid of the lupane series and is found in the bark of some tree species, particularly in birch bark (*Betula*). It is a component of various food additives and cosmetic products. The activity of the betulin derivatives was tested for both *L. donovani* and *L. tropica* species (Table **5**).

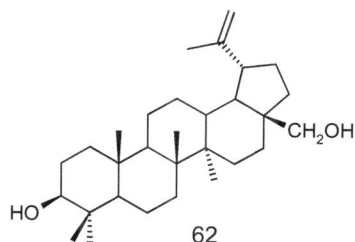

Table 5: Activity *in vivo* of the compounds tested for *L. leishmania*. Initially, the concentration of the compounds was 50μM. Amphotericin B (1μM) was included as positive control. Broth containing DMSO was used as negative control.

Compd	Name	Inhibition of *Leishmania* (%)
62	betulin	35.0
62a	28-acetate of betulonic alcohol	40.6
62b	28-methylester of betulonic acid	40.1
62c	betulinic aldehyde	65.0
62d	betulin 3,28-dioxime	72.4
62e	betulin 28-oxime	66.8
62f	betulonic alcohol	44.0
62g	betulin 3-acetoxyoxime-28-nitrile	66.4
62h	betulin 28-acetic acid methylester	95.3
62i	20,29-hydrobetulonic acid	73.4
62j	2,3-didehydro-3-deoxybetulin	13.2
62k	betulonic acid	97.6
62l	betulinic acid	39.8
62m	28-aspartateamide dimethylester of betulonic acid	69.3
62n	betulonic aldehyde	46.2
62o	betulin 28-*N*-acetylanthranilic acid ester	59.2
62p	betulin 28-chrysanthemate	13.4
62q	betulin 28-carboxymethoxy mentholester	16.6
	positive control: amphotericin B (1 μM)	55.4
	negative control: broth + DMSO	0.0

Kemin Pharma Europe patented the use of bicyclic carbohydrates for the treatment of leishmaniasis [67]. Compound **63** and **64**, bicyclic carbohydrates with halogen containing aryl groups, possessed significant activity against *L. donovani*, with IC$_{50}$s of 1.01 and < 0.98 μM, respectively, compared with 0.47 μM for miltefosine **65**. The use of **65** in *leishmaniasis* therapy should be carefully considered [68].

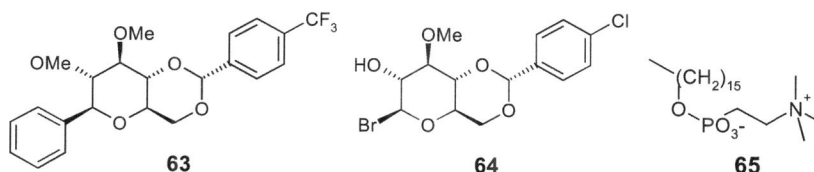

National Institutes of Health reported a novel dinitroaniline sulfonamide based on the herbicide oryzalin **66** [69] preferably less cytotoxic to normal cells than oryzalin [70]. Tubulin, a critical eukaryotic protein responsible for formation of the mitotic spindle has been implicated as the target of dinitroaniline analogs. Compounds **67** and **68** are significantly more potent than compound **66** in blocking leishmanial tubulin assembly (Table **6**). The compound **68** had K$_d$ (dissociation constant) values of 57μM leishmanial tubulin, and the corresponding value for oryzalin were 170μM. These data are consistent with hypothesis that tubulin is the target of the dinitroanilines in Leishmania.

Table 6: Activity *in vivo* of Oryzalin and new dinitroniline compounds against *L. donovani* and *L. Tubulin in vitro.*

Compound	IC$_{50}$ vs. *L. Donovani* Promastigotes (µM)	IC$_{50}$ vs. *L. Donovani* Amastigotes (µM)	% Inhibition of *L. Tubulin* Assembly at 20 µM Compound.
66	44.1	72.5	54
67	17.8	20.1	89
68	14.7	5.41	100

In 2006, De Souza *et al.* related a prophylaxis vaccine for blocking transmission of *Leishmania* infections [71]. It is based on *L. donovani* saponin and promastigotes or amastigotes fractions called Fucose Mannose ligand (FML, an isolated glycoprotein complex). Leishmune® vaccine (FML vaccine) developed 92-95% protection against canine visceral leishmaniasis and showed immunotherapy potential. In 2008, due to high production cost of Leishmune®, Gazzinelli *et al.* developed a vaccine formulation comprising amastigote A2 antigen [72]. The antibodies produced by the dogs in the vaccination process with this antigen are non-reactive to serologic diagnostic infection tests. The formulation is composed of 50 to 200.00 mg/mL of recombinant A2-HIS (rA2) protein of *Leishmania*, produced in *E. coli*, 0.125 to 0.500 mg/mL of Saponin, 1.00 mL of q.s.p tamponade saline solution and 0.01 mL of thimerosal. Therefore, the main innovation of this vaccine formulation is the production of antibodies specifically against the vaccine antigen. The main innovation of this formulation is that the animals vaccinated present serologic reaction against the antigen, but remain seronegativo in the reaction with the total parasite extract, *e.g.* the antibodies do not react with the extract of the promastigote forms of *Leishmania* in the ELISA tests, for example.

Rosalind Franklin University of Medicine & Science disclosed the use of a live mutant *Leishmania* as a suicidal vaccine [73]. The mutant can be selected from natural *Leishmania* species or constructed by genetic engineering. The invention also exploits the virtual absence of heme biosynthesis pathway in trypanosomes to identify or construct the mutant porphyric *Leishmania* as suicidal vaccines. The vaccinated hamsters showed ~10 fold reduction of parasite loads than the control group.

1.8. Treatment of Pneumocistosis

Pneumocystosis is a pneumonia that is characterized by the accumulation of very large numbers of a eucaryotic single-celled organism called *Pneumocystis carinii*, which has not been cultured.

Pneumocystosis also occurs in many other mammalian species. It is not yet established whether *Pneumocystis carinii* is a fungus or a protozoan.

Pneuocistis carinii causes *P. carinii pneumonia* (PCP) in people with depressed immune systems such as AIDS patients, patients undergoing chemotherapy, or transpant patients being treated with immunosuppressants. The infection is treated with pentamidine and the combination of trimethoprim and sulfamethoxazole, however there are side effects and the mortality rate remains high. An alternative treatment is the preparation containing methioninase in combination with an antibiotic, either pentamidine or combination of trimethoprim and sulfamethoxazole related by AntiCancer Inc. [74]. Infection by *P. carinii* can be treated by administering methioninase optionally in combination with additional therapeutic agents, such as antibiotics, either pentamidine or combination of trimethoprim and sulfamethoxazole. The use of methioninase depletes plasma levels of methionine and deprives the parasites of S-adenosylmethionine.

In a futher patent application\development, the preparation of ten carbamate and carbonate prodrugs of bisamidinophenyl furans was patented by The University of North Carolina at Chapel Hill [75]. Compounds were tested against *P. carinii* infected immunosuppressed rats. The more potent compound, **69**, inhibited cyst formation in the lung by 100 and 97.9 % inhibition at doses of 22 and 33 mumol/kg/day, respectively.

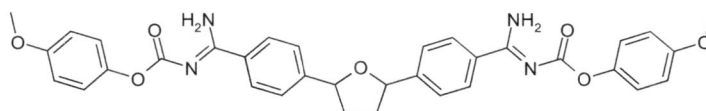

69

A series of primaquine-derived imidazolidin-4-ones were screened for their *in vitro* activity against *P. carinii* by Vale *et al.* [76]. One of the tested imidazolidin-4-ones, **70**, was slightly more active ($IC_{50} = 1.9$ μM) than the parent primaquine ($IC_{50} = 2.5$ μM).

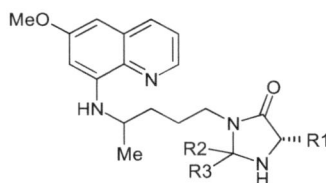

70a R1=Me
70b R2-R3=$(CH_2)_6$

Dihydrofolate reductase (DHFR) is a key enzyme in the treatment of Pneumocistosis. Its role in thymidine biosynthesis (Fig. **1**) is the reduction of 7,8-dihydrofolate (DHF) to 5,6,7,8-tetrahydrofolate (THF) using the cofactor nicotinamide adenine phosphate (NADPH). After reduction, serine hydroxymethyltransferase (SHMT) catalyses the regeneration of 5,10- methylenetetrahydrofolate; and deoxyuridine monophosphate (dUMP) is methylated to give deoxythymidylate (dTMP) in a reaction catalysed by thymidylate synthase (TS). This reaction converts methylene-tetrahydrofolate back to dihydrofolate, completing the cycle. Therefore, the inhibition of DHFR prevents the biosynthesis of thymidine and, as a consequence, DNA biosynthesis [77]. Novel substituted 2,4-Diamino-5-benzylpyrimidine tested as inhibitors of *P. carinii* DHFR, was claimed by Forsch *et al.* (compounds 71-75) [78]. In summary (Table **7**), the results indicated that, 5′-(5-carboxy- 1-pentynyl) substitution is more favorable when the benzyl ring is 3′,4′,5′-trisubstituted than when it is 2′,5′-disubstituted. Compound **73** has the highest specie selectivity for DHFR *P. carinii*. For this reason this novel analogue may be viewed as a novel lead for further structure – activity optimization of DHFR binding. Rosowsky *et al.* reported a concise new route allowing easy access to five 2,4-Diaminopyrido[2,3-d]pyrimidine analogs containing a substituted phenyl or other aromatic ring attached to C6 *via* by a short bridge [79]. The derivatives were tested against *P. carinii* DHFR, however none of the quinazolines or pyridopyrimidines tested (IC$_{50}$ 0.087-72 μM) were more potent against the *P. carinii* enzyme than the structurally related reference compound piritrexim (**77**, IC$_{50}$ = 0.013 μM).

In the patent application, Forsch and Rosowsky, at the Dana-Farber Cancer Institute Inc., synthesized several lipophilic DHFR inhibitors having an aromatic group and a heteroaromatic group linked by a methylene group in the treatment of *P. carinii* [80]. Trimethoprim (**77**) and piritrexim (**76**) are lipid-soluble antifolates that have been used clinically for the prophylaxis and treatment of *P. carinii* infections in patients with AIDS.

In addition, Rosowsky and Forsch, synthesized novel **77** derivatives (Table **8**), compositions comprising them and their use as DHFR inhibitors [81]. The most potent of the O-alkyl derivatives against *P. carinni* DHFR was compound **78c** (n=6). Moreover there was an increase in potency as the length of the 5′-O-alkyl group increased from n=4 to n=6, followed by a decrease in potency as this length increased from n=6 to n=8. The most potent of the 5′-O-(w-carboxyalkyl) analogs was 60g (n=4), 245 times more potent than **77**.

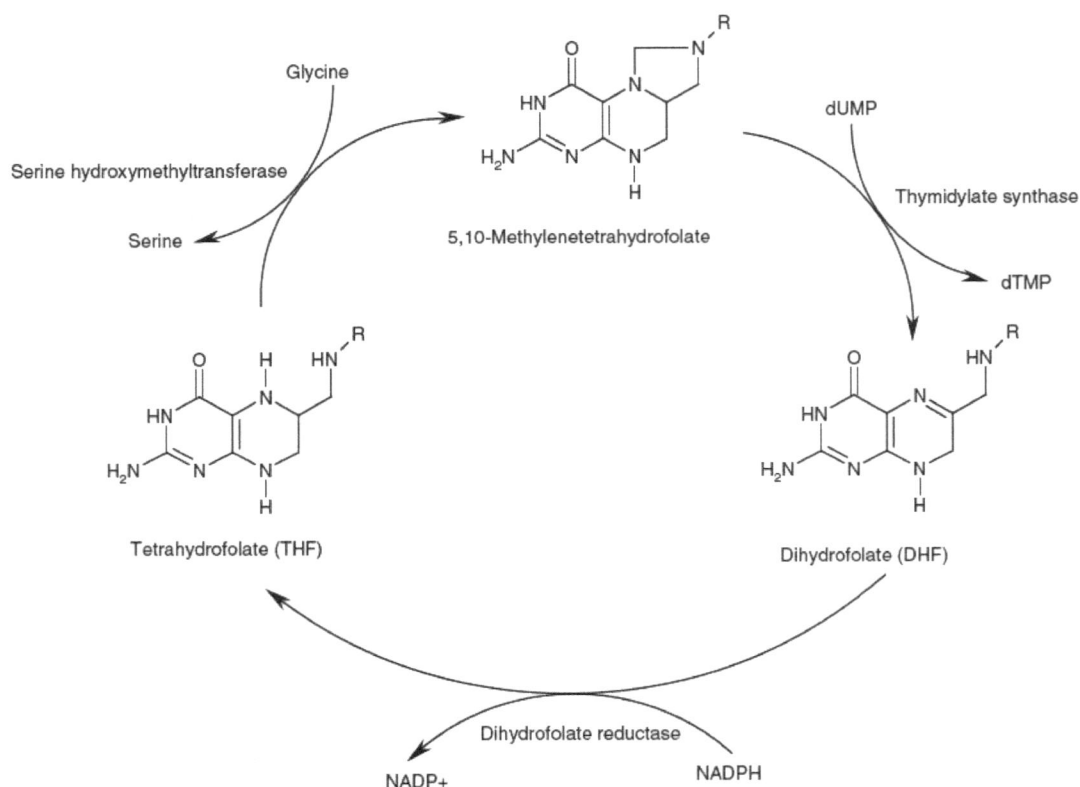

Figure 1: Role of DHFR in the synthesis of thymine.

Table 7: Inhibition of *P. carinii* DHFR by 2,4-diamino-5-(3′,4′-dimethoxy-5′-substituted benzyl] pyrimidines with a carboxyalkynyl or carboxyphenylalkynyl group in the side chain.

Compounds	IC$_{50}$ nM
71	28

Table 7: cont....

	1300
	1.0
	1200
	1.0

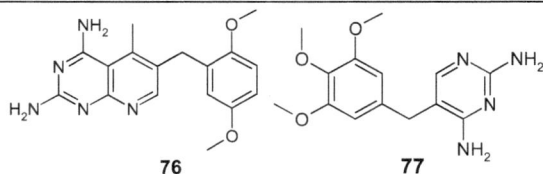

Table 8: Inhibition of *P. carinii* DHFR by trimethoprim derivatives.

Compounds	R	IC$_{50}$ µM
78a	Me, n = 4	19
78b	Me, n = 5	14
78c	Me, n = 6	5.6
78d	Me, n = 7	44
78e	Me, n = 8	51
78f	CO$_2$H, n = 3	0.25

Table 8: cont....

78g	CO_2H, n = 4	0.049
78h	CO_2H, n = 5	0.80
78i	CO_2H, n = 6	2.6
78j	CO_2H, n = 7	7.1
78k	CO_2H, n = 8	4.8
77- trimethoprim		12
76 - piritrexim		0.031

The activity of four triazolyl analogues (**79a-d**) was evaluated as inhibitors of *P. carinii* DHFR by Chan *et al.* [82]. The two most potent compounds are **80a** and **80b**, against the *P. carinii* DHFR. Four compounds with a nitrogen bridge (benzanilides and benzylamines) were also tested for activity. Compounds in the benzanilide series (**80a,b**) are less potent than the benzylamines (**81a,b**; Table **9**). With an IC_{50} value of 0.12 mM, compound (**81a**) is the most potent member of the group as an inhibitor of *P. carinii* DHFR. Interestingly, methylation of the amino bridge (**81b**) showed a 10-fold reduction of activity towards the *P. carinii* enzyme and nearly a 7-fold gain of potency towards the rat liver DHFR.

Table 9: Inhibition of *P. carinii* DHFR by benzanilides and benzylamines derivatives.

Compound	R1	R2	IC_{50} (μM)
79a	H	CH_2Ph	5.18
79b	H	Ph	3.53
79c	H	$SiMe_3$	10.5
79d	H	H	24.8

Compound	R		IC_{50} (μM)
80a	Ph		10.8
80b	$C_6H_2(OMe)_3$		1.3

Compound	R	R1	IC_{50} (μM)
81a	H	CH_2Ph	0.12
81b	Me	CH_2Ph	1.07

Hallberg *et al.* reported a novel series of DHFR inhibitors, where the methylenamino-bridge of non-classical inhibitors was replaced with an ester function [83]. The most potent of the new ester-based DHFR inhibitors, the 1-naphthyl derivative (**82**), exhibits an IC_{50}-value of 110 nM, and is less active than trimetrexate (**83**) (IC_{50} = 42 nM).

82   **83**

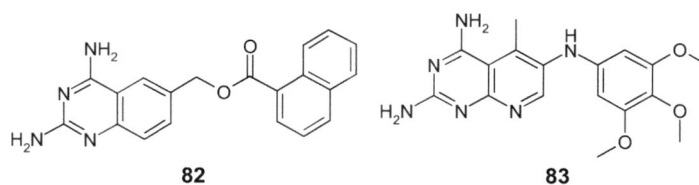

Cushion *et al.* synthesized a series of pentamidine (**2**) congeners and screened for their *in vitro* activity against *P. carinii* [84]. The ATP assay was used to evaluate the effects of this pentamidine on the viability of A549 epithelial lung cell monolayers derived from a human carcinoma. All diamide derivatives (**84a-e**, Table **10**), including the most potent agents **84c** and **84e**, showed no cytotoxicity at 100 times the IC_{50} concentration against *P. carinii*. This study identified the simple small molecules **84c** and **84e** as two very promising therapeutic leads, which deserve further pre-clinical and clinical testing.

Table 10: Inhibition of *P. carinii* DHFR by diamide derivatives.

Compound	Linker	IC_{50} μM (in μg/mL)
84a		5.3 (2.3)
84b		1.2 (0.6)
84c		0.003 (0.0013)
85d		22.8 (0.0013)
84e		0.002 (0.0009)

In a patent application, Walzer *et al.*, University of Cincinnati, disclosed a method of combating infectious agents, such as Pneumocystis pneumonia, and a method of treating a subject in need of such treatment [85]. The method comprises administering a bis-benzamidoxime to the subject. The more active compound of the series, compound **85**, had an IC_{50} value of 0.00087 μg/ml against *P. carinii*. In comparison, pentamidine (**2**) had an IC_{50} value of 0.300 μg/ml.

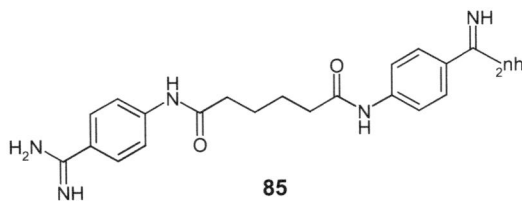

85

1.9. Treatment of Toxoplasmosis

Toxoplasmosis is a disease caused by protozoan *Toxoplasma gondii*. Transmission to man occurs *via* ingestion of meat containing cysts or tachyzoites, ingested oocysts become infective after sporulation or transplacentally.

Toxoplasma gondii (*T. gondii*) is an ubiquitous parasitic protozoan that infects up to one-third of the US population and up to 90% of certain European populations [86]. Risk factors for infection include exposure to infected cats, consumption of rare meats and contact with contaminated soil. Medications that are prescribed for toxoplasmosis are pyrimethamine (**53**) (an antimalarial medication) and sulfadiazine (**86**) (an antibiotic). Thus, the need for new chemotherapeutics active against *T. gondii* is therefore acute. Sanofi-Aventis development herbimycin (**87**) derivatives, a benzoquinone ansamycin antibiotic, as heat shock protein 90 (Hsp90) inhibitors, useful for the treatment of toxoplasmosis [87]. Hsp90 is a molecular chaperone and is one of the most abundant proteins expressed in cells. It is highly conserved and expressed in a variety of different organisms from bacteria to mammals. In an *in vitro* assay, **88** had an IC_{50} of 0.928 μM for the inhibition of the ATPase activity of Hsp82.

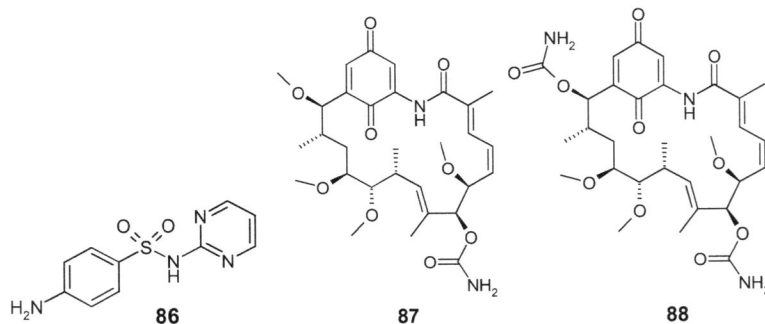

86   **87**   **88**

One feature that distinguishes *T. gondii* from its human host is its inability to synthesize purine "salvage" [88]. Differently from its human host, the *T. gondii* recovers purine precursors from the adenosine kinase, enzyme which converts adenosine in adenosine monophosphate (AMP) (Fig. **2**). Through it, all other purine nucleotides can be synthesized. Thus, the inhibition of the adenosine kinase activity interrupts the purine recovery route ("salvage"), offering a number of potential targets for the antiparasite chemotherapy. It is known in the literature that benzylthioinosine analogues are substrates for the parasite adenosine kinase, but not for human adenosine kinase [89]. Rais *et al.* described a series of 6-benzyladenosine analogues that act as potent and selective substrates for *T. gondii* adenosine kinase [90-92]. 6-Benzylthioinosine analogs (Table **11**) were identified as excellent subversive substrates of *T. gondii* adenosine kinase.

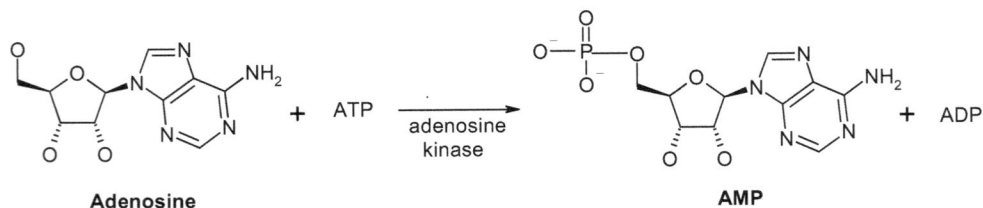

Adenosine   **AMP**

Figure 2: Reaction catalyzed by the adenosine kinase enzyme.

Table 11: Inhibition of *T. gondii* AK by 6-benzyladenosine derivatives.

Compound	Structure	IC$_{50}$ (μM)
89a	*o*-Methyl-Benzylthioinosine	7.7
89b	*o*-Chloro-Benzylthioinosine	6.7
89c	*m*-Methyl-Benzylthioinosine	8.2
89d	*m*-Trifluoromethyl-Benzylthioinosine	8.7
89e	*m*-Nitro-Benzylthioinosine	6.2
89f	*p*-Methyl-Benzylthioinosine	7.8
89g	*p*-Bromo-Benzylthioinosine	14.3
89h	*p*-Methoxy-Benzylthioinosine	3.5
89i	*p*-*tert*-Butyl-Benzylthioinosine	23.3
89j	*p*-Acetoxy-Benzylthioinosine	15.0
89k	*p*-Chloro-Benzylthioinosine	8.11
89l	*p*-Nitro-Benzylthioinosine	12.0
89m	*p*-Cyano-Benzylthioinosine	4.3
89n	*p*-Fluoro-Benzylthioinosine	10.3
89o	2,4-Dichloro-Benzylthioinosine	7.3
89p	2-Chloro-6-fluoro-Benzylthioinosine	8.7
89q	*m,p*-Dichloro-Benzylthioinosine	10.4
89r	*o,m,p*-Trimethyl-Benzylthioinosine	31.1

Currently, there are no tests which can differentiate between oocyst ingestion *versus* tissue cyst ingestion as the infection route [93]. Recombinant proteins have been developed by US Department of Agriculture for the detection of *T. gondii* oocyst proteins in biological fluids for example. Isolates and recombinant *T. gondii*-specific proteins (rDGP5p) can be adsorbed to the surface of microtiter plates or to immunoblotting membranes and used in an ELISA format for detection of antibodies. Recombinant antigens also can be use to prepare monoclonal antibodies which selectively identify *T. gondii* oocyst proteins. In addition, primers directed to *T. gondii*-specific regions of the DNA sequences can be produced for sensitive detection of the parasite by polymerase chain reaction. Using ELISA and MAT, (microscopic agglutination test) assays were performed on 127 human sera known to have resulted from oocyst exposure, five sera from infections resulting from congenital infection, and 76 sera from MAT negative individuals. Of 76 uninfected sera, all were negative using the recombinant ELISA and of 127 oocyst induced infections, the recombinant antigen ELISA detected 119, and did not detect 8. Sera from 40 oocyst infected pigs and 45 tissue cyst infected pigs also were tested using ELISA and MAT. Of the 40 pigs infected with oocysts, 39 were detected using ELISA assay.

In the development of vaccine patents, GlaxoSmithKline Biologicals patented a vaccine composition comprising the toxoplasma protein, SAG3, the major glycosylphosphatidylinositol (GPI)-anchored surface protein of *T. gondii* [94]. SAG1 mediates the host-cell invasion and mono/polyclonal antibodies directed to SAG1 inhibited invasion of cells by tachyzoites, probably by interfering at the parasite attachment level [95]. The binding of SAG1 and SAG3 to CHO K1 cells (Chinese hamster ovary cells) was examined. Just SAG3 displays affinity for sulfated proteoglycans. The presence of alternative cellular receptors used by *T.*

gondii was already suggested, as inhibition of parasite attachment by soluble GAGs or by heparinase treatments of host cells was not complete. The studies show that soluble heparin (glycosaminoglycan) inhibited SAG3 binding by up to 90%. Dermatan sulfate, a glycosaminoglycan that is a structural component of certain body tissues also inhibited the binding of SAG3, but chondroitin sulfate A, a chondrin derivative, did not. In a further patent application on vaccines, Kyushu TLO Co Ltd [Kyushu University] developed a novel fused DNA vaccine produced by constructing a fused gene comprising a gene from an intracellular protozoal parasite, such as toxoplasma, with a ubiquitin gene [96]. This vaccine is said to form an ubiquitinated pathogen antigen which is processed with a proteasome which strongly induces the production of CD8$^+$ T cells.

2. CURRENT AND FUTURE DEVELOPMENT

Neglected tropical diseases are widely related to poverty and disadvantage. The poorest populations often living in remote, rural areas, urban slums or in conflict zones are the most affected. Actually, the neglected tropical diseases show a clear link between health and development.

Adding HIV [97], tuberculosis,[98-100] dengue [101-103], Chagas disease, [104-106] Leishmaniasis [107] and malaria [108-111] to the mix, the diseases kill several million each year, shorten lives and reduce productivity.

Biologists have identified more than 50,000 species of protozoa, of which a fifth are parasitic and some can reach humans by food and water. The majority of food and waterborne infections of parasitic origin are related to poverty, low sanitation, and old food handling habits. Parasitic protozoa do not multiply in foods, but they may survive in or on moist foods for months in damp environments.

Diseases caused by intestinal protozoa are increasing and new treatments are very important because of their association with acute and chronic diarrhea in immunocompromised patients as well as those with a normal immune system.

For African trypanosomiasis, cysteine protease inhibitors were described by Merck for the treatment of trypanosomiasis [12-16] and for American trypanosomiasis, inhibitors of CAC1 cysteine proteases cathepsin K are arising as promising compounds. It is important to mention those compounds were also described in the treatment of Chagas disease [4-10]. SmithKline Beecham Corporation reported a new promising trioxo-[1, 2] thiazepanylamide derivative [26].

Turning to Giardiasis, Suk *et al.* reported the design, synthesis, and activity of a potent and non-toxic second generation anti-giardial agent that was designed as an inhibitor of cyst wall synthase [29].

However, as in the case of G. lamblia, currently (2001-2008), there are no new patents for anti-balantidiasis drugs. Balantidiasis disease is especially dangerous in immunocompromised patients as pulmonary parenchyma involvement is possible. Alternative drugs for the treatment of balantidiasis may be 5-nitroimidazole drugs, such as tinidazole (22) and ornidazole (26) [32].

In this scenario, 5-nitrofuran 2-carboxaldehyde thiosemicarbazones (31) and their subsequent bidentate Cu(II) complexes (32) showed promising results (IC50 value *in vitro*) when compared to the metronidazole (conventional compound used for the Treatment of Amebiasis) [34]. Regarding malaria, two points call special attention: the resistance to currently available antimalarial drugs and the inefficacy of antimalarial vaccines [35]. Thus, new promising Cathepsin k inhibitors (dihydrotriazine derivatives) and inhibitors of a scavenger receptor class protein (ScarB1) were proposed [37,48]. However, in this scenario, a promising way is to use different vaccine formulations against malaria [51-54].

It is interesting to mention that Dihydrofolate reductase (DHFR) is key in the treatment of Pneumocistosis. Novel series of DHFR inhibitors may be viewed as a novel lead for further structure – activity optimization of DHFR binding. In this line, Hallberg and co-workers reported a serie of DHFR inhibitors, where the methylenamino-bridge of non-classical inhibitors was replaced with an ester function [80] leading to interesting results.

The need for new chemotherapeutics active against *T. gondii* is also acute. On this point, herbimycin (87) derivatives, protein 90 (Hsp90) inhibitors, and 6-Benzylthioinosine analogs, which are subversive substrates of *T. gondii* adenosine kinase, were identified as promising compounds in the treatment of toxoplasmosis.

Currently, ultraviolet (UV) irradiation is regarded as being widely effective against all pathogens, bacteria, protozoa and viruses that can be transmitted through drinking water [94]. However, more research is needed to understand the extent to which particles in water may hinder the UV treatment efficacy by interacting with microbial pathogens. UV irradiation is a primary disinfection technology for use in water and wastewater effluents. It has been demonstrated that UV radiation is very effective against (oo)cysts of giardia, a pathogenic microorganism of major importance for the safety of drinking water [112].

In Europe, UV has been widely applied for drinking water disinfection since the 1980s, for the control of incidental contamination of vulnerable groundwater and for reduction of Heterotrophic Plate Counts. Protozoan infections are especially endemic and affect millions of people around the word. In spite of their importance, the development of more efficient drugs against those diseases are still scarce. Furthermore, surprisingly few patents for new drug candidates acting on these diseases have appeared. Observing that scenario, we believe that some advances in anti-protozoan drug candidate research can be expected from the recently published X-ray structures of enzymes, such as Leishmania trypanothione synthetase, Entamoeba histolytica glyceraldehyde-3 phosphate dehydrogenase, glycerol kinase in Plasmodium falciparum, and substrates of *Toxoplasma gondii* adenosine kinase [113-115].

Therefore, we strongly feel that molecular modeling studies, combining docking and molecular dynamic simulations, can improve our understanding of the inhibitor–protein interactions. Furthermore, using these methodologies it is feasible to investigate the structural factors responsible for selectivity of some target enzymes with their inhibitors, and therefore hasten much needed research [100-104].

However, it should be kept in mind that computers are an essential tool in modern medicinal chemistry. Currently, computational approaches and 3D visualization are not used simply to depict pretty pictures of molecules in biological systems; these powerful computational tools allow one to obtain insights on the interaction between enzyme-substrate, mechanism reaction, statistical behavior of molecules and much more, at the molecular level, contributing significantly to the problem solving in biological systems.

ACKNOWLEDGEMENTS

We are grateful to the Brazilian agency FAPEMIG and CNPq for funding part of this work.

REFERENCES

[1]   Renslo AR, Mckerrow JH. Drug discovery and development for neglected parasitic diseases. Nat Chem Biol 2006; 2: 701.

[2]   Landry J, Sutton A, Tafrov ST, Heller RC, Stebbins J, Pillus L, Sternglanz R. The silencing protein SIR2 and its homologs are NAD-dependent protein deacetylases. Proc Natl Acad USA 2000; 97: 5807.

[3]   Schmidt GD, Larry SR. Foundations of Parasitology. 6th ed. McGraw-Hill Higher Education: Dubuque, 2000.

[4]   Quibell M, Watts JP. Tetrahydrofuro[3,2-b]pyrrol-3-one derivatives as inhibitors of cysteine proteinases and their preparation and use in the treatment of diseases. Patent WO 2008007107.

[5]   Quibell M, Watts JP. Tetrahydrofuro[3,2-b]pyrrol-3-ones as cathepsin K inhibitors and their preparation use in the treatment of diseases. Patent WO 2008007112 .

[6]   Quibell M, Watts JP. Tetrahydrofuro [3,2-b]pyrrol-3-ones as cathepsin K inhibitors and their preparation and use in the treatment of diseases. Patent WO 2008007114.

[7]   Quibell M, Watts JP. Furo[3,2-b]pyrrol-3-one derivatives and their use as cysteine proteinase inhibitors and their preparation and pharmaceutical compositions. Patent WO 2008007127.

[8]   Quibell M, Watts JP. Furo[3,2-b]pyrrol-3-one derivatives and their use as cysteinyl proteinase inhibitors and their preparation. Patent WO 2008007109.

[9]   Quibell M, Watts JP. Furo[3,2-b]pyrrol-3-one derivatives as cysteine proteinase inhibitors and their preparation, pharmaceutical compositions and use in the treatment of diseases. Patent WO 2008007130.

[10]   Quibell M, Watts JP. Furo[3, 2-b]pyrrole derivatives as cysteine proteinase inhibitors and their preparation, pharmaceutical compositions and use in the treatment of diseases. Patent WO 2008007103.

[11]   Lecaille F, Weidauer E, Juliano MA, Brömme D, Lalmanach G. Probing cathepsin K activity with a selective substrate spanning its active site. Biochem J 2003; 375: 307.

[12]   Bayly C, Crane SN, Leger S. Cathepsin cysteine protease inhibitors, and use in the treatment of cathepsin-dependent conditions where inhibition of bone resorption is indicated. Patent WO 2005065778.

[13]   Bayly C, Black C, Therien M. reparation of peptides as cathepsin cysteine protease inhibitors. Patent WO 2005066159.

[14]   Black C, Crane S, Oballa R, Robichaud J. 2-(4-Arylpyrazol-3-yl)cyclohexanecarboxamides as cathepsin cysteine protease inhibitors and their preparation, pharmaceutical compositions, and use in the treatment of bone resorption diseases. Patent WO 2007003056.

[15]   Black C, Mellon C, Nicoll-Griffith DA, Oballa R. Papain family cysteine protease inhibitors for the treatment of parasitic diseases. Patent WO 2007012180.

[16]   Bayly C, Black C, Mckay DJ. Preparation of amino acid derivatives as cathepsin inhibitors. Patent WO 2005021487.

[17]   Jeong JU, Yamashita DS. Preparation of 4-amino-3,7-azepinedione amino acid derivatives as protease inhibitors. Patent WO 2003097593.

[18]   Jeong JU, Yamashita, DS. Protease inhibitors for use in treatment of bone loss, excessive cartilage degradation, and parasite infections. Patent WO 2005034838.

[19]   Jeong JU. Preparation of peptidyl 4-amino-3-azepanones as novel cathepsin K inhibitors. Patent WO 2005013909.

[20]   Jeong JU, Yamashita DS. Preparation of 5-amino-1,2-diazepan-4-one amino acid derivs. as protease inhibitors. Patent WO 2003099844.

[21]   Huang TL, Eynde JJV, Mayence A, Bacchi C, Donkor IO, Kode N. Bisbenzamidines and bisbenzamidoximes as parasiticides and their preparation, pharmaceutical compositions and use in the treatment of human african trypanosomiasis. Patent WO 2008070831.

[22]   Bray PG, Barrett MP, Ward SA, de Koning HP. Pentamidine uptake and resistance in pathogenic protozoa: past, present and future. Trends Parasitol 2003; 19: 232.

[23]   Houghton RL, Reed SG, Raychaudhuri S. Fusion proteins comprising *Trypanosoma cruzi* fusion protein TcF and antigenic peptides, their sequences, recombinant production, and use in detecting infections and/or in construction of therapeutic composition. Patent WO 2007056114.

[24]   Reed S. Methods for detection and treatment of *Trypanosoma cruzi* infection using Tc5, Tc48, Tc60 and/or Tc70 antigens. Patent WO 2008057961.

[25]   Wang Y, Benn A, Flinn N, Monk T, Ramjee M, Watts J, Quibell M. *cis*-6-Oxo-hexahydro-2-oxa-1,4-diazapentalene and *cis*-6-oxo-hexahydropyrrolo[3,2-c]pyrazole based scaffolds: design rationale, synthesis and cysteinyl proteinase inhibition. Bioorg Med Chem Lett 2005; 15: 1327.

[26]   Jeong JU, Yamashita DS. Preparation of 1,1,4-1l6-trioxo[1,2]thiazepan-4-ylamides as protease inhibitors. Patent WO 2003104257.

[27]   Amazonas JN, Cosentino-Gomes D, Werneck-Lacerda A, Pinheiro AAS, Lanfredi-Rangel A, Souza W, Meyer-Fernandes JR. *Giardia lamblia*: Characterization of ecto-phosphatase activities. Exp Parasitol 2009; 121: 15.

[28]   Gardner TB, Hill DR. Treatment of Giardiasis. Clin Microbiol Rev 2001; 14: 114.

[29]   Suk DH, Rejman D, Dykstra CC, Pohl R, Pankiewicz KW, Patterson SE. Phosphonoxins: rational design and discovery of a potent nucleotide anti-*Giardia* agent. Bioorg Med Chem Lett 2007; 17: 2811.

[30]   Karr CD, Jarroll EL. Cyst wall synthase: *N*-acetylgalactosaminyltransferase activity is induced to form the novel *N*-acetylgalactosamine polysaccharide in the *Giardia* cyst wall. Microbiol 2004; 150: 1237.

[31]   Dorsch MR, Veal DA. Duncan Adam. Detection of Giardia using probes targeted to its 18S rRNA. Patent WO 200078781.

[32]   Raether W, Hänel H. Nitroheterocyclic drugs with broad spectrum activity. Parasitol Res 2003; 90: S19.

[33]   Cen J, Lu A. Preparation of chiral Ornidazole derivatives as anaerobic bactericides, trichomonacides, and amebicides. Patent WO 2007079653.

[34]   Sharma S, Athar F, Maurya MR, Naqvi F, Azam A. Novel bidentate complexes of Cu(II) derived from 5-nitrofuran-2-carboxaldehyde thiosemicarbazones with antiamoebic activity against *E. histolytica*. Eur J Med Chem 2005; 40: 557.

[35]   França TCC, Pascutti PG, Ramalho TC, Figueroa-Villar JD. A three-dimensional structure of *Plasmodium falciparum* serine hydroxymethyltransferase in complex with glycine and 5-formyl- tetrahydrofolate. Homology modeling and molecular dynamics. Biophys Chem 2005; 115: 1.

[36]   Clark WM, Badham NF, Dai Q, Eldridge AM, Matsuhashi H. A preparation of derivatives of benzofuran-2-carboxylic acid amide, useful as cysteine protease inhibitors. Patent WO 2005069981.

[37]   Jeong JU, Yamashita DS. Preparation of 4-amino-3,7-azepinedione amino acid derivatives as protease inhibitors. Patent WO 2003097593.

[38]   Xie R, Yamashita DS. Preparation of 4-amino-azepan-3-one derivatives as protease inhibitors. Patent WO 2002092563.

[39]   Cummings MD, Marquis RWJr, Ru Y, Thompson SK, Veber DF, Yamashita DS. Preparation of 4-aminoazepan-3-one derivatives as protease inhibitors. Patent WO 2001089451.

[40]   Marquis RWJr, Veber DF, Yamashita DS. Preparation of aminoazepinones as protease inhibitors. Patent WO 2003053331.

[41]   Marquis RWJr, Ru Y, Veber DF, Cummings MD, Thompson SK, Yamashita DS. Patent 2000195911.

[42]   Cummings MD, Marquis RWJr, Ru Y, Thompson SK, Veber DF, Yamashita DS. Patent WO 2000178734.

[43]   Cummings MD, Marquis RWJr, Ru Y, Thompson SK, Veber DF, Yamashita DS. Patent WO 2000170232.

[44]   Ackerman S, Fischer IM. Pantent WO 2003097563.

[45]   Tew DG, Thompson SK, Veber DF. Patent WO 2000217924.

[46]   Lin AJ, Guan J, Kyle DE, Milhous WK. Tricyclic N-(aminoalkyl)-substituted phenothiazines, iminodibenzyls, iminostilbenes, and diphenylamines, active as chemosensitizing agents against chloroquine-resistant *Plasmodium falciparum*, and methods of making and using thereof. Patent WO 2002089810.

[47]   Mueller E, Scheiwe MW. Pharmaceutical combination of artesunate and mefloquine for therapy of malaria. Patent WO 2003075927.

[48]   Hannus M, Martin C, Mota, MM, Prudencio M, Rodrigues CD. Use of inhibitors of scavenger receptor class proteins for the treatment of infectious diseases. Patent WO 2007101710.

[49]   Tripathi R, Puri SK, Srivastava JS, Singh S, Asthana OP, Dwivedi AK. Synergistic combination kit of α,β-arteether, sulfadoxin and pyrimethamine for treatment of severe/multi-drug resistant cerebral malaria. Patent WO 2006070393.

[50]   El Abbadi MS. A new use for scorpion venom as product for malaria treatment and prophylaxis. Patent WO 2004080471.

[51]   Mayer G, Miller LH. *Plasmodium falciparum* erythrocyte binding protein BAEBL for use as vaccine against malarial Plasmodium parasite. Patent WO 2002078603.

[52]   Jensen ATR, Hviid L, Jorgensen L, Lavsten T, Magistrado P, Nielsen MA, Salanti A, Staalso T, Theander TG. Polynucleotide and polypeptide sequences for *Plasmodium falciparum* genes var4-var6 encoding EMP-1 (erythrocyte membrane protein 1) antigens and use thereof for malaria vaccines and diagnosis. Patent WO 2005063804.

[53]   Claudianos C, Crompton TK, Dessens JT, Sinden RE, Trueman HE. Novel modular secreted proteins PfSR, PbSR, PySR and PkSR and antibodies for vaccine and diagnosis. Patent WO 2003004524.

[54]   Schofield L. Immunogenic compositions comprising inositolglycan domain of Plasmodium-derived glycophosphoinositide for diagnosis and therapy against malaria. Patent WO 2004011026.

[55]   Fuertes LL, Jiménez MT. DNA-expression construct for treatment of infections with leishmaniasis. Patent WO 2003031470.

[56]   Duran JJT, Duran ST, Sanchez-Brunete JA, Fernandez FB, Dea-Ayuela MA, Rama-Iniguez S, Rodríguez JMA. Microspheres containing amphotericin B. Patent WO 2004062645.

[57]   Alain-Jacques B. Use of diminazene diaceturate and/or pentamidine for the treatment of malaria. Patent WO 200030631.

[58]   Alain-Jacques B. Patent WO 200030611.

[59]   Bourdichon AJ. Pharmaceutical preparation for treating tropical parasitic diseases containing diminazene diaceturate and/or pentamidine and chloroquine. Patent WO 2004105772.

[60]   Ouaissi A, Sereno D, Vergnes B. Pharmaceutical compositions comprising niacin or niacinamide for the treatment of leishmaniasis. Patent WO 2006008082.

[61]   Jackson MD, Schmidt MT, Oppenheimer NJ, Denu JM. Mechanism of nicotinamide inhibition and transglycosidation by Sir2 histone/protein deacetylases. J Biol Chem 2003; 278: 50985.

[62]   Lemesre JL, Cavaleyra M, Sereno D, Holzmuller P. Antigens from promastigotes or amastigotes of Leishmania for use in Leishmaniasis. Patent WO 2005051989.

[63]   Cordeira da Silva A, Ouaissi A, Sereno D, Vergnes B. Protozoan stocks of attenuated virulence, and their use. Patent WO 2005044844.

[64]   Srebnik M, Takrouri K. Preparation of novel amine-borane compounds and their uses in radioimaging and radiotherapy. Patent WO 2007032005.

[65]   Domb AJ, Polacheck I, Soskolni M, Golenser J. Conjugates of therapeutically active compounds. Patent WO 2007034495.

[66]   Yli-Kauhaluoma J, Alakurtti S, Minkkinen J, Sarcerdoti-Sierra N, Jaffe CL, Heiska T. Betulin derived compounds useful as antiprotozoal agents. Patent WO 2007141391.

[67]   Sas B, Van Hemel J, Vandenkerckhove J, Peys E, Van Der Eycken J, Ruttens B, Van Hoof S, Sas B, Hemel JV, Eycken JVD. Bicyclic carbohydrates as antiprotozoal bioactive for the treatment of infections caused by parasites. Patent WO 2004062590.

[68]   Manna L, Vitale F, Reale S, Picillo E, Neglia G, Vescio F, Gravino AE. Study of efficacy of miltefosine and allopurinol in dogs with leishmaniosis. Vet J 2009; 182: 441. Real-time PCR assay in *Leishmania*-infected dogs treated with meglumine antimoniate and allopurinol. Vet J 2008; 177: 279.

[69]   Bhattacharya G, Salem MM, Werbovetz KA. Antileishmanial dinitroaniline sulfonamides with activity against parasite tubulin. Bioorg Med Chem Lett 2002; 12: 2395.

[70]   Werbovetz K, Bhattacharya G, Sackett D, Salem MM. Antileishmanial dinitroaniline sulfonamides with activity against parasite tubulin. Patent WO 2003090678.

[71]   Palatnik de Souza CB, Chequer Bou-Habib EM. Composition comprising Leishmania promastigotes or Leishmania amastigotes fractions for preparation of Leishmaniasis transmission blocking vaccines. Patent WO 2006122382.

[72]   Fernandes APSM, Abrantes CF, Coelho EAF, Gazzinelli R T. Vaccine composition and immunization method. Patent WO 2008009088.

[73]   Chang K, Kolli BK, Sassa S. Patent WO 2006105044.

[74]   Tan Y, Yang Z, Sun X, Li S, Han Q, Xu M. Methioninase, alone or in combination with other agents, for the treatment of *Pneumocystis carinii* infection. Patent WO 2005000239.

[75]   Boykin DW, Rahmathullah MS, Tidwell RR, Hall JE. Prodrugs for antimicrobial amidines. Patent WO 200103685.

[76]   Vale N, Collins MS, Gut J, Ferraz R, Rosenthal PJ, Cushion MT, Moreira R, Gomes P. Anti-*Pneumocystis carinii* and antiplasmodial activities of primaquine-derived imidazolidin-4-ones. Bioorg Med Chem Lett 2008; 18: 485.

[77]   da Cunha EFF, Ramalho TC, Maia ER, De Alencastro RB. The search for new DHFR inhibitors: a review of patents. Expert Opin Ther Patents 2005; 15: 967.

[78]   Forsch RA, Queener SF, Rosowsky, A. Preliminary *in vitro* studies on two potent, water-soluble trimethoprim analogues with exceptional species selectivity against dihydrofolate reductase from *Pneumocystis carinii* and *Mycobacterium avium*. Bioorg Med Chem Lett 2004; 14: 1811.

[79]   Rosowsky A, Chen H, Fu H, Queener SF. Synthesis of new 2,4-Diaminopyrido[2,3-d]pyrimidine and 2,4-Diaminopyrrolo[2,3-d]pyrimidine inhibitors of *Pneumocystis carinii*, Toxoplasma gondii, and *Mycobacterium avium* dihydrofolate reductase. Bioorg Med Chem 2003; 11: 59.

[80]   Rosowsky A, Forsch RA. Preparation of 2,4-Diamino-5-[5-substituted-benzyl]pyrimidines and 2,4-diamino-6-[5-substituted-benzyl] quinazolines as DHFR inhibitors. Patent WO 2004082613.

[81]   Rosowsky A, Forsch RA. Preparation of benzylpyrimidinamines as dihydrofolate reductase inhibitors. Patent WO 2003043979.

[82]   Chan DCM, Laughton CA, Queener SF, Stevens MFG. Structural Studies on Bioactive Compounds. Part 36: Design, Synthesis and Biological Evaluation of Pyrimethamine-Based Antifolates Against *Pneumocystis carinii*. Bioorg Med Chem 2002; 10: 3001.

[83]   Graffner-Norberg M, Kolmodin K, Åqvist J, Queener SF, Hallberg A. Design, synthesis, and computational affinity prediction of ester soft drugs as inhibitors of dihydrofolate reductase from *Pneumocystis carinii*. Europ J of Pharm Sci 2004; 22: 43.

[84]   Eynde JJV, Mayence A, Huang TL, Collins MS, Rebholz S, Walzer PD, Cushion MT. Novel bisbenzamidines as potential drug candidates for the treatment of *Pneumocystis carinii* pneumonia. Bioorg Med Chem Lett 2004; 14: 4545.

[85]   Walzer PD, Cushion MT, Mayence A, Huang TL, Eynde JJV. Bisbenzamidines for the treatment of *Pneumocystis pneumonia* or other infection. Patent WO 2006021833.

[86]   Schumacher MA, Scott DM, Mathews II, Ealick SE, Roos DS. Crystal Structures of *Toxoplasma gondii* adenosine kinase Reveal a Novel Catalytic Mechanism and Prodrug Binding. J Mol Biol 2000; 298: 875.

[87]   Bellosta V, Bigot A, Canova S, Cossy J, Maillet P, Mignani S, Minoux H. New analog derivatives of herbimycin A, compositions containing them and use. Patent WO 2009004146.

[88]   Krug EC, Man JJ, Berens RL. Purine metabolism in *Toxoplasma gondii*. J Biol Chem 1989; 264: 10601.

[89]   el Kouni MH, Guarcello V, Al Safarjalani ON, Naguib FNM. Metabolism and Selective Toxicity of 6-Nitrobenzylthioinosine in *Toxoplasma gondii*. Antimicro Agents Chem 1999; 43: 2437.

[90]   Yadav V, Chu CK, Rais RH, Al Safarjalani ON, Guarcello VG, Naguib FNM, El Kouni MH. Synthesis, Biological Activity and Molecular Modeling of 6-Benzylthioinosine Analogues as Subversive Substrates of *Toxoplasma gondii* Adenosine Kinase. J Med Chem 2004; 47: 1987.

[91]   Rais RH, Al Safarjalani ON, Yadav V, Guarcello N, Kirk M, Chu CK, Naguib FNM, el Kouni MH. 6-Benzylthionosine analogues as subversive substrate of *Toxoplasma gondii* adenosine kinase: activities and selective toxicities. Biochem Pharmacol 2005; 69: 1409.

[92]   Kim Yah, Sharon A, Chu CK, Rais RH, Al Safarjalania ON, Naguib FNM, El Kouni MH. Synthesis, biological evaluation and molecular modeling studies of N^6-benzyladenosine analogues as potential anti-toxoplasma agents. Biochem Pharmacol 2007; 73: 1558.

[93]   Hill DE, Zarlenga DS, Coss C, Dubey JP. *Toxoplasma gondii* oocyst-specific antigen DGP5p (dense granule protein 6 precursor), recombinant protein and antibody production and diagnostic uses. Patent WO 2008097812.

[94]   Biemans R, Bollen A, De Neve J, Haumont M, Jacquet A. Novel vaccine composition. Patent WO 2001043768.

[95]   Jacques A, Coulon L, De Nève J, Daminet V, Haumont M, Garcia L, Bollen A, Jurado M, Biemans R. The surface antigen SAG3 mediates the attachment of *Toxoplasma gondii* to cell-surface proteoglycans. *Mol* Biochem Parasitol 2001; 116: 35.

[96]   Himeno K, Ishii K. Chimeric gene comprising pathogen-derived antigen gene and ubiquitin gene as vaccine against infectious disease. Patent WO 2004067040.

[97]   da Cunha EFF, Sippl W, Ramalho CT, Antunes OAC, Alencastro RB, Albuquerque MG. 3D-QSAR CoMFA/CoMSIA models based on theoretical active conformers of HOE/BAY-793 analogs derived from HIV-1 protease inhibitor complexes. Eur J Med Chem 2009; 44: 4344.

[98]   Ramalho TC, Caetano MS, da Cunha EFF. Construction and assessment of reaction models of class I EPSP synthase: Molecular docking and density functional theoretical calculations. J Biomol Struct Dyn 2009; 27: 195.

[99]   Ramalho TC, da Cunha EFF, de Alencastro RB. A density functional study on the complexation of ethambutol with divalent cations. J Mol Struct (Theochem) 2004; 676: 149.

[100]   da Cunha EFF, Antunes OAC, Albuquerque MG, de Alencastro RB. n-t-boc-amino acid esters of isomannide - potential inhibitors of serine proteases. Amino Acids 2004; 27: 153.

[101]   da Cunha EFF, Albuquerque MG, Antunes OAC, de Alencastro RB. Pseudo-peptides derived from isomannide as potential inhibitors of serine proteases. Amino Acids 2005; 28: 413.

[102]   da Cunha EFF, Ramalho TC, de Alencastro RB. Theoretical study of adiabatic and vertical electron affinity of radiosensitizers in solution. part 2: analogues of tirapazamine. J Theor Comput Chem 2004; 3: 1.

[103]   Franca TCC, Pascutti PG, Ramalho TC. A three-dimensional structure of *Plasmodium falciparum* serine hydroxymethyltransferase in complex with glycine and 5-formyl- tetrahydrofolate. Homology modeling and molecular dynamics. Biophys Chem 2005; 115: 1.

[104]   Franca TCC, Wilter A, Ramalho TC. Molecular dynamics of the interaction of *Plasmodium falciparum* and human serine hydroxymethyltransferase with 5-formyl-6-hydrofolic acid analogues: design of new potential antimalarials. J Braz Chem Soc 2006; 17: 1383.

[105]   Mamane, H. Impact of particles on UV disinfection of water and wastewater effluents: a review. Rev Chem Eng 2008; 24: 67.

[106]   Hijnen WAM, Beerendonk EF, Medema GJ. Inactivation of UV radiation for viruses, bacteria, and protozoan (oo)cysts in water: a review. Water Res 2006; 40: 3.

[107]   Anthony JP, Fyfe L, Smith H. Plant active components - a resource for antiparasitic agents? Trends Parasitol 2005; 21: 462.

[108]   Fyfe PK, Oza SL, Fairlamb AH, Hunter WN. *Leishmania trypanothione* synthetase-amidase structure reveals a basis for regulation of conflicting synthetic and hydrolytic activities. J Biol Chem 2008; 283: 17672.

[109]   Singh S, Malik BK, Sharma DK. Molecular modeling and docking analysis of *Entamoeba histolytica* glyceraldehyde-3 phosphate dehydrogenase, a potential target enzyme for anti-protozoal drug development. Chem Biol Drug Des 2008; 71: 554.

[110] Schnick C, Polley SD, Fivelman QL, Ranford-Cartwright LC, Wilkinson SR, Brannigan JA, Wilkinson AJ, Baker DA. Structure and non-essential function of glycerol kinase in *Plasmodium falciparum* blood stages. Mol Microbiol 2009; 71: 533.

[111] Yadav V, Chu CK, Rais RH, Al Safarjalani ON, Guarcello VG, Naguib FNM, El Kouni MH. Synthesis, biological activity and molecular modeling of 6-benzylthioinosine analogues as subversive substrates of *Toxoplasma Gondii* adenosine kinase. J Med Chem 2004; 47: 1987.

[112] Souza TCS, Josa D, Ramalho TC, Caetano MS, da Cunha EFF. Molecular modelling of *Mycobacterium tuberculosis* acetolactate synthase catalytic subunit. Mol Sim 2008; 34: 707.

[113] Kundu S, Roy D. Computational Study of Glyceraldehyde-3-phosphate Dehydrogenase of *Entamoeba histolytica*: Implications for Structure-Based Drug Design. J Biomol Struct Dyn 2007; 25: 25.

[114] Josa D, da Cunha EFF, Ramalho TC, Souza TCS, Caetano MS. Hypothesis Paper: Homology Modeling of Wild-type, D516V, and H526L *Mycobacterium tuberculosis* RNA Polymerase and Their Molecular Docking Study with Inhibitors. J Biom Struct Dyn 2008; 25: 373.

[115] Adane L, Bharatam PV. Modelling and informatics in the analysis of *P. falciparum* DHFR enzyme inhibitors. Curr Med Chem 2008; 15: 1552.

CHAPTER 2

Quantitative Structure Activity Analysis of Leishmanicidal Compounds

Alicia Ponte-Sucre*, Emilia Díaz and Maritza Padrón-Nieves

Laboratory of Molecular Physiology, Institute of Experimental Medicine, Faculty of Medicine, Universidad Central de Venezuela, Caracas, Venezuela

Abstract: Several techniques have been used to study the mechanisms by which *receptors* recognize ligands, one of them being quantitative structure activity relationship analysis of compounds. This method facilitates the description of molecular details involved in drug recognition by molecular *receptors*, as well as the molecular mechanism involved. This technique constitutes an essential tool to investigate chemical, electronic, and structural features affecting the leishmanicidal activity of compounds. However, few studies address this topic in *Leishmania*. Efforts should be made to stimulate research in this area and thus describe the characteristics of leishmanicidal drugs and their interaction with molecular *receptors*. The present chapter summarizes progress made recently in quantitative structure activity relationship studies of leishmanicidal compounds in experimental and *in vivo* scenarios. The review highlights possible critical spots in drug design and discusses the potential activity of compounds against different strains of the parasite as a way to optimize the treatment of leishmaniasis.

Keywords: QSAR, leishmaniasis, Lipinski's rule.

1. INTRODUCTION

Arsenites and antimonials were amongst the first synthetic drugs used against infectious diseases at the beginning of the twentieth century, and still organic derivatives of the same heavy metals remain as the drugs of choice for the treatment of diseases caused by Trypanosomatids, including *Leishmania*. Simon Croft [1] wrote this statement in 1999 and unfortunately it still reveals, at least partially, the *state of the art* for chemotherapy against leishmaniasis. Indeed, Glucantime® and Pentostan® remain as the drugs of choice to fight against leishmaniasis in most endemic countries.

These drugs have several limitations like the need of parenteral administration, variable efficacy and high price and toxicity. However, these compounds are essential for control and treatment of the disease since alternative prevention measures, such as pesticide-impregnated bed-nets, fail or prove impractical and no new and efficient drugs exist [2].

Additionally, therapeutic failure attributed to altered drug pharmacokinetics, re-infection, or immunologic compromise of the host is a problem frequently observed in endemic areas [3]. Development of drug resistance further complicates the panorama of the disease, and strong indicators suggest that this may play an important role in therapeutic failure [4]. For this reason, there is an urgent need for markers of chemo-resistance that are easy to monitor in the laboratory and helpful to predict therapeutic prognosis [5-7], and for compounds less prone to induce drug resistance. Of note, despite intensive attempts, there are no effective vaccines for the prevention of leishmaniasis and this will not improve in the near future [8].

For a few years now, identification of the genome sequences of trypanosomatid parasites is either completed or underway (www.genedb.org). Extensive work is being done to characterize the biological function of the encoded proteins and to evaluate their value as antiparasitic drug targets [4]. However, in spite of the efforts made, very few of the identified pharmacophores have successfully entered the clinical pipeline [2, 9]. This means that finding lead drug-like compounds interesting enough to warrant an analysis of their biological activity must be a primary goal to uncover an appropriate cure for leishmaniasis [2].

*Address correspondence to Alicia Ponte-Sucre: Laboratory of Molecular Physiology, Institute of Experimental Medicine, Faculty of Medicine, Universidad Central de Venezuela, Caracas, Venezuela. Phone: +58 212 605 3665. Telefax: +58 212 693 4351; E-mail: aiponte@gmail.com

Teodorico C. Ramalho, Matheus P. Freitas and Elaine F. F. da Cunha (Eds)

The cost of experimentally testing a compound relative to scoring it computationally is very high, and the methods are slow and tedious. In fact, corporate-sized databases, often exceeding 1 million entries, cannot be totally screened at a reasonable cost even by current technologies [10]. For this reason, computer-based methods are useful to suggest which subsets of compounds are most likely to be active. The computational study starts with an interesting biologically active molecule, *e.g.* an already screened inhibitor compound, and a search for the structural requirements responsible for the potency of the compound in a database of chemical structures. The ultimate goal is to find an active molecule different enough from the starting compound(s) that could be considered as a new class of therapeutic agent [10].

2. HOW TO DEFINE STRUCTURE ACTIVITY RELATIONSHIP AND QUANTITATIVE STRUCTURE ACTIVITY RELATIONSHIP

In the last few years, the landscape of drug discovery and development for new anti-parasitic drugs has improved thanks to the financial support from not-for-profit organizations and the involvement of public-private partnerships [11].

For example, catalyzed by the Special Programme for Research and Training in Tropical Diseases (TDR) and with the collaboration of pharmaceutical companies, the search for new leads and anti-parasitic drugs has gained thousands of molecules required for the development of new medicaments against neglected diseases including leishmaniasis [11].

Computational methods and tools have been fundamental and are essential to screen the chemical structure databases and to identify leads to work with. Especially two of them focus on modeling the biological *receptors* and their binding drugs. The main goal of these methods is to optimize drug binding and help in the design of more potent or precise drugs. These methods are 'virtual screening', which is a computational technique that analyzes *in silico* libraries of chemical structures and identifies those most likely to bind to a drug target, and computer-aided design of molecules based on desired properties [12, 13]. This latter method determines which compounds match a specified set of (target) properties; it has a large potential since itpermits designing all kinds of chemical, bio-chemical and material products [13].

For a computational method to be successful, information must be available on theactivity of the compound, and how such activity relates to the molecule's structure, the so called Structure Activity Relationship (SAR). This method is also called structure-property relationship and can be defined as the process by which the chemical structure of molecules is quantitatively correlated with biological activity or chemical reactivity [14]. Although it may happen that different molecules use dissimilar binding modes or trigger diverse mechanisms, the basic assumption for SAR is that analogous molecules may interact with the same *receptor* and might have similar activities. Thus, the SAR analysis is a fundamental requirement for the manufacture of a molecule with specific desired characteristics.

On the other hand, the term Quantitative Structure Activity Relationship (QSAR) (see Fig. **1**) refers to predictive models derived from the application of statistical tools. QSAR correlates biological activity (including desirable therapeutic effect and undesirable side effects) of chemicals, (drugs/toxicants/environmental pollutants) with descriptors representative of molecular structure and/or properties [14]. Success of any QSAR model depends on accuracy of the input data, selection of appropriate descriptors and statistical tools and most importantly, validation of the developed model.

Drugs or ligands vary from the simple to the complex, but the *receptors* they bind to are extremely complex. Predicting the properties of a molecule based on structure is a non-linear and non-intuitive process that requires complex molecular modeling. As a consequence increasingly complicated molecules are designed every day [15]. Drug development is further complicated by the fact that it must function in a bio-molecular system composed of the biological *receptor* and the ligand, and inside a complex organism.

As already mentioned, the central tenet of SAR is that compounds with similar structure act at the same site and with the same mechanism. Unfortunately, many pharmaceutical computer-aided molecular design systems predict the properties of either the ligands that operate on the *receptor* or the *receptor* itself, but usually not both. To overcome this situation, various organizations and web sites foster the validation of anti-infective drugs (http://www.dndi.org/) and of cellular targets (http://www.TDRtargets.org/), especially on diseases that are not a priority for the pharmaceutical industry as is the case for leishmaniasis.

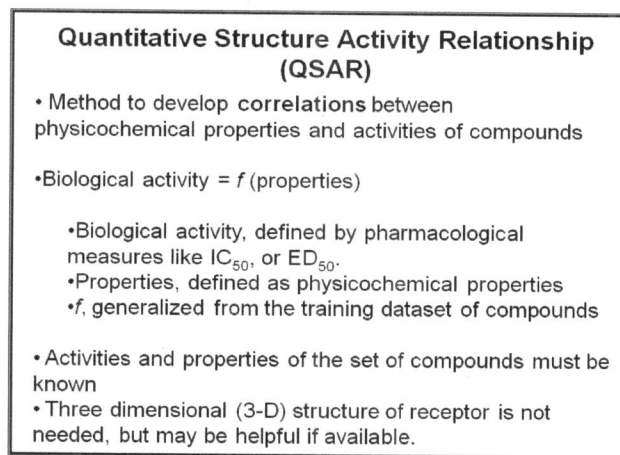

Quantitative Structure Activity Relationship (QSAR)

• Method to develop **correlations** between physicochemical properties and activities of compounds

•Biological activity = f (properties)

   •Biological activity, defined by pharmacological measures like IC_{50}, or ED_{50}.
   •Properties, defined as physicochemical properties
   •f, generalized from the training dataset of compounds

• Activities and properties of the set of compounds must be known
• Three dimensional (3-D) structure of receptor is not needed, but may be helpful if available.

Figure 1: Definition of Quantitative Structure Activity Relationship. For description of the terms, see main text.

3. METHODS AND MODELS USED IN QSAR

The main goal of SAR and QSAR is to uncover correlations between physicochemical properties and molecular function of a group of compounds. To perform this task computational methods are needed and Molecular Descriptors (MD) should be described. According to the web page http://www.molecular-descriptors.eu, a MD is the result of a logic and mathematical procedure, which transforms chemical information encoded within the symbolic representation of a molecule, into a useful number or result of a standardized experiment. The characterization of MD constitutes a scientific field in it self; through it, scientists design strategies to define a feature. In fact, around 1600 MD have been listed and may be used to solve specific situations related to drug discovery and drug-*receptor* interactions [16]. This means that descriptor analysis is an extremely intricate procedure that adds complexity to the models that can be developed for SAR and QSAR.

Web pages (http://www.qsarworld.com/) have been designed to cover all the questions that may arise regarding the complexity of the methods and of the descriptors, and readers are invited to forward their search according to their individual interests. Here in we will remain as simple as possible and will only describe the minimal steps for SAR or QSAR analysis.

In the initial steps of QSAR (see Fig. **2**) a group of compounds, all of which interact in a similar way with the same site in a molecule (*receptor*), must be selected. Afterwards the physicochemical characteristics or MD for each of these compounds must be calculated. Then, the compounds are separated in two sub groups, one to be used for **training** - that is a dataset with known biological values, helpful for designing a model to predict the biological effect of additional molecules- and a second dataset to be used for **testing** in the biological system [17]. This means that a biological attribute has to be measured. These attributes include the molar concentration of an inhibitor that modifies 50 percent the response (IC_{50}), or the amount of a drug that is therapeutic in 50 percent of the persons or animals in which it is tested (ED_{50}) [18].

The three dimensional (3-D) structure of the *receptor* is not an essential parameter; however, the information derived from it may be helpful to perform an integral analysis of the drug-receptor interaction. Once the data are collected, a model is constructed searching for a correlation between the properties and the biological activity. For doing so, regression analysis and statistical methods are chosen [19, 20].

Various automated prediction methods have been developed to annotate biological functions of molecules: molecular docking, molecular packing, Monte Carlo simulated annealing approach, pharmacophore modeling, protein cleavage site prediction, signal peptide prediction and structural bioinformatics, among others. We will not focus on their description but readers are invited to examine the following exhaustive good reviews [19, 21-24].

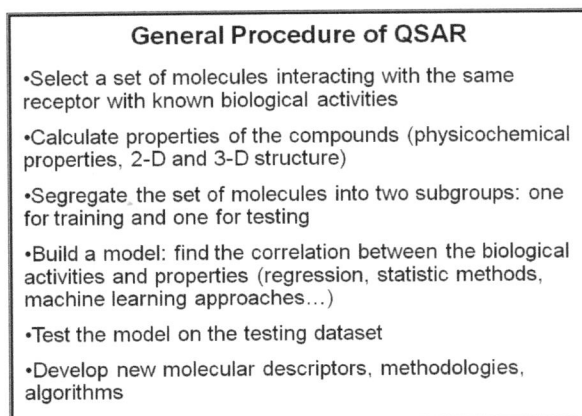

General Procedure of QSAR

•Select a set of molecules interacting with the same receptor with known biological activities

•Calculate properties of the compounds (physicochemical properties, 2-D and 3-D structure)

•Segregate the set of molecules into two subgroups: one for training and one for testing

•Build a model: find the correlation between the biological activities and properties (regression, statistic methods, machine learning approaches...)

•Test the model on the testing dataset

•Develop new molecular descriptors, methodologies, algorithms

Figure 2: Minimal steps for QSAR.

These bioinformatics approaches are based on sequence alignment procedures. However, such methods perform poorly if there is low sequence homology between the examined sequence and the template sequences deposited in the databases. Alternative methods, such as the alignment-free Machine Learning methods have therefore been developed and are useful for predicting the function of the protein and explore its molecular diversity based on structural parameters independently of sequence-sequence similarity [18, 25]. Additional alternatives include graphic methods that provide useful insights into the biological function of proteins [18]. These graphical methods are useful for QSAR studies [17, 26].

On the other hand, in 1977 Chou published his key work about the molecular dynamics of bio-macromolecules [27]. The results obtained using this technique suggests that the low-frequency collective motions that exist in DNA and proteins hold a huge potential for revealing the dynamic mechanisms and the function of drugs in biological systems [18]. Indeed, methods like nuclear magnetic resonance support the certainty of this hypothesis by [28-30] and indicate that an understanding of the molecular interaction of drugs with their *receptors* must consider their static structures and their dynamical interactions. Thus, Molecular Dynamics has become the foremost computational technique to investigate structure and function of peptides (*receptors*) [18] and to predict activity against microbial species, growth inhibition of cancer lines and, in general mechanisms of action of any drug or pharmacophore.

4. LIPINSKI'S RULE OF 5

We have already mentioned that QSAR is a method that correlates the biological function (or activity) of a series of compounds with physicochemical and/or structural properties. For that reason it can be defined in terms of these simple equations:

$$\text{Activity} = f(\text{physicochemical properties}),$$

or,

$$\text{Activity} = f(\text{structural properties})$$

From these equations, it is easy to find out which physical and chemical characteristics should be fulfilled by the compounds to determine their activity. Lipinski's rule of five, described initially in 1997, lists which regulations the physicochemical properties should follow (see Fig. **3**) [31].

Most medically important drugs are relatively small and lipophilic molecules. Based on this observation, Lipinski's rule of 5 evaluates if a biologically active compound holds properties that would make it an efficient, orally active drug in humans. This has been called "drug likeness" of the compound; of note, this concept does not predict its pharmacological activity [31].

Lipinski's rule of 5[1]

An orally active drug has no more than one violation of the following criteria[2]:
- Not more than 5 hydrogen bond donors (nitrogen or oxygen atoms with one or more hydrogen atoms)
- Not more than 10 hydrogen bond acceptors (nitrogen or oxygen atoms)
- A molecular weight under 500 Daltons (160 to 480[3])
- An octanol-water partition coefficient log P of less than 5 (-0.4 to +5.6[3])
  - Molar refractivity from 40 to 130[3]
  - Number of atoms from 20 to 70[3]

- [1] This rule and its validation is a hope for a better starting point in compound discovery to save time and cost.
- [2] All numbers are multiples of five, which is the origin of the rule's name.
- [3] Lipinski's profiling tool for drug likeness has led to further investigations to extend profiling tools to lead-like properties of compounds Ghose et al., 1999 [76].

Figure 3: Minimal requirements for designing an orally active drug. Lipinski's rule of 5.

The key to drug discovery is to identify chemical structures with good inhibitory effects on specific targets, without or with minimal toxicity (non-specific activity). Modification of the molecular structure of compounds results in drugs with higher molecular weight, more rings and bonds with free rotation, and a higher lipophilicity. These features do not guaranty a stepwise increase in activity and selectivity, and in drug-like properties [31]. For this reason Lipinski's rule of 5 must be strictly followed during the active lead structure optimization and is fundamental for drug development and consequently for QSAR.

To help scientists to design the best compounds, a web page has been developed (http://www.chemaxon.com/marvin/sketch/index.jsp) to calculate the relevant physicochemical parameters that should be followed to design a drug according to Lipinski's rule of 5.

5. ADVANTAGES AND DISADVANTAGES OF QSAR

Drug discovery both in the pharmaceutical industry and in the academia is a very challenging process; therefore, the preparation of virtual libraries of compounds for high through-screening is often desired [10, 18]. The design and preparation of such libraries is time consuming and challenging and the characterization of drug-like properties of compounds within the libraries must correlate two-dimensional (2-D) and three-dimensional (3-D) descriptors with drug action; otherwise these properties will be of limited use [18].

As already mentioned, QSARs, the basis of rational drug design, assume a unique mechanism for all compounds that belong to a data set, and are therefore fundamental for developing predictive relationships to improve the profile of virtual libraries. As the main goal of SAR and QSAR is to uncover correlations between physicochemical properties and molecular function of a group of compounds, at least two advantages derive from this quantification: 1) An understanding of the relationship that exists between structure and activity. 2) An understanding of the effect that a given molecular structure may have on activity. This information is extremely helpful for predictions leading to the synthesis of novel and effective analogues. Last but not least, QSAR reinforce the identification of active molecules for a selected target when compared with random selection or other traditional methods (http://www.qsarworld.com).

However, systems that depend on the structure of a biological *receptor* face extreme difficulties: 1) Membrane-bound *receptors* are elusive for the determination of their 3-D structure, either by nuclear magnetic resonance (NMR) or X-ray crystallography. 2) The data collections describe only a few thousand receptors; these structures are artificially modeled in water or organic solvents, far away from the weak salt solutions where the *receptors* function. 3) *Receptors* suffer conformational changes when they accept the ligand; this implies an even more complicated binding kinetics, and makes the unbound *receptor* not necessarily relevant for QSAR. 4) The determination of 3-D structures of ligand bound *receptors* adds extra complexity to the procedure. 5) QSAR techniques are accurate on a small scale when determining the properties of specific regions, such as the active site in an enzyme; however, they do not produce an accurate global description of the whole molecule. 6) False correlations may arise because biological data are subject to considerable experimental error (noisy data). 7) If the training dataset is not large enough, the data collected may not reflect the complete molecular space (http://www.qsarworld.com).

We can thus conclude that although the results can seldom be used to predict the most likely compounds with best activity, the information obtained through 3-D QSAR has led to better characterization of molecules and compounds and improved calculation of their properties. The main drawback is that QSAR methods are unable to model steric interactions accurately, particularly when these interactions involve large regions of the molecular surface [32]. This is especially relevant when functional areas of the drug that participate in binding easily identified with QSAR, heavily contribute in determining the molecule's activity and defining the "pharmacophore" areas of the molecule [33].

6. QSAR IN *LEISHMANIA*

The first initiatives to perform SAR in *Leishmania* used antifolates as leishmanicidal drugs, and measured the capacity of triazine derivatives to inhibit cell growth by targeting the enzyme dihydrofolate reductase [34]. Linear free energy descriptors through Molecular Shape Analysis (MSA) were used to develop the QSARs. The MSA-QSAR results suggested that the activity of the antifolates was independent of the conformation assigned to the large flexible substituents [34, 35]. Since then, different types of compounds have been screened against *Leishmania*. The structure of the isolated (natural) and designed (synthetic) molecules include acridines, phenothiazines, purines, pyrimidines and quinolines that contain nitrogen heterocycles [36] (see Fig. **4**), although certain amino acids, amides and esters, aniline derivatives, flavonoids and quinones have also been reported to be leishmanicidal [37]. In spite of many efforts done, few substances have reached the pre-clinic and the clinic stage. Here in we will present a summary of the leishmanicidal compounds SAR analyses published until now. In many cases the mechanism of action of the compounds is unknown; we will address this issue when the information is available.

Lead compounds used for QSAR in *Leishmania*	
Arylisoquinolines	[54, 55]
Aurones	[77]
Azatrypanthrin derivatives	[56]
Bisphosphonate analogs	[69, 70]
Chalcones	[38, 40, 41, 42]
Dinitroaniline sulfonamide derivatives	[62, 63]
Folate Antagonists	[34, 35]
Lactones	[48]
Lignans benzoflurans	[53, 78]
Miltefosine analogs	[68]
Nifuroxazide	[58]
Phloroglucinol	[47]
Pyrazol pyridines	[59]
Sesquiterpenes	[45, 46]
Trypanothione antagonists	[60]

Figure 4: Examples of compounds used for QSAR in *Leishmania*. Brackets, main references.

7. NATURAL PRODUCTS AND DERIVATIVES

Lidochalcone A, isolated from Chinese licorice, efficiently inhibits the proliferation of *Leishmania* (*L.*) (*L. major* and *L. donovani*). It does so by interfering with the function of parasite's mitochondria [38]. This compound has a dual -maybe unwanted- effect since it also inhibits the phytohemmaglutinin A-induced proliferation of human lymphocytes [39]. However, changes in the substitution pattern of lidochalcone A modulate the activity against parasites and macrophages in an opposite way, making it possible to prepare chalcones with high selectivity against *Leishmania* [38].

Therefore, various groups have synthesized numerous chalcones and their derivatives, based on statistical and rational design methods, and have performed QSAR analysis [38-42]. For example, the group of Nielsen [38] developed a method to correlate the biological activity of the synthesized chalcones to physicochemical parameters like molar refractivity, lipophilicity and changes in the type of aromatic reactant substitutes (Hammet electronic effect) [43]. They concluded that positions 4 and 4' in the molecule are crucial for the biological activity of the derivatives and that steric interactions between the chalcones and the target molecules are of major importance for the potencies of the compounds; additionally, these results suggest that substitutions in positions 4', 2, 3 and 4 of the parent compound decrease the toxicity towards lymphocytes [38]. Unfortunately, these substitutions interfere at the same time with the solubility of the molecules and decrease the activities against *Leishmania* [38]. A fundamental conclusion essential for the development of leishmanicidal compounds aroused from this work: it is possible to separate the leishmanicidal activity from the antilymphocite properties of the molecules. Similar conclusions were obtained by the group of Lunardi [40] that designed chalcones by substitutions in the ring structures; indeed, their best compound was active *in vitro* against *L. braziliensis* at 13.7 μM, without affecting macrophages even at concentrations of 300 μM.

Chalcones have also been synthesized as a set of 4-methoxychalcone derivatives to identify if sulfonamide and methoxy substitutions could be promising adding-groups for the design of lead antiparasitic compounds. SAR analysis of these derivatives was devoted to determining structural and stereo-electronic features that could direct the leishmanicidal activity to the sulfonamide moiety [41]. This study evaluated *in vitro* the activity of the best molecules against promastigotes and their capacity to decrease the infection rate of *Leishmania*-infected macrophages. Additionally, the study evaluated the cytotoxicity of the compounds against mouse peritoneal macrophages to determine the selectivity index [41]. The MD used to compare with the activity of the compounds were the dipole, the Highest Occupied Molecular Orbital (HOMO), the Lowest Occupied Molecular Orbital (LUMO), the calculated log *P* (octanol/water partition), the molecular weight and the molecular volume [44]. The SAR analysis of these sulfonamide 4-methoxychalcone derivatives suggests that the molecular volume, the HOMO density concentrated in the chalcone moiety, and the conformational configuration of the compounds are fundamental structural and stereo-electronic features for the leishmanicidal activity. Of note, these compounds were designed according to Lipinski's rule of 5 and their "drug likeness" was similar to that of classical leishmanicidal drugs. Interestingly, there were very minor structural differences among the best and worst compounds and these differences were related to only one additional spacer (carbon residue) present in the structure of the best compound [41]. This key work for SAR in *Leishmania* demonstrated an enhanced leishmanicidal activity of derivatives due to the combination of two different pharmacophoric groups (*i.e.* chalcone and sulfonamide).

Natural products isolated from plants of the order Celastraceae have also been useful in the discovery of leishmanicidal compounds. These plants produce active secondary metabolites, the so called sesquiterpenes, of variously polyoxygenated tricyclic scaffolds, all based on a core C15 skeleton known as dihydro-â-agarofuran[5,11-epoxy-5â, 10R-eusdesm-4(14)-ene]. Many sesquiterpenes isolated from Celastraceae are lipophilic and cross cell membranes have a wide range of biological actions, and probably interact with a variety of cellular targets. Interestingly they have been described as useful in reversing multidrug resistance (MDR) and as efficient antitumor-promoting compounds [45, 46], two properties which make them clinically interesting.

The biological evaluation of sesquiterpenes as modulators of the MDR phenotype in *L. tropica* included a 3-D-QSAR both for natural sesquiterpenes from *Maytenus cuzcoina* and semi synthetic derivatives. The 3-D-QSAR was performed using the Comparative Molecular Similarity Indices Analysis (COMSIA), an extension of the Comparative Molecular Field Analysis Methodology (COMFA). This technique permitted the comparison of various sesquiterpenes bearing substituents at different positions of the dihydro-â-agarofuran skeleton, and allowed the characterization of the steric, electrostatic, hydrophobic, and hydrogen-bond-donor and acceptor requirements needed at the active sites of the *receptors* for ligand recognition. The 3-D-QSAR showed that the electrostatic features represented 58.9 percent of the total requirements needed to act as modulators at the P-glycoprotein-like transporter, being the most salient features those related to the H-bond interaction between the substituents at the C-2 and C-6 positions of the molecule with the receptor [45, 46]. Of note, these compounds selectively decreased the phenotype of drug-resistant parasites without significant toxicity.

Agelasine D, benzofuranes, flavonoids, lactones and phloroglucinol compounds are active antiparasitic compounds [47-51]. Many of these molecules are isolated from bacteria, plants or sponges and constitute lead compounds that when modified at their side chains display good selectivity against parasites. In fact, analogs of the O-prenylated phloroglucinol derivative based O-alkylated and formylated acylphloroglucinols, have a demonstrated leishmanicidal activity against *L. donovani* promastigotes (IC$_{50}$ ~ 5 µg ml^{-1}). In these compounds the prenyl, methyl and isovaleryl side chains were replaced with allyl, isopentyl, formyl, acetyl and isopentyl groups. The O-alkylated phloroglucinols bearing formyl/isovaleryl moieties are the most active molecules against *L. donovani* compared to compounds bearing formyl/acetyl, acetyl/acetyl or isovaleryl/isovaleryl moieties. These results suggest that the activity of the acylphloroglucinol molecules depends on the length of the acyl side chain and that substitution of the acyl group with the alkyl group results in loss of activity [47].

Synthetic aliphatic and aromatic lactones and dimers are active against *L. panamensis* at 0.8 µg ml^{-1} but are still very cytotoxic. The SAR demonstrated that the aliphatic side chain enhances the biological activity and decreases the cytotoxicity. This means that a reduced size of the lactone ring increases the selectivity index, although it also decreases the activity of the compound [48].

Inhibitors of tubulin polymerization at the colchicine binding site constitute an example worth to mention. A series of synthetic dihydrobenzofuran-lignans and related benzofurans have thus been tested against *Leishmania* [50]. The results suggest that compound 2g may function as a promising leishmanicidal dihydrobenzofuran derivative lead compound. Interestingly the concept of quasi-atomistic receptor surface modeling was used to develop a 4D-QSAR study of this interaction (http://www.biograf.ch) [52, 53].

A final example we would like to address is the SAR analysis of the naturally occurring plant-derived naphthylisoquinoline alkaloids. This is an interesting model since, besides studying the *in vitro* activity of the compounds, the decrease in the infection rate of infected macrophages and the cytotoxicity of the compounds against different mammalian cell types were investigated. The initial results and the SAR studies, where the aryl portion of the lead substrate was modified, demonstrated that the alkaloids ancistrocladinium A and B and the synthetic isoquinolinium salt decrease the macrophage infection rate by acting directly on the intracellular amastigotes [54, 55]. Additionally, the new isoquinolines act synergistically with amphotericin B and do not interact with cytochrome P450 enzymes involved in the metabolism of leishmanicidal drugs, thus making naphthylisoquinoline alkaloid derivatives promising candidates to be considered as leishmanicidal pharmacophores [55].

8. RANDOM AND RATIONAL DESIGN OF COMPOUNDS

The mode of action of many active leishmanicidal compounds is yet unknown and information on the molecular mechanisms is very scarce to support a rational design or selection of active molecules. However, since normally steric and electronic properties are responsible for the recognition of drugs by *receptors*, analysis of these properties in series of compounds should help obtain better insights on the mechanism of action and design more efficient analogues [56]. Within this aim, many groups have devoted their studies to finding out which electronic structure of specific molecules and derivatives should guide the most efficient activity and receptor-drug interaction to obtain lead compounds against *Leishmania*.

For example, the indolo[2,1-b]quinazoline-6,12-dione derivatives or azatrypanthrin derivatives [57] have been analyzed and tested for *in vitro* leishmanicidal activity. Their molecular characteristics in a 3-D space generated a satisfactory model that correlates experimental and estimated activity of the trypanthrins and confirms the stereo-electronic and redox potential analyses of SAR. The results are interesting and extend to structurally diverse classes of compounds, especially since the drugs are less toxic to mammalian cell lines than to *Leishmania*, making tryptanthrins suitable for further studies as potential leishmanicidal candidates [56].

In another study fifty-three nitro molecules were designed based on the antibacterial agent nifuroxazide backbone. The core structure permitted different substitution patterns in the aromatic ring, thus generating several analogues with diverse physicochemical properties but with similar steric characteristics. The side chain groups were chosen based on the relative electronic and lipophilic contributions of an appropriate substituent to the global structure. The results of using mono- and di-substituted analogues and 2-D QSAR studies suggest that there are compounds that exhibit IC_{50} values lower than the standard drugs pentamidine and amphotericin and are good candidates for further *in vivo* leishmanicidal assays. Additionally the results indicate that nitrothiophene derivatives were more active than the nitrofuran analogues [58].

Our final example relates to fused heterocyclic systems containing a pyrazole ring, for example 1H-pyrazolo[3,4-b]pyridine. These compounds are considered among the most versatile bioactive molecules. Pyrazolo[3,4-b]pyridines are potential specific antagonists of nucleic acid metabolism, and derivatives of this heterocyclic ring system are substrate inhibitors of purine-requiring enzymes [59]. The pyrazolo-pyridine is considered an aminoquinoline analogue, whose derivatives such as chloroquine and amodiaquine are antimalarial. Amodiaquine had never been used against leishmaniasis and de Mello and collaborators synthesized three series of 4-anilino-1H-pyrazolo[3,4-b]pyridine-5-carboxylic esters to study their potential leishmanicidal activity. The compounds were tested against promastigote forms of *L. amazonensis* and the activities correlated with the octanol/water partition parameter, log *P*. The results demonstrated that 3'-diethylaminomethyl-substituted compounds are the most active, with IC_{50} between 0.39 and 0.12 μM. Molecular modeling predicted the log *P* and steric parameters as the most significant contributions to biological activity [59].

9. TARGET ORIENTED DESIGN OF COMPOUNDS

The design of selective toxic agents should exploit key differences in metabolism between the host and pathogen, to exploit these differences in the design of selective toxic agents. We will now comment on a few examples that have used rational design to devise compounds directed towards known validated targets and new molecular targets in *Leishmania* pathogens [18].

For example, the thiol metabolism of trypanosomatids depends on the hexapeptide trypanothione (N1,N8-bis(glutathio-nyl)spermidine) (T(SH)2), an antioxidant replacing glutathione, the major antioxidant in other eukaryotic cells. Trypanothione may thus be an important antiprotozoal drug target and in fact some trypanocidal drugs, notably the arsenicals (melarsoprol) and difluoromethylornithine (eflornithine), may work by interfering with the metabolism or synthesis of this hexapeptide. The central role of trypanothione in *Trypanosoma* and *Leishmania* parasites make thiol-dependent enzymes potential targets for the development of chemotherapeutic drugs. Daunes and collaborators synthesized a series of N-S-blocked glutathione monoester and diester derivatives based on N-benzyloxy-carbonyl-S-(2,4-dinitrophenyl)glutathione and evaluated them for activity against *Trypanosoma* (*T.*) *brucei* (*b.*) brucei, *T. cruzi*, and *L. donovani in vitro*. Only monoesters with a log *P* value > 2.7 inhibited the growth of *T. b. brucei* bloodstream form trypomastigotes. Diester compounds were better inhibitors of *T. b. brucei* growth than monoester compounds, and some displayed high activity against *T. cruzi* and *L. donovani* [60]. Again, analysis of the inhibition data *vs.* calculated log *P* values provided evidence to support membrane penetration and steric factors as the key components in the activity of these compounds.

Dyneins are important proteins of *Leishmania* governing fundamental processes such as cilia and flagella motion, nuclear migration, organization of the mitotic spindle, and chromosome separation during mitosis. The first QSAR for dynein proteins in *Leishmania* contained 411 protein sequences of different species. This analysis discriminated more than 92 % of dyneins and other proteins in four different training and cross-validation datasets [61]. The analysis allowed the description of a new dynein sequence through a combined experimental and theoretic analysis and illustrated the usefulness of the model to search for potential drug targets. This was possible by the use of a combined strategy of 2D-electrophoresis analysis of *L. infantum* biological samples, followed by the excision from 2D-E gels of the spot of interest - unknown protein or protein fragment in the region M < 20,200 and isoelectric point < 4-. Afterwards, and through the Modular Approach to Software Construction Operation and Test (MASCOT) search engine (http://www.matrixscience.com), which uses mass spectrometry data to identify proteins from primary sequence databases, a *L. infantum* protein containing a dynein heavy chain was successfully identified [61].

Dinitroaniline sulfonamide compounds are known for their leishmanicidal activity. A virtual screen analysis of drug-like compounds belonging to the Maybridge® database led to the identification of the compound **BTB 06237** (2-[(2,4-dichloro-5-methylphenyl)sulfanyl]-1,3-dini-tro-5-(trifluoromethyl)benzene, 1), with potent activity against *L. donovani* axenic amastigotes (IC_{50} = 0.52 ± 0.20 μM) [62]. The leishmanicidal activity of **BTB 06237** correlated with the production of reactive oxygen species and the dissipation of the mitochondrial membrane potential; at the end, the results of SAR studies suggested that **BTB 06237** is a promising leishmanicidal pharmacophore [63].

The phosphocoline analogue miltefosine was approved in 2000 as an orally active drug for improving the classical therapies against leishmaniasis, *i.e.*, meglumine antimoniate and sodium stibogluconate. Still, its side effects remain challenging, especially teratogenicity. However, the fact that its cellular target is known represents an advantage in the effort to design that are active against leishmaniasis but lack the above mentioned drawback. With this aim, series of compounds have been designed as ring-substituted ether phospholipids carrying *N,N,N*-trimethylammonium, N-methylpiperidino, or N-methylmorpholino head groups [64]. Interestingly, the leishmanicidal activity of the different series demonstrated that some analogues selectively act against the promastigote forms of *L. donovani* or *L. infantum*. Additionally, some compounds were even more potent than miltefosine (hexadecylphosphocholine) against either *L. donovani* and *L. infantum* or both. Interestingly, Differential Scanning Calorimetry (DSC) demonstrated that the active compounds affect the thermotropic properties of a model membrane bilayer to a lesser extent than the less active ones [64]. Miltefosine is a ring-substituted ether phospholipid that challenges the use of SAR analysis since it is a rigid molecule and normally the analogues have increased structural flexibility [65]. This means that to perform a 3-D-QSAR the optimized 3-D-conformations of molecules are needed [66-68]. This challenging situation has been successfully addressed and a systematic procedure correlating the leishmanicidal activity of ring-substituted ether phospholipids against *L. infantum* and *L. donovani* has been performed and correlated with electrostatic, hydrophobic and steric parameters [65]. As a result, good COMFA and COMSIA activity models with similar results were obtained and suggest that the steric effect determines the activity of the miltefosine type ether phospholipid analogues. Interestingly this work demonstrated that the high leishmanicidal activity of these compounds can be achieved by a combination of an adamantylidene group in the lipohilic part of the choline at the head group of specific compounds [65].

Using the mevalonate/isoprene biosynthesis pathway enzyme, farnesyl pyrophosphate synthase, the activity of 62 bisphosphonates has been evaluated for inhibition of the enzyme from *L. major*. The investigated compounds exhibit activities (IC_{50} values) ranging from 100 nM to 80 μM (corresponding to Ki values as low as 10 nM). Bisphosphonates containing longer or multiple (N,N-) alkyl substitutions were inactive, as were aromatic species lacking an ortho- or meta-nitrogen atom in the ring, or possessing multiple halogen substitutions or a para-amino group. The active compounds were the bisphosphonates containing short (n) 4, 5) alkyl chains [69, 70]. The COMSIA fields were useful for investigating which structural features correlated with high activity, and indicated that a positive charge in the bisphosphonate side chain and a hydrophobic prospect significantly contributed to the activity of the compounds. These results represent the first detailed

QSAR of the inhibition of an expressed farnesyl pyrophosphate synthase enzyme by bisphosphonate inhibitors, and demonstrate that the activity of these inhibitors can be predicted using 3-D-QSAR techniques [70].

10. BEYOND STRUCTURE-BASED DRUG DISCOVERY: FRAGMENT APPROACHES, MULTITASKING QSAR AND POLYPHARMACOLOGY

QSAR models predict the biological activity of drugs against only one target. However, the interaction of a compound with multiple targets is more common than previously recognized. In fact the terms promiscuous targets, dirty drugs, and complex pharmacology are in common use now days. Consequently, the prediction of drug activity against different targets is an interesting challenge, usually difficult to achieve. Fragment approach drug design, multitasking QSAR and pharmacological networks are approaches that go beyond QSAR. When successful these approaches may yield opportunities to construct and map the contribution of sub-structures that function for multiple targets and species [52, 71, 72]. This would be especially interesting when dealing with a parasite like *Leishmania* where at least 20 species (with different isoforms of proteins or targets) could be pathogenic for humans.

Although efforts to deal with this situation have already begun, the obtained information is still scarce and difficult to understand. However, if successful, these methods would allow identifying compounds according to their profile of biological activity [17, 73, 74] and will constitute a fundamental methodology for designing drugs, including those for leishmaniasis.

11. FINAL COMMENTS

Parasitic diseases are still a threat to mankind in the 21st Century. The lack of effective chemotherapy for tropical diseases, and the alarming decrease in our arsenal of effective antiparasitic drugs due to the development of drug resistance emphasize the need for new therapeutic agents. In determining the best drug candidates, it is critical to keep in mind that any new drug must be affordable to those affected by these diseases, of course, including leishmaniasis [75].

New leishmanicidal products should be structurally simple enough to serve as leads for easy analogue synthesis or be available in sufficient quantities to permit economical access by affected populations.

Several techniques have been used to study the mechanisms by which *receptors* recognize their ligands, and QSAR facilitates the description of the molecular details involved in the mechanism of action and/or behavior of a specified interaction and recognition model. Given the tremendous chemical diversity, the successful development of antiparasitic products based in this method could have a dramatically positive impact on the treatment of leishmaniasis [76]. In the case of *Leishmania* QSAR constitute an essential tool to investigate chemical, electronic, and structural features affecting the leishmanicidal activity of compound libraries either with known mechanism of action or chosen at random. However, the information regarding *Leishmania* QSAR is rather scarce and major efforts are needed to stimulate this type of research in order to collect information that is fundamental for describing the determinants of leishmanicidal drugs.

Table **1** summarizes efforts made in this direction. Although many of the studies include an evaluation of the cytotoxicity of compounds against mammalian cells, only two describe *in vitro* the decrease in percentage of macrophage infection, and unfortunately only one has gone as far as *in vivo* testing of compounds. This means that a critical spot in leishmanicidal drug design is to encourage researchers to perform *in vivo* experiments and go into pre-clinical and clinical phases of research to reach the patient as soon as possible. We can never forget that our ultimate goal is to cure those suffering from the disease.

ACKNOWLEDGEMENTS

The authors are grateful for the financing support received from the Coordination for Research, Faculty of Medicine, UCV; and the Council for Scientific and Humanistic Research (CDCH), Universidad Central de

Venezuela. Likewise they are grateful for the support conferred by CDCH to Maritza Padrón Nieves to finish her PhD and the support conferred by the Alexander von Humboldt Foundation, Germany, to Alicia Ponte-Sucre.

Table 1: *In vitro* and *In vivo* activity of leishmanicidal compounds in selected SAR analysis.

Strain	Compound Type	*In Vitro* Activity µg / ml		*In Vitro* Activity µM		Cytotoxicity		*In Vivo*	References*
		Promastigotes	Percentage ▼ of Infection	Promastigotes	Percentage ▼ of Infection	µg / ml	µM	Percentage ▼ of Infection	
L. amazonensis	pyrazol pyridines	0.44	ND	1.23	ND	ND	ND	ND	de Mello *et al.*,2004
L. braziliensis	chalcones	ND	ND	13.7 (9.9-18.9)	ND	ND	> 300	ND	Lunardi *et al.*, 2003
L. braziliensis	chalcones	ND	ND	3.5 ± 0.6	ND	ND	69.0 ± 3.7	ND	Andriguetti-Fröhner *et al.*, 2009
L. donovani	aurones	0.45	1.40			<2.32>25	ND	ND	Kayser *et al.*, 1999
L. donovani	chalcones	ND	ND	3.4 ± 0.5	ND	ND	3.7 ± 0.2	ND	Nielsen *et al.*, 1998
L. donovani	trypanothione antagonists	ND	ND	7.8	ND	ND	269	ND	Daunes et al, 2001
L. donovani	dinitroaniline sulfonamide	ND	ND	0.67 ± 0.24	2.6 ± 1.2	ND	11.0 ± 2.9	28	Delfín *et al.*, 2009
L. donovani	Pentamidine analogs	ND	ND	1.17 [1, 2]	ND	ND	6.40 [1]	ND	Bakunova *et al.*, 2009
L. major	bisphosphonates	ND	ND	0.11	ND	ND	ND	ND	Sanders *et al.*, 2003
L. major	arylisoquinolines	ND	ND	2.65 ± 1.51	0.092	ND	12.67 ± 3.07	ND	Ponte-Sucre *et al.*, 2009
L. panamensis	lactones	2.8 ± 0.8	33.9 ± 1.4	ND	ND	ND	ND	ND	Castaño *et al.*, 2009

* de Mello *et al.*,2004, [59]; Lunardi *et al.*, 2003, [40]; Andriguetti-Fröhner *et al.*, 2009, [41]; Kayser *et al.*, 1999, [77] ; Nielsen *et al.*, 1998, [38] ; Daunes et al, 2001, [60]; Delfín *et al.*, 2009, [63]; Sanders *et al.*, 2003, [69]; Ponte-Sucre *et al.*, 2009, [55]; Castaño *et al.*, 2009, [48] Bakunova *et al.*, 2009 [79]. ND= not determined. [1] average of two determinations. [2] axenic amastigotes.

REFERENCES

[1]   Croft SL. Pharmacological approaches to antitrypanosomal chemotherapy. Mem Inst Oswaldo Cruz 1999; 94: 215.

[2]   Croft SL, Seifert K, Yardley V. Current scenario of drug development for leishmaniasis. Indian J Med Res 2006; 123: 399.

[3]   Handman E, Kedzierski L, Uboldi AD, Goding JW. Fishing for anti-Leishmania drugs: principles and problems. Adv Exp Med Biol 2008; 625: 48.

[4]   Ouellette M, Drummelsmith J, Papadopoulou B. Leishmaniasis: drugs in the clinic, resistance and new developments. Drug Resist Updat 2004; 7: 257.

[5]   Croft SL, Yardley V, Kendrick H. Drug sensitivity of Leishmania species: some unresolved problems. Trans R Soc Trop Med Hyg 2002; 96 S1: 127.

[6]   Croft SL, Sundar S, Fairlamb AH. Drug resistance in leishmaniasis. Clin Microbiol Rev 2006b; 19: 111-26.

[7]   Natera S, Machuca C, Padrón-Nieves M, Romero A, Díaz E, Ponte-Sucre A. Leishmania spp.: proficiency of drug-resistant parasites. Int J Antimicrob Agents 2007; 29: 637.

[8]   Cruz AK, de Toledo JS, Falade M, Terrão MC, Kamchonwongpaisan S, Kyle DE, Uthaipibull C. Current treatment and drug discovery against *Leishmania spp.* and *Plasmodium spp.*: a review. Curr Drug Targets 2009; 10: 178.

[9]   Mishra J, Saxena A, Singh S. Chemotherapy of Leishmaniasis: past, present and future. Curr Med Chem 2007; 14: 1153.

[10]   Sheridan RP, Kearsley SK. Why do we need so many chemical similarity search methods? Drug Discov Today 2002; 7: 903.

[11]   Dardonville C, Fernández-Fernández C, Gibbons SL, Jagerovic N, Nieto L, Ryan G, Kaiser M, Brun R. Antiprotozoal activity of 1-phenethyl-4-aminopiperidine derivatives. Antimicrob Agents Chemother 2009 [in press].

[12]   Stiefl N, Bringmann G, Rummey C, Baumann K. Evaluation of extended parameter sets for the 3D-QSAR technique MaP: implications for interpretability and model quality exemplified by antimalarially active naphthylisoquinoline alkaloids. J Comput Aided Mol Des 2003; 17: 347.

[13]   de Azevedo WF Jr, Dias R, Timmers LF, Pauli I, Caceres RA, Soares MB. Bioinformatics tools for screening of antiparasitic drugs. Curr Drug Targets 2009; 10: 232.

[14]   Tong W, Hong H, Xie Q, Shi L, Fang H, Perkins R . "Assessing QSAR Limitations - A Regulatory Perspective". Curr Comput-Aid Drug Des 2005; 1: 195.

[15]   Leonard JT, Roy K. On selection of training and test sets for the development of predictive QSAR models. QSAR Comb Sci 2006; 25: 235.

[16]   Prado-Prado FJ, González-Díaz H, de la Vega OM, Ubeira FM, Chou KC. Unified QSAR approach to antimicrobials. Part 3: first-tasking QSAR model for input-coded prediction, structural back-projection, and complex networks clustering of antiprotozoal compounds. Bioorg Med Chem 2008; 16: 5871.

[17]   Prado-Prado FJ, Uriarte E, Borges F, González-Díaz H. Multi-target spectral moments for QSAR and Complex Networks study of antibacterial drugs. Eur J Med Chem 2009 [in press].

[18]   González-Díaz H, Dea-Ayuela MA, Pérez-Montoto LG, Prado-Prado FJ, Agüero-Chapín G, Bolas-Fernández F, Vazquez-Padrón RI, Ubeira F. QSAR for RNases and theoretic-experimental study of molecular diversity on peptide mass fingerprints of a new Leishmania infantum protein. Mol Divers 2009 [in press].

[19]   González-Díaz H, Sánchez-González A, González-Díaz Y. 3D-QSAR study for DNA cleavage proteins with a potential anti-tumor ATCUN-like motif. J Inorg Biochem 2006; 100: 1290.

[20]   Concu R, Dea-Ayuela MA, Perez-Montoto LG, Bolas-Fernández F, Prado-Prado FJ, Podda G, Uriarte E, Ubeira FM, González-Díaz H. Prediction of enzyme classes from 3d structure: a general model and examples of experimental-theoretic scoring of peptide mass fingerprints of Leishmania proteins. J Proteome Res 2009; 8: 4372.

[21]   Masimirembwa CM, Bredberg U, Andersson TB. Metabolic stability for drug discovery and development: pharmacokinetic and biochemical challenges. Clin Pharmacol 2003; 42: 515.

[22]   Livingstone DJ. Pattern recognition methods in rational drug design. Methods Enzymol 1991; 203: 613-38.

[23]   Eriksson L, Jaworska J, Worth AP, Cronin MT, McDowell RM, Gramatica P. Methods for reliability and uncertainty assessment and for applicability evaluations of classification- and regression-based QSARs. Environ Health Perspect 2003; 111: 1361.

[24]   Baranczewski P, Stańczak A, Sundberg K, Svensson R, Wallin A, Jansson J, Garberg P, Postlind H. Introduction to *in vitro* estimation of metabolic stability and drug interactions of new chemical entities in drug discovery and development. Pharmacol Rep 2006; 58: 453.

[25]   Mohr JA, Jain BJ, Obermayer K. Molecule kernels: a descriptor- and alignment-free quantitative structure-activity relationship approach. J Chem Inf Model 2008; 48: 1868.

[26]   González-Díaz H, Bonet I, Teran C, De Clercq E, Bello R, García MM, Santana L, Uriarte E. ANN-QSAR model for selection of anticancer leads from structurally heterogeneous series of compounds. Eur J Med Chem 2007; 42: 580.

[27]   Chou KC, Chen NY. The biological functions of low-frequency phonons. Sci Sinica 1977; 20: 447-57.

[28]   Chou JJ, Li S, Klee CB, Bax A. Solution structure of Ca^{2+}- calmodulin reveals flexible hand-like properties of its domains. Nat Struct Biol 2001; 8: 990.

[29]   Gordon G. Designed electromagnetic pulsed therapy: clinical applications. J Cell Physiol 2007; 212: 579-82.

[30]   Gordon G. Extrinsic electromagnetic fields, low frequency (phonon) vibrations, and control of cell function: a non-linear resonance system. J Biomed Sci Eng 2008; 1: 152.

[31]   Lipinski CA, Lombardo F, Dominy BW, Feeney PJ. Experimental and computational approaches to estimate solubility and permeability in drug discovery and development settings. Adv Drug Del Rev 1997; 23: 3.

[32]   Ahlers J, Stock F, Werschkun B. Integrated testing and intelligent assessment-new challenges under REACH. Environ Sci Pollut Res Int 2008; 15: 565.

[33]   González-Díaz H, Vilar S, Santana L, Uriarte E. Medicinal chemistry and bioinformatics-current trends in drugs discovery with networks topological indices. Curr Top Med Chem 2007; 7: 1015.

[34]   Booth RG, Selassie CD, Hansch C, Santi DV. Quantitative structure-activity relationship of triazine-antifolate inhibition of Leishmania dihydrofolate reductase and cell growth. J Med Chem 1987; 30: 1218.

[35]   Koehler MG, Rowberg-Schaefer K, Hopfinger AJ. A molecular shape analysis and quantitative structure-activity relationship investigation of some triazine-antifolate inhibitors of Leishmania dihydrofolate reductase. Arch Biochem Biophys 1988; 266: 152.

[36]   del Olmo E, Alves M, López JL, Inchaustti A, Yaluff G, Rojas de Arias A, San Feliciano A. Leishmanicidal activity of some aliphatic diamines and amino-alcohols. Bioorg Med Chem Lett 2002; 12: 659.

[37]   Jeong JM, Choi CH, Kang SK, Lee IH, Lee JY, Jung H. Antioxidant and chemosensitizing effects of flavonoids with hydroxy and/or methoxy groups and structure-activity relationship. J Pharm Sci 2007; 10: 537.

[38]   Nielsen SF, Christensen SB, Cruciani G, Kharazmi A, Liljefors T. Antileishmanial chalcones: statistical design, synthesis, and three-dimensional quantitative structure-activity relationship analysis. J Med Chem. 1998 19; 41: 4819.

[39]   Chen M, Christensen SB, Blom J, Lemmich E, Nadelmann L, Fich K, Theander TG, Kharazmi A. Licochalcone A. Novel antiparasitic agent with potent activity against human pathogenic protozoan species of Leishmania. Antimicrob Agents Chemother 1993; 37: 2550.

[40]   Lunardi F, Guzela M, Rodrigues AT, Corrêa R, Eger-Mangrich I, Steindel M, Grisard EC, Assreuy J, Calixto JB, Santos AR.Trypanocidal and leishmanicidal properties of substitution-containing chalcones. Antimicrob Agents Chemother 2003; 47: 1449.

[41]   Andrighetti-Fröhner CR, de Oliveira KN, Gaspar-Silva D, Pacheco LK, Joussef AC, Steindel M, Simões CM, de Souza AM, Magalhaes UO, Afonso IF, Rodrigues CR, Nunes RJ, Castro HC. Synthesis, biological evaluation and SAR of sulfonamide 4-methoxychalcone derivatives with potential antileishmanial activity. Eur J Med Chem 2009; 44: 755.

[42]   Souza AM, Castro HC, Brito MA, Andrighetti-Fröhner CR, Magalhães U, Oliveira KN, Gaspar-Silva D, Pacheco LK, Joussef AC, Steindel M, Simões CM, Santos DO, Albuquerque MG, Rodrigues CR, Nunes RJ. Leishmania amazonensis Growth Inhibitors: Biological and Theoretical Features of Sulfonamide 4-Methoxychalcone Derivatives. Curr Microbiol 2009 [in press].

[43]   Hammett LP. Physical organic chemistry. New York: McGraw-Hill Book Co., Inc. 1940.

[44]   Costa MS, Boechat N, Rangel EA, da Silva Fde C, de Souza AM, Rodrigues CR, Castro HC, Junior IN, Lourenço MC, Wardell SM, Ferreira VF. Synthesis, tuberculosis inhibitory activity, and SAR study of N-substituted-phenyl-1,2,3-triazole derivatives. Bioorg Med Chem 2006; 14: 8644.

[45]   Cortés-Selva F, Campillo M, Reyes CP, Jiménez IA, Castanys S, Bazzocchi IL, Pardo L, Gamarro F, Ravelo AG. SAR studies of dihydro-beta-agarofuran sesquiterpenes as inhibitors of the multidrug-resistance phenotype in a Leishmania tropica line overexpressing a P-glycoprotein-like transporter. J Med Chem 2004; 47: 576.

[46]   Cortés-Selva F, Jiménez IA, Muñoz-Martínez F, Campillo M, Bazzocchi IL, Pardo L, Ravelo AG, Castanys S, Gamarro F. Dihydro-beta-agarofuran sesquiterpenes: a new class of reversal agents of the multidrug resistance phenotype mediated by P-glycoprotein in the protozoan parasite Leishmania. Curr Pharm Des 2005; 11: 3125.

[47]   Bharate SB, Khan SI, Yunus NA, Chauthe SK, Jacob MR, Tekwani BL, Khan IA, Singh IP. Antiprotozoal and antimicrobial activities of O-alkylated and formylated acylphloroglucinols. Bioorg Med Chem 2007; 15: 87-96.

[48]   Castaño M, Cardona W, Quiñones W, Robledo S, Echeverri F. Leishmanicidal activity of aliphatic and aromatic lactones: correlation structure-activity. Molecules 2009; 14: 2491.

[49]   Tasdemir D, Kaiser M, Brun R, Yardley V, Schmidt TJ, Tosun F, Rüedi P. Antitrypanosomal and antileishmanial activities of flavonoids and their analogues: *in vitro, in vivo*, structure-activity relationship, and quantitative structure-activity relationship studies. Antimicrob Agents Chemother 2006; 50: 1352.

[50]   Van Miert S, Van Dyck S, Schmidt TJ, Brun R, Vlietinck A, Lemière G, Pieters L. Antileishmanial activity, cytotoxicity and QSAR analysis of synthetic dihydrobenzofuran lignans and related benzofurans. Bioorg Med Chem 2005; 13: 661.

[51]   Vik A, Proszenyák A, Vermeersch M, Cos P, Maes L, Gundersen LL. Screening of agelasine D and analogs for inhibitory activity against pathogenic protozoa; identification of hits for visceral leishmaniasis and Chagas disease. Molecules 2009; 14: 279.

[52]   Vedani A, Dobler M. Multi-dimensional QSAR in drug research. Predicting binding affinities, toxicity and pharmacokinetic parameters. Prog Drug Res 2000; 55: 105.

[53]   Vedani A, Dobler M. 5D-QSAR: the key for simulating induced fit? J Med Chem 2002; 45: 2139.

[54]   Ponte-Sucre A, Faber JH, Gulder T, Kajahn I, Pedersen SE, Schultheis M, Bringmann G, Moll H. Activities of naphthylisoquinoline alkaloids and synthetic analogs against *Leishmania major*. Antimicrob Agents Chemother 2007; 51: 188.

[55]   Ponte-Sucre A, Gulder T, Wegehaupt A, Albert C, Rikanović C, Schaeflein L, Frank A, Schultheis M, Unger M, Holzgrabe U, Bringmann G, Moll H. Structure-activity relationship and studies on the molecular mechanism of leishmanicidal N,C-coupled arylisoquinolinium salts. J Med Chem 2009; 52: 626.

[56]   Bhattacharjee AK, Skanchy DJ, Jennings B, Hudson TH, Brendle JJ, Werbovetz KA. Analysis of stereoelectronic properties, mechanism of action and pharmacophore of synthetic indolo[2,1-b]quinazoline-6,12-dione derivatives in relation to antileishmanial activity using quantum chemical, cyclic voltammetry and 3-D-QSAR CATALYST procedures. Bioorg Med Chem 2002; 10: 1979.

[57]   Scovill J, Blank E, Konnick M, Nenortas E, Shapiro T. Antitrypanosomal activities of tryptanthrins. Antimicrob Agents Chemother 2002; 46: 882.

[58]   Rando DG, Avery MA, Tekwani BL, Khan SI, Ferreira EI. Antileishmanial activity screening of 5-nitro-2-heterocyclic benzylidene hydrazides. Bioorg Med Chem 2008; 16: 6724.

[59]   de Mello H, Echevarria A, Bernardino AM, Canto-Cavalheiro M, Leon LL. Antileishmanial pyrazolopyridine derivatives: synthesis and structure-activity relationship analysis. J Med Chem 2004; 47: 5427.

[60]   Daunes S, D'Silva C, Kendrick H, Yardley V, Croft SL. QSAR study on the contribution of log P and E(s) to the *in vitro* antiprotozoal activity of glutathione derivatives. J Med Chem 2001; 44: 2976.

[61]   Dea-Ayuela MA, Pérez-Castillo Y, Meneses-Marcel A, Ubeira FM, Bolas-Fernández F, Chou KC, González-Díaz H. HP-Lattice QSAR for dynein proteins: experimental proteomics (2D-electrophoresis, mass spectrometry) and theoretic study of a Leishmania infantum sequence. Bioorg Med Chem 2008; 16: 7770.

[62]   Delfín DA, Bhattacharjee AK, Yakovich AJ, Werbovetz KA. Activity of and initial mechanistic studies on a novel antileishmanial agent identified through *in silico* pharmacophore development and database searching. J Med Chem 2006; 49: 4196.

[63]   Delfín DA, Morgan RE, Zhu X, Werbovetz KA. Redox-active dinitrodiphenylthioethers against Leishmania: synthesis, structure-activity relationships and mechanism of action studies. Bioorg Med Chem 2009; 17: 820.

[64]   Avlonitis N, Lekka E, Detsi A, Koufaki M, Calogeropoulou T, Scoulica E, Siapi E, Kyrikou I, Mavromoustakos T, Tsotinis A, Grdadolnik SG, Makriyannis A. Antileishmanial ring-substituted ether phospholipids. J Med Chem 2003; 46: 755.

[65]   Kapou A, Benetis NP, Avlonitis N, Calogeropoulou T, Koufaki M, Scoulica E, Nikolaropoulos SS, Mavromoustakos T. 3D-Quantitative structure-activity relationships of synthetic antileishmanial ring-substituted ether phospholipids. Bioorg Med Chem 2007;15: 1252.

[66]   Nicklaus MC, Milne GW, Burke TR Jr. QSAR of conformationally flexible molecules: comparative molecular field analysis of protein-tyrosine kinase inhibitors. J Comput Aid-Mol Des 1992; 6: 487.

[67]   Demeter DA, Weintraub HJ, Knittel JJ. The local minima method (LMM) of pharmacophore determination: a protocol for predicting the bioactive conformation of small, conformationally flexible molecules. J Chem Inf Comput Sci 1998; 38: 1125.

[68]   Radwan AA, Gouda H, Yamaotsu N, Torigoe H, Hirono S. Rational procedure for 3D-QSAR analysis using TRNOE experiments and computational methods: application to thermolysin inhibitors. Drug Des Discov 2001; 17: 265.

[69]   Sanders JM, Gómez AO, Mao J, Meints GA, Van Brussel EM, Burzynska A, Kafarski P, González-Pacanowska D, Oldfield E. 3-D QSAR investigations of the inhibition of *Leishmania major* farnesyl pyrophosphate synthase by bisphosphonates. J Med Chem 2003; 46: 5171.

[70]   Sanders JM, Song Y, Chan JM, Zhang Y, Jennings S, Kosztowski T, Odeh S, Flessner R, Schwerdtfeger C, Kotsikorou E, Meints GA, Gómez AO, González-Pacanowska D, Raker AM, Wang H, van Beek ER, Papapoulos SE, Morita CT, Oldfield E. Pyridinium-1-yl bisphosphonates are potent inhibitors of farnesyl diphosphate synthase and bone resorption. J Med Chem 2005; 48: 2957.

[71]   Mestres J, Gregori-Puigjané E, Valverde S, Solé RV. The topology of drug-target interaction networks: implicit dependence on drug properties and target families. Mol Biosyst 2009; 5: 1051.

[72]   Hubbard RE. Fragment approaches in structure-based drug discovery. J Synchrotron Radiat 2008; 15: 227.

[73]   Hemmateenejad B, Miri R, Niroomand U, Foroumadi A, Shafiee A. A mechanistic QSAR study on the leishmanicidal activity of some 5-substituted-1,3,4-thiadiazole derivatives. Chem Biol Drug Des 2007; 69: 435.

[74]   González-Díaz H, Prado-Prado F, Ubeira FM. Predicting antimicrobial drugs and targets with the MARCH-INSIDE approach. Curr Top Med Chem 2008; 8: 1676.

[75]   Salem MM, Werbovetz KA. Natural products from plants as drug candidates and lead compounds against Leishmaniasis and trypanosomiasis. Curr Med Chem 2006; 13: 2571.

[76]   Ghose AK, Viswanadhan VN, Wendoloski JJ. A knowledge-based approach in designing combinatorial or medicinal chemistry libraries for drug discovery. 1. A qualitative and quantitative characterization of known drug databases. J Comb Chem 1999; 1: 55.

[77]   Kayser O, Kiderlen AF, Folkens U, Kolodziej H. *in vitro* leishmanicidal activity of aurones. Planta Med 1999; 65: 316.

[78]   Pieters L, Van Dyck S, Gao M, Bai R, Hamel E, Vlietinck A, Lemière G. Synthesis and biological evaluation of dihydrobenzofuran lignans and related compounds as potential antitumor agents that inhibit tubulin polymerization. J Med Chem 1999; 42: 5475.

[79]   Bakunova SM, Bakunov SA, Patrick DA, Kumar EV, Ohemeng KA, Bridges AS, Wenzler T, Barszcz T, Jones SK, Werbovetz KA, Brun R, Tidwell RR. Structure-activity study of pentamidine analogues as antiprotozoal agents. J Med Chem 2009; 52: 2016.

CHAPTER 3

Antimalarial Agents: Homology Modeling Studies

Tanos C.C. França[1,*] and Magdalena N. Rennó[2]

[1]*Chemical Engineering Department, Military Institute of Engineering, Pça General Tibúrcio, 80 - Urca, 22290-270, Rio de Janeiro/RJ, Brazil and* [2]*Pharmacy Course, Federal University of Rio de Janeiro, Campus Macaé, Rua Aloísio da Silva Gomes, 50, Granja dos Cavaleiros, 27930-560, Macaé/RJ, Brazil*

Abstract: Homology modeling may be a very useful tool to obtain theoretical 3D structures of molecular targets when their experimental structures are still unknown. If 3D structures of the appropriate templates are available, it is possible to build a very consistent model using one of several softwares available today for this purpose and, further, use it to analyze the overall target structure, its active site residues and possible interactions with potential ligands. These models can also be used for further MD simulations, docking and QSAR studies in order to afford additional information toward the rational design of inhibitors to the molecular target in focus. Literature has reported some interesting studies using this approach on neglected diseases, especially malaria. Those studies have afforded very useful models for the design of more selective and powerful antimalarial agents.

Keywords: Homology modeling, molecular dynamics, antimalarial agents.

1. INTRODUCTION

1.1. Building 3D Proteins Models by Homology Modeling

According to Higgins [1], the genome sequencing facilities available today have doubled the number of protein sequences discovered each year, increasing quickly the gap between the amount of structural information in the protein databanks and the number of sequences waiting for elucidation of their 3D structures. This happens because experimental methods for 3D elucidation are much more time consuming, restricted and usually not automatic when compared to the genomic facilities. The development of practical methods for prediction of protein structures from their sequences is, therefore, very important in the biologic field [1].

Several different approaches have been applied to predict protein structures from their sequences, with varied degrees of success. Most of these methods are based on known experimental structures, used as templates, and are strongly dependent on the Identity Degree (ID), or the percentage of identical amino acid residues present among the target sequence and their templates [1,2].

When there is no template available with a significant ID to the target sequence, one can use *ab initio* methods for the prediction of 3D structures. These methods include theretical methodologies to calculate the coordinates of a protein from the first principles, *i.e.*, with no reference to known structures. They have been of limited success and have produced more theory than really useful methodology [1].

If the ID between a target protein, whose structure is intended to be predicted, and its homologues are too low (< 25 %) for the identification of an unequivocal template, the best way to face the problem is the principle of fold recognition in order to find a protein with a more suitable folding for the modeling of the target protein [1-6]. This method evaluates the compatibility sequence-structure by means of an empirical potential, like Boltzman, derived from a table of contacts of the residues observed in proteins with known structures [1, 7].

Despite not having the same precision of other homology modeling techniques, the fold recognition could eventually become a powerful tool for the prediction of protein structures [8] considering the fact that only 10 different foldings (the supercoils) contribute to 50% of all structural similarities known among the

*Address correspondence to Tanos C.C. França: Chemical Engineering Department, Military Institute of Engineering, Pça General Tibúrcio, 80 - Urca, 22290-270, Rio de Janeiro/RJ, Brazil; Tel: +00552125467195; E-mail: tanos@ime.eb.br

protein super families [9]. So, instead of trying to find the correct structure for a protein among numerous possibilities of conformations for polypeptidic chains, a more practical approach would be to query first if the correct conformation has not been previously observed in another protein.

The homology modeling, or comparative modeling, tries to predict the protein structure related to another known structure (the template) following the idea that similar sequences imply in similar structures. This method has been quite successful, but presents some constraints like dependence on the alignment quality among the target sequence and the templates and the need for a good similarity with templates (above 35 % identity) [1]. The protocol for model building by homology modeling today involves template identification, alignment of the sequences, generation of the model coordinates, optimization and validation of the model [1]. These steps are quickly discussed below.

1.1.1. Template Identification, Alignment of the Sequences and Generation of the Model Coordinates

The homology modeling process starts with the identification of at least one protein with known 3D structure as template to the target protein [7]. Here, the protein family of the target protein can or can not be known.

If this family is known, the search for the best template should be done in a unique protein family and the chances of major accuracy in the final model are increased [2]. If the goal is to model a Serine Hydroxymethyltransferase (SHMT), for example, the logical way would be to search for templates inside the group of SHMTs with 3D structures available in the Protein Data Bank (PDB) [10, 11] or another protein data bank.

When the target protein family is not known, it is necessary to look for structures in the data banks of aminoacid sequences that could be compared to the target using the algorithms like BLAST [2, 12] and FastA [2, 13]. Aminoacid sequence data banks, like *GenBank*, [14], *Protein Identification Resource* (PIR) [15] and *SWISS-PROT* [16] can also be used [2].

The chosen templates must be aligned to the target sequence in order to establish the spatial correspondence between the target amino acid residues and those of the templates [2, 7]. This task can be performed by softwares like *BLAST* [12], *FastA*, [13], *Multalign* [17] and SWISSMODEL [18].

For sequences with more than 80 residues, if the sequential identity between the target and one of the templates is ≥ 25 %, the probability of their 3D structures being similar is high and, consequently, the 3D structure of the template could be used for modeling [2, 19]. In this case, most of the incertitude depends on the loop regions. Also, if the sequential identity is > 60%, the accuracy level of the resulting model is similar to the level of an experimental structure [2, 19].

When there are several potential templates, it is possible to choose between multiple and single alignments. The multiple alignments offer more options for modeling poorly aligned regions in the protein and afford a model reflecting the mean values among all templates. However, this necessarily implies in a deviation of the main protein chain from the more sequentially similar template. In the single alignment, on the other hand, despite the possibility of some regions being poorly modeled, the final model is more similar to the template with higher sequential identity.

Once the templates and the type of alignment to be used are defined, the next step is to generate the model coordinates. For this, it is necessary to perform, in sequence, the modeling of the structurally conserved regions, the loops and the side chains.

The alignment is of fundamental importance for the spatial location of most atoms in the target protein. In multiple alignments, each template contributes to the modeling of a specific segment in the target [7] according to the alignment obtained. The secondary structure (α-helix and β-sheets) model elements are usually defined according to the structure of the most similar template in the multiple alignment, similar to the single alignment. The crude model of the protein is obtained changing the different residues in the template for their corresponding residues in the target protein [2].

It is also possible to build an average structure for the target elements of secondary structure using all templates in this process and, further, use this structure when building the global model [2, 20].

In order to complete the model, it is necessary to build the loops and the elements of secondary structure not available in the templates [7]. The loops can be modeled with algorithms like those proposed by Greer *et al.* [21, 22]. Each loop is defined based on its length (number of residues) and the geometry of the anchor atoms, *i.e.*, the C_α of the four residues before and after the loop, and the fragment corresponding to it is extracted from structures available on PDB [2, 10, 11]. This method, however, does not always generates convincing solutions, mainly when dealing with large loops (more than 8 residues). In this case, it is necessary to complement the procedure using conformational analysis techniques in which the loops are filtered according to criteria like exposition of hydrophobic groups on the surface and relative conformational energies [2, 23].

The modeling of the side chain conformations is performed through libraries of rotamers allowed for side chains (χ_i) (Fig. **1**), deduced from high resolution PDB structures [2, 7, 24] and developed by Bower *et al.* [25]. The library affords the probabilities of fitting to side chains rotamers (χ_i) dependent on the conformational angle values of the main chain, ψ and ϕ (see Fig. **1**) as well as the residue type [2].

Besides the library of rotamers dependent on the main chain [25], Bower and Cohen applied a statistical Bayesian analysis to side chain conformations [26]. This kind of analysis combines previous information on measurable amounts with experimental data (usually limited). As a consequence, it affords a better estimative for a parameter of interest than the use of only experimental data [2, 26]. For previous distributions of rotamers χ_2, χ_3, and χ_4, it is assumed that the probability of each type of rotamer depends only on the previous rotamer in the main chain. For rotamers χ_1, depending on the main chain, previous distributions from probability products dependent on the angles ϕ and ψ are derived [26].

1.2. Optimization

Despite the final model presenting a conformation very similar to their templates, its structure still needs to be optimized, in order to remove or minimize unfavorable interactions between non covalently bonded atoms, by means of calculations using Molecular Dynamics (MD) simulation force fields [27, 28]. However, such calculations have to be restricted to the necessary minimum. Optimizations too extensive could excessively deviate the system from the original template causing the loss of the model experimental configuration [2] *i.e.*, the active conformation of the enzyme in nature. An excess of minimization could destroy this conformation creating a false model.

1.3. Validation

The last step in the process of homology modeling is the analysis of model consistency in order to validate the final model. Because crystallographic structures of the templates can have experimental and interpretation errors [2, 29-31], their quality needs to be evaluated by the resolution, the R-factor and temperature factor or B factor [32-34]. The better the protein resolution, the better the number of different experimental observations derived from the diffraction data and, consequently, the better the accuracy of the protein structure [2, 32]. The R factor is the measurement of the agreement between the 3D structure and the real crystalline structure of the protein [2, 33], and can be determined by comparing the amplitudes of experimental X-ray reflections and the amplitudes calculated for the protein structure with the best agreement to the electronic density map. The better the agreement between these amplitudes (lower R factor), the better the agreement between the crystalline and the derived structures [2, 33], considering the structures with resolution equal or better than 2.0 Å and R factor bellow 20 % as those most reliable.

The B factor of the protein [2, 34] is a measurement of the extension of the degree of the thermal and static disorder in each region of structure [2, 35]. The higher the B factor in a specific region of the protein, the higher its dynamic delocalization and, consequently, lower the degree of assurance of the spatial coordinates in that region. Higher B factor values are usually found in the loops and lower values are found in the α-helix and β-sheets regions [2]. The stereochemistry of the model can be verified by the analysis of parameters like bond lengths and angles, torsional angles and quirality of residues using software like PROCHECK [36], WHATCHECK [37] and PROSA [38].

An important indicator of the stereochemistry of a protein, among the several checks performed by the software above, is the distribution of the torsional angles φ and ψ from the main chain. This distribution can be examined through the Ramachandran plot [2, 39] (Fig. **1**). From the analysis of crystallographic data of a huge amount of proteins, Ramachandran *et al.* [39] mapped regions to the values of the angles φ and ψ of the residues that would be acceptable for functional protein tridimensional composition. Besides, this information also allows the identification of regions corresponding to the secondary structures. If a protein presents residues with stereochemical problems, they will fall in non-permitted regions of the Ramachamdran plot [2]. Only residues Pro (with a cycled side chain) and Gly (with no side chain at all) can occupy large areas of the Ramachamdran plot and should be analyzed separately [2]. The percentage of residues in favorable regions of the plot is one of the best ways to evaluate the stereochemical quality of a protein model [2, 33]. The ideal model should present more than 90% of the residues inside these regions [2, 35].

A similar analysis can be applied for the rotamers of side chains. The distribution of these angles to all residues of the model can be inspected in plots of χ_1 *versus* χ_2 [2, 35]. This kind of plot indicates which regions are more favorable to these angles based on a databank of well resolved crystallographic structures [2].

Some protein structure parameters considered constant should also be analyzed. These parameters are the planarity of the peptidic bond, the quirality of the Cα and the bond lengths and angles in the main chain [2, 33, 35, 40].

Also, the planarity of aromatic systems (Phe, Tyr, Trp and Hys) and terminal groups with sp^2 hybridization (Arg, Asn, Asp, Glu and Gln), the inner packing of globular proteins and the elements of secondary structure (in order to verify if the α-helix and β-strands found in the template are preserved during the model building) should be verified [33, 41, 42]. Besides, the distribution of hydrophobic and hydrophilic residues in the protein structure and the distribution of non polar and charged residues on the protein surface [43] can also be used to estimate the reliability of a model [2, 36].

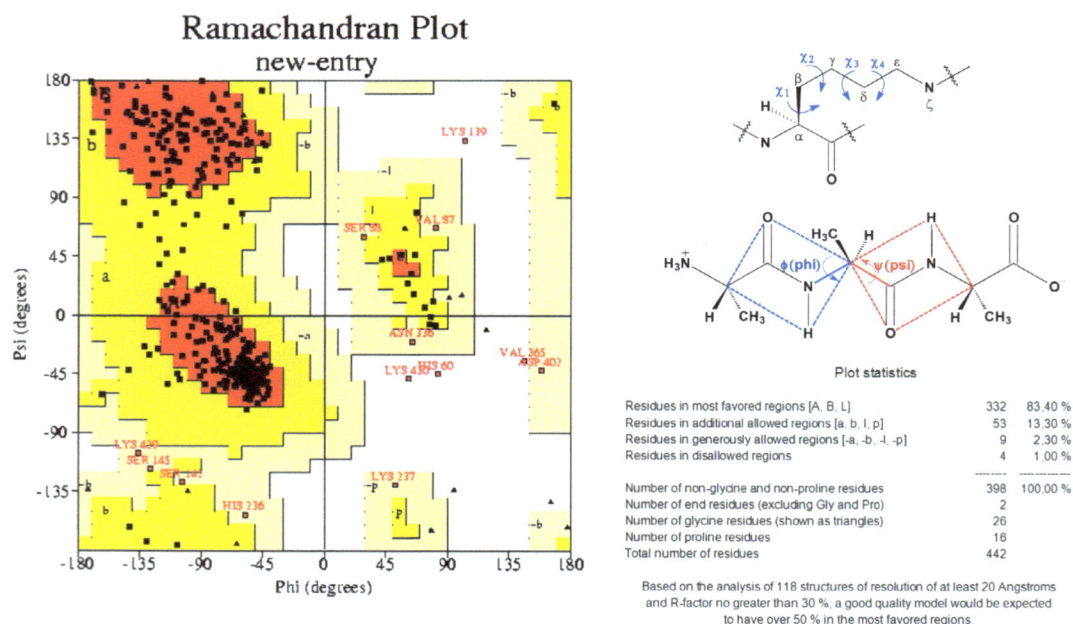

Figure 1: General representation of a Ramachandram plot [39] for a protein containing 442 residues. It is possible to identify the regions of the mean types of secondary structures. The plot was generated in the PDB validation server: http://deposit.rcsb.org/cgi-bin/validate/adit-session-driver. The angles φ, ψ and χ_i are illustrated beside the plot.

2. HOMOLOGY MODELING APPLIED TO OBTAIN AND STUDY MOLECULAR TARGETS AGAINST MALARIA

Today, homology modeling techniques are well established and widely employed by theoretical medicinal chemists to obtain molecular targets further used in drug design. However, there are still few studies in the literature comprising theses techniques focused on molecular targets in the parasites responsible for neglected diseases. Some examples of these studies on malaria are the homology modeling studies of the folato cycle enzymes from *P. falciparum* by Santos-Filho *et al.* [44], Delfino *et al.* [45] and França *et al.* [46, 47], and the work by Sabnis *et al.* [48] on the enzyme falcipain-2 from *P. falciparum*. Also, it is worth mentioning the works by Yuan Ping Pang [49], who modeled the enzyme acethylcolinesterase from *Anopheles gambiae*, the vector of malaria, and proposed it as a potential molecular target for antimalarial vectors; the work by Banerjee *et al.* [50] who modeled the enzyme Thioredoxin reductase from *P. falciparum*; and the studies of Cheng *et al.* [51], on the modeling of the *N*-myristoyltransferase from *P. falciparum* (*Pf*NMT). The more relevant aspects of these works will be discussed in the next topic.

2.1. Homology Modeling Studies on the Enzymes of *P. falciparum* Folate Cycle

Two of the first homology modeling studies on the enzymes of the folate cycle from *P. falciparum* are those works by Santos-filho *et al.* [44] and Delfino *et al.* [45], based on wild and mutated models of the Dihydrofolate reductase (DHFR) domain of the bifunctional enzyme Dihydrofolate reductase - Thymidilate synthase from *P. falciparum* (*Pf*DHFRTS), in order to check the effect of mutations on this enzyme active site on the resistance observed in some *P. falciparum* strains facing DHFR inhibitors.

In the first work, Santos-Filho *et al.* [44] proposed homology models for the wild-type *P. falciparum* DHFR (*Pf*DHFR) and the cicloguanyl (CYC) and pyrimetamine (PYR) cross-resistant mutant *Pf*DHFR that presents the following mutations: N51I, C59R, S108N and I164L. These models were further docked with CYC and PYR and submitted to steps of MD simulations in order to obtain information on the binding process and the probable resistance mechanism.1

Santos-Filho *et al.* [44] used the alignment between the crystallographic structure of chicken (*Gallus gallus*) liver DHFR [52] (*Gg*DHFR), PDB [10, 11] entry 8DFR, and the aminoacid sequence of *Pf*DHFR reported by Bzik and co-workers [53] (Fig. **2**) to build the wild-type model of *Pf*DHFR, with the Swiss-PdbViewer program [18] and the SWISS MODEL server [18, 54, 55], and further validated it with PROCHECK [36] and WHAT IF [56]. From specific mutations on the wilt-type model, Santos-Filho *et al.* [44] built the mutant model and, then, docked both the wild-type and mutant models with CYC and PYR for further studies by MD simulations. These studies made the evaluation of the interactions between CYC and PYR with the active site residues of both models possible, giving a clear view of the importance of residue mutations regarding resistance mechanisms, contributing with very useful information to the design of new potential antimalarial drugs.

Delfino *et al.* [45] have refined the studies by Santos-Filho *et al.* [44] by performing some modifications, using SPDBViewer [18], on the original alignment of *Pf*DHFR with *Gg*DHFR first proposed by Santos-Filho *et al.* [44], since the first model presented some drawbacks, like the incorrect representation of some interactions among enzyme, ligands [Dihydrofolate (DHF) or drug] and cofactor (NADPH).

```
PfDHFR    1    MMEQVCDVFD IYAICACCKV ESKNEGKKNE VFNNYTFRGL GNKGVLPWKC
GgDHFR    1      VRSLNS IVAVCQNM-- ---------- --------GI GKDGNLPWP-

PfDHFR   51    NSLDMKYFCA VTTYVNESKY EKLKYKRCKY LNKETVDNVN DMPNSKKLQN
GgDHFR   26    -PLRNEYKYF QR------- ---------- --MTSTSHVE GK------QN

PfDHFR  101    VVVMGRTSWE SIPKKFKPLS NRINVILSRT LKKEDFDEDV YIINKVEDLI
GgDHFR   49    AVIMGKKTWF SIPEKNRPLK DRINIVLSRE LKEAP-KGAH YLSKSLDDAL

PfDHFR  151    VLLG----KL NYYKCFIIGG SVVYQEFLEK KLIKKIYFTR INSTYECDVF
GgDHFR   98    ALLDSPELKS KVDMVWIVGG TAVYKAAMEK PINHRLFVTR ILHEFESDTF

PfDHFR  197    FPEINENEYQ IIS-----VS DVYTSNNTTL DFIIYKK
GgDHFR  148    FPEIDYKDFK LLTEYPGVPA DIQEEDGIQY KFEVYQKSV
```

Figure 2: Reproduction of the alignment between *Pf*DHFR and *Gg*DHFR proposed by Santos-Filho *et al.* [44]. Matching residues are shown in red. Similar residues are shown in blue.

The modified alignment was used to build a new and more representative model for wild-type *Pf*DHFR using the same procedure adopted by Santos-Filho *et al.* [44]. From this new model, Delfino *et al.* [45] have built models for 14 *Pf*DHFR mutants by replacing the appropriate residues in the model, for single, double, triple and quadruple mutants according to the evolutionary tree proposed, for *Pf*DHFR, by Sirawaraporn *et al.* [57]. The models of wild-type and mutant *Pf*DHFR were validated using the software PROCHECK [36] and WHAT IF [56] and further energy minimized with the AMBER force field [58] in the Insight-discovery® package [59].

These new models were used to build complexes by docking NADPH and the antimalarial drugs CYC, PYR, and WR99210 (formerly constructed and minimized with the Gaussian software [60]) inside the active sites of the models in order to build the ternary complexes *Pf*DHFR-NADPH-Drug. These complexes where further minimized with the AMBER force field [58] and used to propose possible reasons for the antifolate resistance.

Based on structural and functional considerations reported earlier in the literature [61-65], Delfino *et al.* [45] performed the following modifications on the alignment first proposed by Santos-Filho *et al.* [44] (see Figs. **2** and **3**): 1) Asn51 in *Pf*DHFR was aligned to Leu27 in *Gg*DHFR; 2). All residues from Pro26 to Arg36 were manually moved two positions in order to align Asp54 in *Pf*DHFR with Glu30 in *Gg*DHFR; and 3) all residues from Val1 to Met14 were manually moved three positions to the right in order to align Ile-14 of *Pf*DHFR with Ile-7 of *Gg*DHFR and approximate residues Ile14 and Ala16 of *Pf*DHFR to the active site.

```
PfDHFR    1    MMEQVCDVFD IYAICACCKV ESKNEGKKNE VFNNYTFRGL GNKGVLPWKC
GgDHFR    1           VRS LNSIVAVCQN M--------- --------GI GKDGNLPWPP

PfDHFR   51    NSLDMKYFCA VTTYVNESKY EKLKYKRCKY LNKETVDNVN DMPNSKKLQN
GgDHFR   26    LRNEYKYFQR ---------- ---------- --MTSTSHVE GK------QN

PfDHFR  101    VVVMGRTSWE SIPKKFKPLS NRINVILSRT LKKEDFDEDV YIINKVEDLI
GgDHFR   49    AVIMGKKTWF SIPEKNRPLK DRINIVLSRE LKEAP-KGAH YLSKSLDDAL

PfDHFR  151    VLLG----KL NYYKCFIIGG SVVYQEFLEK KLIKKIYFTR INSTYECDVF
GgDHFR   98    ALLDSPELKS KVDMVWIVGG TAVYKAAMEK PINHRLFVTR ILHEFESDTF

PfDHFR  197    FPEINENEYQ IIS-----VS DVYTSNNTTL DFIIYKK
GgDHFR  148    FPEIDYKDFK LLTEYPGVPA DIQEEDGIQY KFEVYQKSV
```

Figure 3: Reproduction of the alignment between *Pf*DHFR and *Gg*DHFR proposed by Delfino *et. al.* [45]. Matching residues are shown in red. Similar residues are shown in blue.

The refined models proposed by Delfino *et al.* [45] contemplated all the interactions usually found in the active sites of DHFR from other species [66] and, also, found in other theoretical models of *Pf*DHFR [62, 63]. Some of these interactions were not observed in the model proposed by Santos-Filho *et al.* [44]. Besides it was probably the first application of the homology modeling technique to an extensive study of *Pf*DHFR mutants. Similar earlier studies found in the literature were limited to analyze some few mutations [44, 62, 63]. The analysis of superimposition of the active sites of mutant complexes with the equivalent active site of the wild-type *Pf*DHFR performed by Delfino *et al.* [45] allowed the proposition of hypotheses that could explain the mechanisms by which the studied mutations act on PYR and CYC resistance, improving the results first obtained by Santos-Filho *et al.* [44].

In two other interesting works involving the folate cycle enzymes, França *et al.* [46, 47] have proposed two homology modeling models for the enzymes *Pf*DHFR-TS [46] and SHMT from *P. falciparum* (*Pf*SHMT) [47]. In the first one [46], they proposed a theoretical model for *Pf*DHFRTS, which includes the 55 aminoacid residues not visible in the crystallographic structure first published by Yuvaniyama *et al.* [67] and deposited on PDB [10, 11] under the code 1J3I. The model proposed by França *et al.* [46] exhibits the electrostatic potential on its calculated surface and revealed a continuous region of positive potential between the two active sites, suggesting an optimized mechanism for dihydrofolate transport. This achievement was used to rationalize some experimental observations concerning *Pf*DHFRTS, especially the existence of an optimized substrate transport process from one active site to another, as earlier proposed by Liang and Anderson [68]. The template used by França *et al.* [47] was the crystallographic structure of *Pf*DHFRTS deposited in PDB [10, 11] by Yuvaniyama *et al.* [67].

The complete sequence of *Pf*DHFRTS was manually aligned, in the Swiss-PdbViewer program, [18] to the incomplete crystallographic structure [67] in order to match all residues. This alignment was further submitted to the Optimize mode of the Swiss-Model server [18, 54, 55] in order to generate the complete model by searching its loops database. Therefore, this model was validated using the software PROCHECK [36] and WHATCHECK [37]. França *et al.* [46] also performed energy minimization steps using the GROMACS 3.1.4 package [69, 70] as additional steps for model validation and calculated the potential surface of *Pf*DHFRTS using the Swiss-PdbViewer software [18] and the method of Poisson-Boltzmann with a medium dielectric constant of 80 and a protein dielectric constant of 4.

The 3D structure of Yuvaniyama *et al.* [67] missed 55 aminoacids located from Val-86 to Ser-95 (09 residues) and between Leu-234 and Asp-281, corresponding to 46 residues of the junction region between DHFR and TS domains. According to Shallom *et al.* [71], there are important interactions between the DHFR and TS domains which are necessary for the activation of the TS active site. Besides, the knowledge of this junction region is of fundamental importance to understand the mechanism of substrate transport from one active site to another.

After analysis of the model, França *et al.* [46] noticed that the region located between residues Val-86 and Ser-95 was modeled by Swiss model based on the twenty fifth loop of the three dimensional structure of transketolase from *Sacharomices cerevisiae* (*Sc*TK) (PDB [10, 11] code: 1TRK), while the junction region was modeled based on the initial loop of the three-dimensional structure of human carbonic anhydrase II (*Hs*CA2) (PDB [10, 11] code: 2CBA) (Fig. **4**).

```
            Loop1      84      TVDNVND MPNSK
            ScTK      453      TLAHFRS LPNIQ

Loop2      233      MLNEQNCI KGEEKNNDMP LKNDDKDTCH MKKLTEFYKN VDKYKINYEN DDD
HsCA2       25      GERQSPVD IDTHTAKYDP SLKPLSVSYD QATSLRILNN GHAFNVEFDD SQD
```

Figure 4: Reproduction of the alignments used by França *et al.* [46] to build the *Pf*DHFRTS complete model: Loop 1 (Thr85 - Lys96) was aligned to the 25ᵗʰ loop of ScTK. Loop 2 (Met233 - Asp282) was aligned to the 1ˢᵗ loop of *Hs*CA2. Similar residues are shown in blue, while the identical ones are shown in red.

After building the complete model of *Pf*DHFRTS (Fig. **5**) França *et al.* [46] also calculated its potential surface using the program Swiss PDB viewer [18] in order to investigate the possibility of the existence of an optimized substrate transport process from one active site to another. Results showed the existence of an electrostatic potential path that may serve to conduct dihydrofolate from the TS active site to the DHFR active site (Fig. **6**), in agreement with similar findings in *Leishmania major* DHFRTS (*Lm*DHFRTS) [72] and other literature reports [73]. This optimized substrate transport mechanism may be a potential target for new antimalarial drugs, as it may be important for the constant maintenance of thymidylate production by the *Plasmodium* cell.

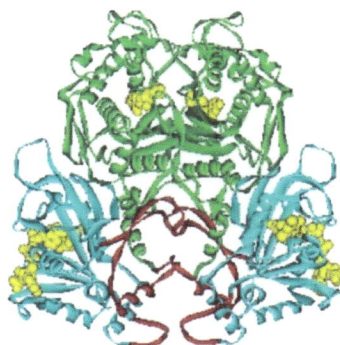

Figrue 5: Model of dimeric *Pf*DHFRTS proposed by França *et al.* [46]. DHFR domains are shown in blue, TS domains in green, the modeled regions in red and the substrates in the active sites in yellow CPK. Available in: http://jbcs.sbq.org.br/jbcs/2004/vol15_n3/18-187-03.pdf. Copyright of the Brazilian Chemical Society.

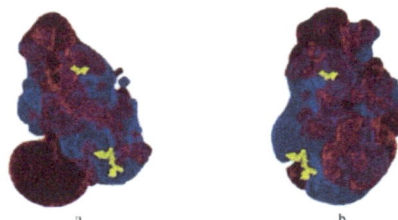

Figure 6: Electrostatic isopotential contours around *Pf*DHFRTS model. a) front view of the monomeric structure; b) backview of the monomeric structure. In blue are the regions with positive potential and in red the ones with negative potential. The substrates in the active sites are shown in yellow CPK. Available in: http://jbcs.sbq.org.br/jbcs/2004/vol15_n3/18-187-03.pdf. Copyright of the Brazilian Chemical Society.

The complete *Pf*DHFRTS model is a useful starting point for further studies, by MD simulations, of the possibility of existence of an optimized substrate mechanism transfer between the sites.

In their second study involving the enzymes of the folate cycle, França *et al.* [47] proposed cytosolic *Pf*SHMT as a potential target for antimalarial chemotherapy and built a multiple alignment 3D model for this enzyme in complex with its cofactor *N*-glycine-[3-hydroxy-2-methyl-5-phosphonooxymethyl-pyridin-4-yl-methane] (PLG) and the substrate 5-formyl tetrahydrofolate (5-FTHF) or 5-formyl-6-hydrofolic acid (FFO). This model was further compared to human SHMT in a search for key differences that could be useful in the design of selective *Pf*SHMT inhibitors.

The templates used by França *et al.* [47] were the crystallographic structures of SHMT from *E. coli* (*Ec*SHMT; PDB [10, 11] entry 1DF0) [74], Human (*Hs*SHMT; PDB [10, 11] entry 1BJ4) [75], Murine (*Mm*SHMT; PDB [10, 11] entry 1EJI) [76], Rabbit (*Oc*SHMT; PDB [10, 11] entry 1CJ0) [77] and *Bacillus stearothermophyllus* (*Bs*SHMT; PDB [10, 11] entry 1KKJ) [78]. The template sequences were aligned with the sequence of *Pf*SHMT and manually refined by comparison with the *Pf*SHMT alignment with other sequences available in the literature [77, 79]. This refined multiple alignment (Fig. **7**) was then submitted to the SWISS Model server [18, 54, 55] in the optimize mode to generate the model further validated with the PROCHECK [36] and WHATIF [56] programs. PLG and FFO, after optimization with the Gaussian98 package, [60] were docked into the *Pf*SHMT and *Hs*SHMT active sites in order to build the structures of the respective holoenzymes. These structures were submitted to rounds of energy minimization followed by MD simulations in order to investigate the dynamic behavior of PLG and FFO inside each active site in a search for key differences that could be useful in the design of selective *Pf*SHMT inhibitors.

Figure 7: Reproduction of the multiple alignments used by França *et al.* [47] to build the *Pf*SHMT model. Similar residues are shown in blue while the identical ones are shown in red.

Analysis of the MD simulations performed on the model proposed by França *et al.* [47] and crystallographic *Hs*SHMT, pointed out residues Glu137 and Asp136 in *Pf*SHMT active site that could be taken into consideration in the design of selective inhibitors for *Pf*SHMT. As these residues are involved in the interactions of the FFO tail, França *et al.* [47] suggested that compounds analogous to FFO, with either a positively charged tail instead of a glutamate tail or with bulkier substituents at the tail, could be able to interact more strongly with *Pf*SHMT than *Hs*SHMT active site.

In two subsequent MD studies França *et al.* [80] and da Silva *et al.* [81] proposed and refined structures of analogues of FFO as potential selective inhibitors for *Pf*SHMT. These works would not be possible without the *Pf*SHMT homology model earlier built by França *et al.* [47].

2.2. Homology Modeling of Falcipain-2 from *P. falciparum*

Also in an early homology modeling study on *P. falciparum* molecular targets, Sabnis *et al.* [48] have proposed a multiple alignment model for Falcipain-2 further used to investigate binding requirements on this enzyme crucial to the design and synthesis of isoquinoline inhibitors which exhibited *in vitro* enzyme inhibition at micromolar concentrations.

Falcipain-2 (FP-2) is a papain-family (C1A) cysteine protease that plays an important role in the parasite life cycle by degrading its principal source of aminoacids: the erythrocyte proteins, most notably hemoglobin [82]. Inhibition of FP-2 and its paralogues (aspartic and metaloproteases) prevents parasite maturation, suggesting these proteins as potential targets for the design of novel antimalarial drugs. Despite Hogg *et al.* [82] having elucidated the 3D structure of FP-2 in 2006, no 3D FP-2 structure was available in PDB [10, 11] when Sabnis *et al.* [48] developed their work.

The homology modeling studies by Sabnis *et al.* [48] were performed within the COMPOSER module of SYBYL 6.7® and the FP-2 sequence used was obtained from SWISS-PROT and TrEMBL databases of ExPASy Molecular Biology Server [16]. Only the mature sequence (Q9N6S8) was considered for deriving the homology model. The templates used by Sabnis *et al.* [48] were the crystallographic structures of: cruzain from *T. cruzi* (*Tc*Kr) inhibited by benzoyl-tyrosine-alanine, fluoromethylketone (PDB [10, 11] code 1AIM) proposed by Gillmor *et al.* [83]; Human protakepsin K (*Hs*PrK) (PDB [10, 11] code 1BY8) proposed by Lalonde *et al.* [84]; cruzain from *T. cruzi* bound to MOR-LEU-HPQ also proposed by McGrath *et al.* [85] (PDB [10, 11] code 1EWP), caricain D158E from *Carica papaya* (*Cp*CarD158E) mutant in complex with E-64 (PDB [10, 11] code 1MEG) proposed by Katerelos *et al.* [86] and a papaya protease omega from Carica papaya (*Cp*PO) proposed by Pickersgill *et al.* [87] (PDB [10, 11] code 1PPO). The alignment proposed by Sabnis *et al.* [48], after manual adjustments, to build the model, is reproduced in Fig. **8**.

The work by Sabnis *et al.* [48] also describes the procedure used to derive the Structurally Conserved Regions (SCRs) and the Structurally Variable Regions (SVRs), whose parameters used are reproduced in Table **1**.

```
PfFP2
TcKr
HsPrK        5                          EILDT HWELWKKTHR KQYNNKVDEI SRRLIWEKNL KYISIHNLEA SLGVHTYELA MNHLGDMTSE EVVQKMTGLK
TcKr2
CpCarD158E
CpPO


PfFP2        1                          YDWRLH SGVTPVKDQK NCGSCWAFSS IGSVESQYAI RQNKLITLSE QELVDCSFKN YGCNGGLINN AFEDMIELGG
TcKr         1             APAAVDWRAR GAVTAVKDQG QCGSCWAFSA IGNVECQWFL AGHPLTNLSE QMLVSCDKTD SGCSGGLMNN AFEWIVQENN
HsPrK        76  VPLSHSRSND APDSVDYRKK GYVTPVKNQG QCGSCWAFSS VGALEGQLKK KTGKLLNLSP QNLVDCVSEN DGCGGGYMTN APQYVQKN--
TcKr2        1             APAAVDWRAR GAVTAVKDQG SCGSCWAFSA IGNVECQWFL AGHPLTNLSE QMLVSCDKTD SGCSGGLMNN AFEWIVQENN
CpCarD158E   1             LPENVDWRKK GAVTPVRHQG SCGSCWAFSA VATVEGINKI RTGKLVELSE QELVDCERRS HGCKGGYPPY ALEYVAKN--
CpPO         1             LPENVDWRKK GAVTPVRHQG QCGSCWAFSA VATVEGINKI RTGKLVELSE QELVDCERRS HGCKGGYPPY ALEYVAKN--


PfFP2        78  ICPDGDYFYV SDIAPNLGNI DRCTEKYGIK NYLSVPDNKL KEALRFLGPI SISVAIVSDD FAFYKEGIFD GECGDQLNHA VMLVGPGMKE
TcKr         81  GAVYTEDSYP YASGEGISPP CTTSGHTVGA TITGHVELPQ DEAQIAAWLA VNGPVAVAVD ----ASSWMT YTGVMTSCV SEALDHGVLL
HsPrK        154 RGIDSEDAYP YVGQ--EESC MYNPTGKAAK CRGYREIPEG NEKALKRAVA RVGPVSVAID ASLTSFQFYS KGVYYDESCN SDNLNHAVLA
TcKr2        81  GAVYTEDSYP YASGEGISPP CTTSGHTVGA TITGHVELPQ DEAQIAAWLA VNGPVAVAVD ----ASSWMT YTGVMTSCV SEQLDHGVLL
CpCarD158E   79  -GIHLRSKYP YKAKQGTCRA KQVGGPIVKT SGVGRVQ-PN NEGNL-LNAI AKQPVSVVVE SKGRPFQL-- YKGGIFEGPC GTKVEHAVTA
CpPO         79  -GIHLRSKYP YKAKQGTCRA KQVGGPIVKT SGVGRVQ-PN NEGNL-LNAI AKQPVSVVVE SKGRPFQL-- YKGGIFEGPC GTKVDHAVTA


PfFP2        169 IVNPLTKKGE KHYYYIIKNS WGQQWGERGP INIETDESGL MRKCGLGTDA FIPLIE
TcKr         168 VGYND----- -----SAAVP YWIIKNSWTT QWGEEGYIRI AKGSNQCLVK EEASSAVVG
HsPrK        243 VGYGI----- -----QKGNK HWIIKNSWGE NWGNKGYILM ARNK---NNA CGIANLASFP KM
TcKr2        168 VGYND----- -----SAAVP YWIIKNSWTT QWGEEGYIRI AKGSNQCLVK EEASSAVVGL
CpCarD158E   165 VGYGK----- -----SGGKG YILIKNSWGT AWGEKGYIRI KRAPGNSPGV CGLYKSSYYP TKN
CpPO         165 VGYGK----- -----SGGKG YILIKNSWGT AWGEKGYIRI KRAPGNSPGV CGLYKSSYYP TKN
```

Figure 8: Representation of the multiple alignment of falcipain 2 with homologs of the long-chain subfamily of cysteine proteases proposed by Sabnis *et al.* [48]. Residues stressed in bold represent the Structurally Conserved Regions (SCRs).

Table 1: Parameters used by Sabnis *et al.* [48] for searching Structurally Conserved Regions (SCRs) and Structurally Variable Regions (SVRs) on *Pf*FP-2.

For Structurally Conserved Regions (SCRs)	
Minimum number of residues in SCR	4
Maximum number of iterations before updating SCRs	50
Maximum times to update SCRs	20
Maximum distance between equivalent $C_{\alpha s}$	3,5 Å
RMS difference considered significant	0.00001 Å
Residual difference considered significant	0.00001 Å
For Structurally Variable Regions (SVRs)	
End to end threshold	1.5 Å
Inter-C_α RMS threshold	1.0 Å
RMS fit to anchor threshold	1.0 Å
Minimum distance between loop and SCR $C_{\alpha s}$	3.5 Å

Cruzain presented an overall identity percentage of 55% and more than 90% in the SCRs with falcipain 2, being chosen to derive coordinates for building the SCRs of falcipain 2. They used the Tweak Loop approach [88, 89] to build loops onto the SCRs with the parameters described in Table **1**. The model was further refined by energy minimization until a gradient 0.01 kcal/mol.Å and MD simulations assisted by Simulated Annealing (SA) to further resolve the structure [90, 91].

Sabnis *et al.* [48] were able to propose a good quality 3D structure for *Pf*FP-2 by multiple alignment homology modeling. The model was reasonably further validated by docking studies of known vinyl sulfone inhibitors performed with the program DOCK 4 [92] and provided, at that time, the best alternative for insights into binding specificities of *Pf*FP-2, being useful for structure-based inhibitor design. Based on the information obtained from the model active site analysis and the docking studies, Sabnis *et al.* [48] designed, synthesized and evaluated the two isoquinoline compounds 1 and 2 presented in Fig. **9** as potential FP-2 inhibitors. These compounds presented *in vitro* IC$_{50}$ against FP-2 of 8 and 10 μM respectively. Such a result confirms the power of homology modeling as a powerful tool for the elucidation of 3D structures in drug design.

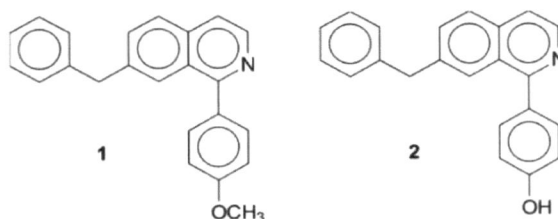

Figure 9: Compounds designed and evaluated by Sabnis *et al.* [48] as FP-2 inhibitors based on information obtained from the model.

2.3. Homology Modeling of Acethylcolinesterase from *Anopheles gambiae*

In an unconventional approach for the search of drug targets to the design of antimalarials, Pang [49] has built a homology model for the enzyme Acetylcholinesterase from *Anopheles gambiae* (*Ag*AChE) and compared this model sequence to another 73 AChE sequences. Results showed that two residues of *Ag*AChE are conserved at the opening of the active site of AChEs in 17 invertebrate and four insect species, respectively. Both residues are absent in the active sites of AChEs from human and other mammals. Such achievement could be very useful for the design of safer pesticides targeting the residues present only in mosquitoes and, also, turned *Ag*AChE into a promising target for the design of vector antimalarials.

AChE is a serine hydrolase vital to the regulation of the neurotransmitter acetylcholine in mammals and insects. This enzyme is a target for pesticides to control pests, including the malaria-carrying mosquito (*A. gambiae*). However, pesticides targeting insect AChE also inhibit mammal AChE being toxic to humans and animals. Also, the use of anticholinesterase pesticides has been limited by resistance problems caused by mosquitoes possessing AChE mutants insusceptible to current pesticides [48, 93].

The homology model proposed by Pang [49] was generated by the SWISS-MODEL program [18], according to the multiple sequence alignment reproduced in Fig. **10**, and using, as templates, the crystallographic structures of: mouse (PDB [10, 11] codes 1J07 and 1N5R) [94] and electric eel (PDB [10, 11] code 1C2O) [95]. According to Pang [49], these crystallographic structures were automatically identified by the SWISSMODEL [18] program and have the highest sequence identity (46%) to *Ag*AChE.

Pang [49] constructed the substrate-bound *Ag*AChE model by manually docking acetylcholine from *Torpedo californica* AChE (PDB [10, 11] code 1C2O) [96] into the active site of the model. The homology complex model was then refined by Multiple MD Simulations (MMDSs) in order to ensure the correct conformation of some loops in the structure.

After building the model, Pang [49] used the CLUSTALW program [97] to perform a sequence analysis of AChE from 73 species, including *Ag*AChE, currently available at the GenBank, and presented in Table **2**.

Results of the analysis performed by Pang [49] pointed towards two residues, C286 and R339, that are conserved in only four insect species and absent in AChEs from all other species listed in Table **2**.

```
AgAChE         1      DNDPLVV NTDKGRIRGI TVDAPSGKKV DVWLGIPYAQ PPVGPLRFRH PRPAEKWTGV LNTTTPPNSC VQIVDTVFGD
MmAChE(1J07)   4      EDPQLLV RVRGGQLRGI RLKAPGGP-V SAFLGIPFAE PPVGSRRFMP PEPKRPWSGV LDATTFQNVC YQYVDTLYPG
MmAChE(1N5R)   4      EDPQLLV RVRGGQLRGI RLKAPGGP-V SAFLGIPFAE PPVGSRRFMP PEPKRPWSGV LDATTFQNVC YQYVDTLYPG
TcAChE         5      DPQLLV RVRGGQLRGI RLKAPGGP-V SAFLGIPFAE PPVGSRRFMP PEPKRPWSGV LDATTFQNVC YQYVDTLYPG

AgAChE         78     FPGATMWNPN TPLSEDCLYI NVVAPRPRPK NAA-VMLWIF GGGFYSGTAT LDVYDHRALA SEENVIVVSL QYRVASLGFL
MmAChE(1J07)   80     FEGTEMWNPN RELSEDCLYL NVWTPYPRPA SPTPVLIWIY GGGFYSGAAS LDVYDGRFLA QVEGAVLVSM NYRVGTFGFL
MmAChE(1N5R)   80     FEGTEMWNPN RELSEDCLYL NVWTPYPRPA SPTPVLIWIY GGGFYSGAAS LDVYDGRFLA QVEGAVLVSM NYRVGTFGFL
TcAChE         80     FEGTEMWNPN RELSEDCLYL NVWTPYPRPA SPTPVLIWIY GGGFYSGAAS LDVYDGRFLA QVEGAVLVSM NYRVGTFGFL

AgAChE         157    FLG-TPEAPG NAGLFDQNLA LRWVRDNIHR FGGDPSRVTL FGESAGAVSV SLHLLSALSR DLFQRAILQS GSPTAPWALV
MmAChE(1J07)   160    ALPGSREAPG NVGLLDQRLA LQWVQENIAA FGGDPMSVTL FGESAGAASV GMHILSLPSR SLFHRAVLQS GTPNGPWATV
MmAChE(1N5R)   160    ALPGSREAPG NVGLLDQRLA LQWVQENIAA FGGDPMSVTL FGESAGAASV GMHILSLPSR SLFHRAVLQS GTPNGPWATV
TcAChE         160    ALPGSREAPG NVGLLDQRLA LQWVQENIAA FGGDPMSVTL FGESAGAASV GMHILSLPSR SLFHRAVLQS GTPNGPWATV

AgAChE         236    SREEATLRAL RLAEAVGC-- ----PHEPSK LSDAVECLRG KDPHVLVNNE WGTL---GIC EFPFVPVVDG AFLDETPQRS
MmAChE(1J07)   240    SAGEARRRAT LLARLVGCP- -----ND--- -TELIACLRT RPAQDLVDHE WHVLPQESIF RFSFVPVVDG DFLSDTPEAL
MmAChE(1N5R)   240    SAGEARRRAT LLARLVGCP- -----ND--- -TELIACLRT RPAQDLVDHE WHVLPQESIF RFSFVPVPDG DFLSDTPEAL
TcAChE         240    SAGEARRRAT LLARLVGCPP GGAGGND--- -TELIACLRT RPAQDLVDHE WHVLPQESIF RFSFVPVVDG DFLSDTPEAL

AgAChE         307    LASGRFKKTE ILTGSNTEEG YYFIIYYLTE LLRKEEGVTV TREEFLQAVR ELNPYVNGAA RQAIVFEYTD WTEPDNPNSN
MmAChE(1J07)   311    INTGDFQDLQ VLVGVVKDEG SYFLVYGVPG FSKDNESL-I SRAQFLAGVR IGVPQASDLA AEAVVLHYTD WLHPEDPTHL
MmAChE(1N5R)   311    INTGDFQDLQ VLVGVVKDEG SYFLVYGVPG FSKDNESL-I SRAQFLAGVR IGVPQASDLA AEAVVLHYTD WLHPEDPTHL
TcAChE         316    INTGDFQDLQ VLVGVVKDEG SYFLVYGVPG FSKDNESL-I SRAQFLAGVR IGVPQASDLA AEAVVLHYTD WLHPEDPTHL

AgAChE         387    RDALDKMVGD YHFTCNVNEF AQYAEEGNNV YMYLYTHRSK GNPWPRWTGV MHGDEINYVF GEPLNPTLGY TEDEKDFSRK
MmAChE(1J07)   390    RDAMSAVVGD HNVVCPVAQL AGLAAQGARV YAYIFEHRAS TLTWPLWMGV PHGYEIEFIF GLPLDPSLNY TTEERIFAQR
MmAChE(1N5R)   390    RDAMSAVVGD HNVVCPVAQL AGLAAQGARV YAYIFEHRAS TLTWPLWMGV PHGYEIEFIF GLPLDPSLNY TTEERIFAQR
TcAChE         395    RDAMSAVVGD HNVVCPVAQL AGLAAQGARV YAYIFEHRAS TLTWPLWMGV PHGYEIEFIF GLPLDPSLNY TTEERIFAQR

AgAChE         467    IMRYWSNFAK TGNPNPNTAS SEFPEWPKHT AHGRHYLELG LNTSFVGRGP RLRQCAFWKK YL
MmAChE(1J07)   470    LMKYWTNFAR TGDPNDPRDS KS-PQWPPYT TAAQQYVSLN LKPLEVRRGL RAQTCAFWNR FL
MmAChE(1N5R)   470    LMKYWTNFAR TGDPNDPRDS KS-PQWPPYT TAAQQYVSLN LKPLEVRRGL RAQTCAFWNR FL
TcAChE         475    LMKYWTNFAR TGDPNDPRDS KS-PQWPPYT TAAQQYVSLN LKPLEVRRGL RAQTCAFWNR FL
```

Figure 10: Reproduction of the SwissModel-generated multiple sequence alignments of Anopheles gambiae proposed by Pang [49] to built the *Ag*AChE model.

The four insects are house mosquito (*Culex pipiens*), Japanese encephalitis-carrying mosquito (*Culex tritaeniorhynchus*), German cockroach (*Blattella germanica*) and the African malaria-carrying mosquito (*A. gambiae*) including the one that is resistant to current pesticides (the G119S mutant, GenBank ID: AJ515149 [98]), and species listed in Table **2**. Pang [49] also observed that C286 is present in AChE of 17 invertebrate species and absent in all other species studied. These 17 invertebrates species are stressed in bold in Table **2**.

Table 2: Names of the 73 species which Acetylcholinesterases were used in the studies of Pang Y. [49].

	Official Name	**Common Name**
1	*Aedes aegypti*	Yellow fever mosquito
2	*Anopheles gambiae*	African malaria mosquito
3	*Anopheles stephensi*	Urban malaria mosquito on the Indian subcontinent
4	*Culex pipiens*	House mosquito
5	*Culex tritaeniorhynchus*	Japanese Encephalitis mosquito
6	*Aphis gossypii cotton aphid;*	Melon aphid
7	*Myzus persicae green peach aphid;*	Peach-potato aphid;
8	*Rhopalosiphum padi oat aphid; wheat aphid;*	Bird cherry-oat aphid
9	*Schizaphis graminum*	Greenbug
10	*Sitobion avenae English grain aphid;*	Grain aphid
11	*Bactrocera dorsalis*	Oriental fruit fly
12	*Bactrocera oleae olive fruit fly;*	Olive fly
13	*Bemisia tabaci sweetpotato whitefly;*	Silverleaf whitefly
14	*Drosophila melanogaster*	Fruit fly
15	*Haematobia irritans*	Horn fly
16	*Lucilia cuprina*	Australian sheep blowfly
17	*Musca domestica*	House fly
18	*Trialeurodes vaporariorum*	Greenhouse whitefly
19	*Blattella germanica*	German cockroach
20	*Bombyx mori*	Domestic silkworm; silk moth
21	*Apis mellifera*	Honey bee
22	*Cydia pomonella*	Codling moth
23	*Plutella xylostella*	Diamondback moth
24	*Helicoverpa armigera*	Cotton bollworm; tobacco budworm; corn ear
Worm		
25	*Helicoverpa assulta*	Oriental tobacco budworm; Cape gooseberry budworm
26	*Leptinotarsa decemlineata*	Colorado potato beetle
27	*Oulema oryzae*	Rice leaf beetle
28	*Nephotettix cincticeps*	Green rice leafhopper
29	*Nilaparvata lugens*	Brown planthopper
30	*Spodoptera exigua*	Beet armyworm
Mammals		
31	*Bos Taurus*	Cattle; domestic cow
32	*Canis familiaris*	Dog
33	*Felis catus*	Domestic cat
34	*Homo sapiens*	Human
35	*Macaca mulatta*	Rhesus monkey; rhesus macaque
36	*Mus musculus*	House mouse; mouse
37	*Oryctolagus cuniculus*	Rabbit; domestic rabbit
38	*Rattus norvegicus*	Norway rat; brown rat; rat
Other		
39	*Boophilus decoloratus*	Blue tick; type of tick
40	*Boophilus microplus*	Southern cattle tick; cattle tick
41	*Branchiostoma floridae*	Florida lancelet

	Official Name	**Common Name**
42	*Branchiostoma lanceolatum*	Common lancelet; amphioxus
43	*Bungarus fasciatus*	Banded krait
44	*Caenorhabditis briggsae*	Free-living nematode, bacterivore Clade V
45	*Caenorhabditis elegans*	Nematode;''C. elegans''; the worm
46	*Carassius auratus*	Goldfish
47	*Ciona intestinalis*	Sea vase
48	*Ciona savignyi a tunicate;*	Aquatic Invertebrate from the United States
49	*Danio rerio*	Zebrafish; zebra fish; zebra danio
50	*Dermacentor variabilis*	American dog tick
51	*Dictyocaulus viviparus*	Lungworm of cattle; bovine lungworm
52	*Electrophorus electricus*	Electric eel; electric knifefish
53	*Fugu rubripes*	Torafugu; tiger puffer; Japanese pufferfish
54	*Gallus gallus*	Chicken
55	*Loligo opalescens*	California market squid
56	*Meloidogyne incognita*	Southern root-knot nematode; cotton root-knotnematode
57	*Meloidogyne javanica*	Root knot nematode; root-knot nematode
58	*Myxine glutinosa*	Atlantic hagfish
59	*Necator americanus*	New world hookworm of humans, the American killer
60	*Nippostrongylus brasiliensis*	A common intestinal nematode of rats worldwide
61	*Oryzias latipes*	Japanese medaka; Japanese rice fish
62	*Rhipicephalus appendiculatus*	Brown ear tick
63	*Rhipicephalus sanguineus*	Brown dog tick
64	*Schistosoma bovis*	Blood-fluke in cattle
65	*Schistosoma haematobium*	Trematode; blood-flukes; human blood-fluke
66	*Schistosoma mansoni*	Trematode, human parasite
67	*Tetranychus cinnabarinus*	Carmine spider mite
68	*Tetranychus kanzawai*	Kanzawa spider mite
69	*Tetranychus urticae*	Two-spotted spider mite; red spider mite
70	*Tetraodon nigroviridis*	Puffer fish, Green spotted puffer
71	*Torpedo californica*	Pacific electric ray
72	*Torpedo marmorata*	Marbled electric ray; marbled torpedo ray
73	*Xenopus tropicalis*	Western clawed frog

After a profound analysis of the active site residues important to the catalytic activity of AChE and, also, the relevant interactions of R339 and C286 with important neighboring residues, Pang [49] have proposed that C286 and R339 can be used as species markers for developing effective and safer pesticides that can covalently bind to C286 of *Ag*AChE. These results suggest a conceptually new paradigm for pesticide design, thus offering a potential, effective control of malaria mosquitoes.

2.4. Homology Modeling of Thioredoxin from *P. falciparum*

In the search for new and promising targets against malaria, Banerjee *et al.* [50] have built a homology model for *P. falciparum* Thioredoxin reductase (*Pf*TrxR) and further refined it by MD simulations. This model has provided insight into the structure of the *Pf*TrxR from malarial parasite and aided in the rational drug designing.

*Pf*TrxR belongs to the flavoenzymes family like lipoamide dehydrogenase, glutathione reductase and mercuric ion reductase and reduces thioredoxin [98]. *Pf*TrxR inhibition could affect the parasite in several ways like enhancing oxidative stress, reducing the efficiency of DNA synthesis and cell division and, also, disturbing redox regulatory process. Besides there are significant mechanistic and structural difference between *Pf*TrxR and human TrxR to make *Pf*TrxR a promising drug target [99]. However, the lack of a 3D structure available had avoided effective studies on this important target [50].

The homology model proposed by Banerjee *et al.* [50] was generated in the software MODELLER9v327 [100] using as template the crystallographic structure of TrxR type 2 of mouse (PDB [10, 11] code 1ZDL), that presented 43.1% identity with *Pf*TrxR, according to the alignment reproduced in Fig. **11**. According to Banerjee *et al.* [50] the primary sequence of *Pf*TrxR [99] was obtained from the public domain protein sequence database of NCBI (*http://www.ncbi.nlm.nih.gov*).

```
PfTrxR    1   MCKDKNEKKN YEHVNANEKN GYLASEKNEL TKNKVEEHTY DYDYVVIGGG PGGMASAKEA
MmTrxR    1   ---------- ---------- ---------- ---------Q SFDLLVIGGG SGGLACAKEA

PfTrxR   70   AAHGARVLLF DYVKPSSQGT KWGIGGTCVN VGCVPKKLMH YAGHMGSIFK LDSKAYGWKF
MmTrxR   22   AQLGKKVAVA DYVEPSPRGT KWGLGGTCVN VGCIPKKLMH QAALLGGMIR -DAHHYGWEV

PfTrxR  130   DN-LKHDWKK LVTTVQSHIR SLNFSYMTGL RSSKVKYING LAKLKDKNTV SYYLKGDLSK
MmTrxR   81   AQPVQHNWKT MAEAVQNHVK SLNWGHRVQL QDRKVKYFNI KASFVDEHTV RGVDKG--GK

PfTrxR  189   EETVTGKYIL IATGCRPHIP DDVEGAKELS ITSDDIFSLK KDPGKTLVVG ASYVALECSG
MmTrxR  139   ATLLSAEHIV IATGGRPRYP TQVKGALEYG ITSDDIFWLK ESPGKTLVVG ASYVALECAG

PfTrxR  249   FLNSLGYDVT VAVRSIVLRG FDQQCAVKVK LYMEEQGVMF KNGILPKKLT KMDD-KILVE
MmTrxR  219   FLTGIGLDTT VMMRSIPLRG FDQQMSSLVT EHMESHGTQF LKGCVPSHIK KLPTNQLQVT

PfTrxR  308   FSDKTS---- -ELYDTVLYA IGRKGDIDGL NLESLNMNVN KSNNKIIADH LSCTNIPSIF
MmTrxR  279   WEDHASGKED TGTFDTVLWA IGRVPETRTL NLEKAGISTN PKNQKIIVDA QEATSVPHIY

PfTrxR  363   AVGDVAENVP ELAPVAIKAG EILARRLFKD SDEIMDYSYI PTSIYTPIEY GACGYSEEKA
MmTrxR  339   AIGDVAEGRP ELTPTAIKAG KLLAQRLFGK SSTLMDYSNV PTTVFTPLEY GCVGLSEEEA

PfTrxR  423   YELYGKSNVE VFLQEFNNLE ISAVHRQKHI RAQKDEYDLD VSSTCLAKLV CLKNEDNRVI
MmTrxR  399   VALHGQEHVE VYHAYYKPLE FTVADRD--- ---------- ---ASQCYIKMV CMREPPQLVL

PfTrxR  483   GFHYVGPNAG EVTQGMALAL RLKVKKKDFD NCIGIHPTDA ESFMNLFVTI SSGLSYAAKG
MmTrxR  443   GLHFLGPNAG EVTQGFALGI KCGASYAQVM QTVGIHPTCS EEVVKLHISK RSGLEPTVTG

PfTrxR  543   GGCGGGKCG
MmTrxR  ---   ---------
```

Figure 11: Reproduction of the sequence alignment proposed by Banerjee *et al.* [50], between *Pf*TrxR and the template *Mm*TrxR.

For prediction of the secondary structure, Banerjee *et al.* [50] employed the methods: Double Prediction Method (DPM) [101], Discrimination of protein secondary structure class (DSC) [102], GOR4 [103], Hierarchical Neural Network (HNN) [104], PHD [105], Predator [106], SIMPA96 [107], Self-Optimized Prediction Method with Alignment (SOPMA) [108] and Sec.Cons [109]. In order to identify regions of higher flexibility, they used the tools DisEMBL [110], Globplot [111], Regional Order Neural Network (RONN) [112] and Protein disorder prediction system.

After building the model, Banerjee *et al.* [50] refined it with steps of MD simulation using CHARMM force field [113] and validated it with the programs PROCHECK [36], WHATCHECK [37], WHATIF [56], PROSA [38], VERIFY 3D [114] and ERRAT [115].

In the absence of experimental structure, the final model built and extensively validated by Banerjee *et al.* [50] opened possibility to additional molecular modeling studies on such an important molecular target, providing a foundation for elucidating structure function relationship and paved the way for the rational drug design on this important target.

2.5. Homology Modeling of *N*-Myristoyltransferase from *P. falciparum*

In a classical application of homology modeling studies, Sheng *et al.* [51] have performed an interesting work in which they constructed models for the enzyme Myristoyl-CoA:protein *N*-myristoyltransferase (NMT) from *P. falciparum* (*Pf*NMT), *L. major* (*Lm*NMT) and *Trypanosoma brucei* (*Tb*NMT) on the basis of the crystal structures of fungal NMTs using a homology modeling method. The models were further refined by energy minimization and MD simulations and the respective active sites characterized by Multiple Copy Simultaneous Search (MCSS) revealing that *Pf*NMT, *Lm*NMT and *Tb*NMT share a similar active site topology, which is defined by two hydrophobic pockets, a Hydrogen-Bonding (HB) pocket, a negatively-charged HB pocket and a positively-charged HB pocket. They have further used the models to perform flexible docking of known inhibitors into the active site of *Pf*NMT and investigated in detail the binding mode, structure-activity relationships and selectivity of inhibitors. Results suggested key residues responsible for inhibitor binding, providing insights for the design of novel inhibitors of parasitic NMTs [51].

NMT is a cytosolic monomeric enzyme that catalyzes the transfer of the myristoyl group from myristoyl-CoA to the *N*-terminal glycine of a number of eukaryotic cellular and viral proteins [116, 117] and is important for diverse biological processes including signal transduction cascades and apoptosis [118-120]. Besides, myristoylation of the proteins can result in increased lipophilicity, facilitating their association with cellular/subcellular membranes and mediating protein-protein interactions [121]. These many cellular functions have turned NMT into a therapeutic target for the development of anticancer [122], antiviral [123], and antifungal [124] agents. Additionally, comparative biochemical studies of NMT from *Pf*NMT with human NMT (*Hs*NMT) highlighted the potential of the enzyme for the development of selective antiparasitic compounds [125] and a few protozoan parasite NMT inhibitors, previously discovered as fungal NMT inhibitors, have been reported [126, 127]. The achievement of 3D structures for parasitc NMT could accelerate this process and lead to potential new antimalarial structures.

The homology models proposed by Sheng *et al.* [51] were generated in the Align 123 module of the program InsightII® using as templates the crystallographic structures of NMTs from *Candida albicans* (PDB [10, 11] code 1IYL) and *Sacharomices cerevisae* (PDB [10, 11] code 2P6G). They obtained the primary sequences of *Pf*NMT, *Lm*NMT and *Hs*NMT from the SwissProt database and the *Tb*NMT sequence from Genome Survey Sequences (GSS) Database. The final models were subjected to energy minimization, using the builder module of InsightII® until a RMS energy gradient lower than 0.001 kcal/mol Å.

The validations of the models were performed by PROCHECK program [36] and the Profile-3D program [128] and InsightII® was used to evaluate the fitness of the model sequences in their current 3D environment. The best quality models of *Pf*NMT, *Lm*NMT and *Tb*NMT obtained by Sheng *et al.* [51] were chosen for further calculations and molecular modeling studies. MCSS program were used to explore the key regions in the active site that are necessary for ligand binding, by calculating the energetically favorable positions and orientations of given functional groups in the active sites of *Pf*NMT, *Lm*NMT and *Tb*NMT. Also, Sheng *et al.* [51] performed docking studies of the known NMT inhibitors 1-10 illustrated in Fig. **12** in the active site of *Pf*NMT in order to check the binding mode and selectivity of *Pf*NMT inhibitors.

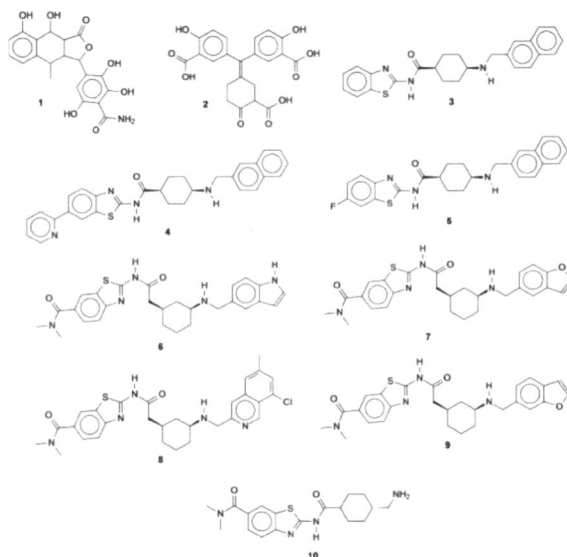

Figure 12: Structures of the NMT inhibitors used by Sheng *et al.* [51] in flexible docking studies on the *Pf*NMT model.

The 3D structures of *Pf*NMT, *Lm*NMT and *Tb*NMT proposed by Sheng *et al.* [51] permitted additional studies by MCSS and flexible docking that revealed key features of the active sites of these enzymes. Flexible docking corroborated the experimental values and showed that all the inhibitors studied are stabilized by hydrophobic and HB interactions. Also, the models have shown that the selectivity of inhibitors for either *Pf*NMT or *Hs*NMT was mainly attributed to HB interaction with residues Lys389 and Lys25. The structural models reported in the study by Sheng *et al.* [51] can be used in virtual screening and/or *de novo* inhibitor design for discovery of new lead compounds.

REFERENCES

[1] Higgins D, Taylor WR. Bioinformatics sequence, structure, and databanks. Higgins D and Taylor WR (eds),1st ed. Oxford, Oxford University Press, 2001.

[2] Santos Filho AO, Alencastro RB. Modelagem de proteínas por homologia. Quim Nova 2003; 26: 253.

[3] Finkelstein AV, Reva BA. A search for the most stable folds of protein chains. Nature 1991; 351: 497.

[4] Jones DT, Taylort WR, Thornton, JM. A new approach to protein fold recognition. Nature 1992; 358: 86.

[5] Bryant SH, Lawrence CE. An empirical energy function for threading protein sequence through the folding motif. Proteins Struct Funct Genet 1993; 16: 92.

[6] Sippl MJ, Weitckus S. Detection of native-like models for amino acid sequences of unknown three-dimensional structure in a data base of known protein conformations. Proteins Struct Funct Genet 1992; 13: 258.

[7] Peitsch MC. Practical Application of Computer-Aided Drug Design. In: Charifson PS, Ed., Marcel Dekker, New York, 1997.

[8] Box GEP, Hunter WG, Hunter JS. Statistics for experimenters: An introduction to design, data analysis, and model building.1st ed. Princeton: John Wiley & Sons, Inc., 1978.

[9] Orengo CA, Jones DT, Thornton JM. Protein superfamilles and domain superfolds. Nature 1994; 372: 631.

[10] Bernstein FC, Koetzle TF, Williams GJB, Meyer Jr EF, Brice MD, Rodgers JR, Kennard O, Schimanouchi T. Tasumi M. The protein data bank: A computer-based archival file for macromolecular structures. J Mol Biol 1977; 112: 535.

[11] Berman HM, Westbrook J, Feng Z, Gilliland G, Bhat TN, Weissig H, Shindyalov IN, Bourne PE. The protein data bank. Nucleic Acids Res 2000; 28: 235.

[12] Altschul SF, Gish W, Miller W, Myers EW, Lipman DJ. Basic local alignment search tool. J Mol Biol 1990 215: 403-410.

[13] Pearson WR, Lipman DJ. Improved tools for biological sequence comparison. Proc Natl Acad Sci USA 1988; 85: 2444.

[14] Benson DA, Boguski M, Lipman DJ, Ostell J. GenBank. Nucleic Acids Res 1994; 22: 3441.

[15] Barker WC, Garavelli JS, Huang H, McGarvey PB, Orcutt BC, Srinivasarao GY, Xiao C, Yeh LSL, Ledley RS, Janda JF, Pfeiffer F, Mewes HW, Tsugita A, Wu C. The protein information resource (PIR). Nucleic Acids Res 2000; 28: 41.

[16] Bairoch A, Apweiller R. The SWISS-PROT protein sequence database and its supplement TrEMBL in 2000. Nucleic Acids Res 2000; 28: 45.

[17] Barton GJ. Protein multiple sequence alignment and flexible pattern matching. Methods Enzymol 1990; 183: 403.

[18] Guex N, Peitsch MC. SWISS-MODEL and the Swiss-PdbViewer: An environment for comparative protein modeling; Electrophoresis 1997; 18: 2714.

[19] Rost B, Sander C. Bridging the protein sequence-structure GaP by structure predictions. Annu Rev Biophys Biomol Struct 1996; 25: 113.

[20] Taylor WR, Flores TP, Orengo CA. Multiple protein structure alignment. Protein Sci 1994; 3: 1858.

[21] Jones DT, Thirup S. Using known substructures in protein model building and crystallography. EMBO J 1986; 5: 819.

[22] Greer J. Comparative modeling: Application to the family of the mammalian serine proteases. Proteins Struct Funct Genet 1990; 7: 317.

[23] Bruccoleri RE, Karplus M. Prediction of the folding of short polypeptide segments by uniform conformational sampling. Biopolymers 1987; 26: 137.

[24] Sánchez R, Šali A. Advances in comparative protein-structure modeling. Curr Opin Struct Biol 1997; 7: 206.

[25] Bower MJ, Cohein FE, Dunbrack Jr., RL. Prediction of protein side-chain rotamers from a backbone-dependent rotamer library: a new homology modeling tool. J Mol Biol 1997; 267: 1268.

[26]   Dunbrack Jr., RL, Cohein FE. Bayesian statistical analysis of protein side-chain rotamer preferences. Protein Sci 1997; 6: 1661.

[27]   Jensen F. Introduction to Computational Chemistry 1st ed. Chichester, John Wiley & Sons, 1999.

[28]   Leach AR, Molecular Modelling: Principles and Applications; 2nd ed. London: Prentice Hall, 2001.

[29]   Bränden CI, Jones TA. Between objectivity and subjectivity. Nature 1990; 343: 687.

[30]   Jones TA, Zou JY, Cowan SW, Kjeldgaard M. Improved methods for building protein models in electron density maps and the location of errors in these models. Acta Cryst 1991; A47(Part 2): 110.

[31]   Engh RA, Huber R. Accurate bond and angle parameters for X-ray protein structure refinement. Acta Cryst 1991; A47(Part4): 392.

[32]   Hubbard TJP; Blundell TL. Comparison of solvent inaccessible cores of homologous proteins: Definitions useful for protein modeling. Protein Eng 1987; 1: 159.

[33]   Höltje HD, Folkers G. Molecular Modeling: Basic Principles and Applications; Mannhold R, Kubinyi H, Timmerman H. Eds. Weinheim, 1997.

[34]   Lumry R. Protein-Solvent Interactions; Gregory RB Ed. New York, Marcel Dekker. 1995; pp. 10.

[35]   Johnson MS, Srinivasan N, Sowdhamini R, Blundell TL. Knowledge-based protein modeling. Crit Rev Biochem Mol Biol 1994; 29: 1.

[36]   Laskowski RA, MacArthur MW, Moss DS, Thornton JM. PROCHECK: a program to check the stereochemical quality of protein structures. J Appl Cryst 1993; 26(Part 2): 283.

[37]   Hooft RWW, Vriend G, Sander C, Abola EF. Errors in protein structures. Nature 1996; 381: 272.

[38]   Sippl MJ. Recognition of errors in three-dimensional structures of proteins. Proteins 1993; 17: 355.

[39]   Ramachandram GN, Sasisekharan V. Conformation of polypeptides and proteins. Adv Prot Chem 1968; 23: 283.

[40]   Brandén C, Tooze J. Introduction to Protein Structure. 1st ed. New York, Garland, 1991.

[41]   Kabsch W, Sander C. Dictionary of protein secondary structure: Pattern recognition of hydrogen-bonded and geometrical features. Biopolymers 1983; 22: 2577.

[42]   Frishman D, Argos P. Knowledge-based protein secondary structure assignment. Proteins Struct Funct Genet 1995; 23: 556.

[43]   Creighton TE. Proteins: Structures and Molecular Properties. 2nd ed. New York, Freeman WH and Company, 1993.

[44]   Santos-Filho OA, Alencastro RB, Figueroa-Villar JD. Homology modeling of wild type and pyrimethamine/cycloguanil-cross resistant mutant type *Plasmodium falciparum* dihydrofolate reductase. A model for antimalarial chemotherapy resistance. Biophys Chem 2001; 91: 305.

[45]   Delfino RT, Santos-Filho OA, Figueroa-Villar JD. Molecular modeling of wild-type and antifolate resistant mutant *Plasmodium falciparum* DHFR. Biophys Chem 2002; 98: 287.

[46]   França TCC, Medeiros ALR, Santos ECP, Santos-Filho OA, Figueroa-Villar JD. A Complete Model of the *Plasmodium falciparum* Bifunctional Enzyme Dihydrofolate Reductase-Thymidylate Synthase. A Model to Design New Antimalarials. J Braz Chem Soc 2004; 15: 450.

[47]   França TCC, Pascutti PG, Ramalho TC, J Figueroa-Villar JD. A three-dimensional structure of *Plasmodium falciparum* serine hydroxymethyltransferase in complex with glycine and 5-formyl-tetrahydrofolate. Homology modeling and molecular dynamics. Bioph Chem 2005; 115: 1.

[48]   Sabnis Y, Rosenthal PJ, Desai P, Avery MA. Homology modeling of falcipain-2: validation, *de novo* ligand design and synthesis of novel inhibitors. J Biomol Struc Dyn 2002; 19: 765.

[49]   Pang YP. Novel acetylcholinesterase target site for malaria mosquito control; PLoS ONE 2006; 1(1): 1-8.

[50]   Banerjee AK, Arora N, Murty USN. Structural model of the *Plasmodium falciparum* thioredoxin reductase: a novel target for antimalarial drugs. J Vector Borne Dis 2009; 46: 171.

[51]   Sheng C, Ji H, Miao Z, Che X, Yao J, Wang W, Dong G, Guo W, Lu J, Zhang W. Homology modeling and molecular dynamics simulation of N-myristoyltransferase from protozoan parasites: active site characterization and insights into rational inhibitor design. J Comput Aided Mol Des 2009; 23: 375.

[53]   Bzik DJ, Li W, Horii T, Inselburg J. Molecular cloning and sequence analysis of the *Plasmodium falciparum* dihydrofolate reductase-thymidylate synthase gene. Proc Natl Acad Sci USA 1987; 84: 8360.

[54]   Peitsch MC. Protein modeling by E-mail. BioTechnology 1995; 13: 658.

[55]   Peitsch MC. ProMod and Swiss-Model: internet-based tools for automated comparative protein modeling. Biochem Soc Trans 1996; 24: 274.

[56]   Vriend G. J. WHAT IF: a molecular modeling and drug design program. J Mol Graph 1990; 8: 52.

[57]   Sirawaraporn W, Sathitkul T, Sirawaraporn R, Yuthavong Y, Santi DV. Antifolate-resistant mutants of *Plasmodium falciparum* dihydrofolate reductase. Proc Natl Acad Sci USA 1997; 94: 1124.

[58] Weiner PK, Kollman PA. AMBER: assisted model building with energy refinement. A general program for modeling molecules and their interactions. J Comp Chem 1981; 2: 287.

[59] Accelrys Inc. 9685 Scranton Road, San Diego, CA 92121-3752, USA.

[60] Gaussian Inc., Carnegie Office Park, Building 6, Suite 230, Carnegie, PA 15106, USA.

[61] Sirawaraporn W. Dihydrofolate reductase and antifolate resistance in malaria. Drug Res Update 1998; 1: 397.

[62] Lemcke T, Christensen IT, Jorgensen FS. Towards an understanding of drug resistance in Malaria: Three-dimensional structure of *Plasmodium falciparum* dihydrofolate reductase by homology building. Bioorg Med Chem 1999; 7: 1003.

[63] Rastelli G, Sirawaraporn W, Sompornpisut P. Vilaivan T, Kamchonwongpaisan S, Thebtaranonth Y, Yuthavong Y, Lowe G, Quarrel R. Interaction of pyrimethamine, cycloguanil, WR99210 and their analogues with *Plasmodium falciparum* dihydrofolate reductase: structural basis of antifolate resistance. Bioorg Med Chem 2000; 8: 1117.

[64] Warhurst DC. Antimalarial drug discovery: development of inhibitors of reductase active in drug-resistance. Drug Disc Today 1998; 3: 538.

[65] Matthews DA, Bolin JT, Burridge JM, Filman DJ, Volz KW, Kraut J. Dihydrofolate reductase. The stereochemistry of inhibitor selectivity. J Biol Chem 1985; 260: 392.

[66] Blakley RL, Folates and Pterins; Blakley RL, Benkovic SJ. (eds). John Wiley & Sons, New York, 1984.

[67] Yuvaniyama J, Chitnumsub P, Kamchonwongpaisan S, Vanichtanankul J, Sirawaraporn W, Taylor P, Walkinshaw MD, Yuthavong Y. Insights into antifolate resistance from malarial DHFR-TS structures. Nat Struct Biol 2003; 10: 357.

[68] Liang PH, Anderson KS. Substrate channeling and domain-domain interactions in bifunctional thymidylate synthase-dihydrofolate reductase. Biochemistry 1998; 37: 12195.

[69] Lindahl E, Hess B, Van der Spoel D. GROMACS 3.0: A Package for molecular simulation and trajetory analysis. J Mol Mod 2001; 7: 306.

[70] Berendsen HJC, Van der Spoel D, Van Drunen R. GROMACS: A message-passing parallel molecular dynamics implementation. Comp Physic Comm 1995; 91: 43.

[71] Shallom S, Zhang K, Jiang L, Rathod PK. Essential Protein-protein interactions between *P. falciparum* thymidilate synthase and dihydrofolate reductase domains. J Biol Chem 1999; 274: 37781.

[72] Knighton DR, Kan CC, Howland E, Janson CA, Hostomska Z, Welsh KM, Matthews DA. Structure of and kinetic channelling in bifunctional dihydrofolate reductase–thymidylate synthase. Nat Struct Biol 1994; 1: 186.

[73] Ivanetich KM, Santi DV. Bifunctional thymidylate synthase-dihydrofolate reductase in protozoa. The FASEB J 1990; 4: 1591.

[74] Scarsdale JN, Radaev S, Kazanina G, Schirch V, Wright HT. Crystal structure at 2.4 Å resolution of E. coli serine hydroxymethyltransferase in complex with glycine substrate and 5-formyl tetrahydrofolate. J Mol Biol 2000; 296: 155.

[75] Renwick SB, Snell K, Baumann U. The crystal structure of human cytosolic serine hydroxymethyltransferase a target for cancer chemotherapy. Structure 1998; 6: 1105.

[76] Szebenyi DME, Liu X, Kriksunov IA, Stover PJ, Thiel DJ. Structure of a murine cytoplasmic serine hydroxymethyltransferase quinonoid ternary complex: evidence for asymmetric obligate dimers. Biochemistry 2000; 39: 13313.

[77] Scarsdale JN, Kazanina G, Radaev S, Schirch V, Wright HT. Crystal Structure of Rabbit Cytosolic Serine Hydroxymethyltransferase at 2.8 Å Resolution: Mechanistic Implications. Biochemistry 1999; 38: 8347.

[78] Trivedi V, Gupta A, Jal VR, Saravanan P, Rao GSJ, Rao NA, Savithri HS, Subramanya HS. Crystal structure of binary and ternary complexes of serine hydroxymethyltransferase from *Bacillus stearothermophilus*: insights into the catalytic mechanism. J Biol Chem 2002; 10: 17161.

[79] Alfadhli S, Rathod PK. Gene organization of a *Plasmodium falciparum* erine hydroxymethyltransferase and its functional expression in Escherichia coli. Mol Biochem Parasitol 2000; 110: 283.

[80] França TCC, Wilter A, Ramalho TC, Pascutti PG, Villar JDF. Molecular dynamics of the interaction of *plasmodium falciparum* and human erine hydroxymethyltransferase with 5-formyl-6-hydrofolic acid analogues: design of new potential antimalarials. J Braz Chem Soc 2006; 17: 1383.

[81] Silva ML, Gonçalves AS, Batista PR, Villar JDF, Pascutti PG, França TCC. Design, docking studies and molecular dynamics of new potential selective inhibitors of *Plasmodium falciparum* serine hydroxymethyltransferase. Mol Sim 2010; 36: 5.

[82] Hogg T, Nagarajan K. Herzberg S, Chen L, Shen X. Jiang H, Wecke M. Blohmke C, Hilgenfeld R, Schmidt CL. Structural and functional characterization of falcipain-2, a hemoglobinase from the malarial parasite *Plasmodium falciparum*. Structural J Biol Chem 2006; 281: 25425.

[83]   Gillmor SA, Craik CS, Fletterick RJ. Structural determinants of specificity in the cysteine protease cruzain. Protein Sci 1997; 6: 1603.

[84]   LaLonde JM, Zhao B, Janson CA, D`Alessio KJ, McQueney MS, Orsini MJ, Debouck CM, Smith WW. The crystal structure of human procathepsin k. Biochemistry 1999; 38: 862.

[85]   McGrath ME, Eakin AE, Engel JC, McKerrow JH, Craik CS, Fletterick RJ. The crystal structure of Cruzain: a therapeutic target for chagas disease. J Mol Biol 1995; 247: 251.

[86]   Katerelos NA, Taylor MA, Scott M, Goodenough PW, Pickersgill RW. Crystal structure of a caricain D158E mutant in complex with E-64. FEBS Lett 1996; 392: 35.

[87]   Pickersgill RW, Rizkallah P, Harris GW, Goodenough PW. Determination of the structure of papaya protease omega. Acta Crystallogr Sect B 1991; 47: 766.

[88]   Shenkin PS, Yarmush DL, Fine RM, Wang H, Levinthal C. Predicting antibody hypervariable loop conformation. I Esembles of random conformations for ring like structures. Biopolymers 1987; 26: 2053.

[89]   Fine RM, Wang H, Shenkin PS, Yarmush DL, Levinthal C. Predicting antibody hypervariable loop conformations. II: Minimization and molecular dynamics studies of MCPC603 from many randomly generated loop conformations. Proteins 1986; 1: 342.

[90]   Stillinger FH. Theory and molecular models for water. Adv. in Chem. Phys. ed. I. Prigogine and S. Rice, J. Wiley and Sons, Inc. New York 1975; 31: 1.

[91]   Berendsen HJC, Postma JPM, van Gunsteren WF, Dinola A, Haak JR. Molecular dynamics with coupling to an external bath. J Chem Phys 1984; 81: 3684.

[92]   Kuntz ID, Blaney JM, Oatley SJ, Langridge RL. Ferrin TE. A geometric approach to macromolecule-ligand interactions. J Mol Biol 1982; 161: 269.

[93]   Weill M, Lutfalla G, Mogensen K, Chandre F, Berthomieu A, Berticat C, Pasteur N, Philips A, Fort P, Raymond M. Comparative genomics: Insecticide resistance in mosquito vectors. Nature 2003; 423: 136.

[94]   Bourne Y, Taylor P, Radic Z, Marchot P. Structural insights into ligand interactions at the acetylcholi-nesterase peripheral anionic site. EMBO J 2003; 22: 1.

[95]   Bourne Y, Grassi J, Bougis PE, Marchot P. Conformational flexibility of the acetylcholinesterase tetramer suggested by x-ray crystallography. J Biol Chem 1999; 274: 30370.

[96]   Raves ML, Harel M, Pang YP, Silman I, Kozikowski AP, Sussman JL. Structure of acetylcholinesterase complexed with the nootropic alkaloid, (-)-huperzine A. Nat Struct Biol 1997; 4: 57.

[97]   Chenna R, Sugawara H, Koike T, Lopez R, Gibson TJ, Higgins DG, Thompson JD. Multiple sequence alignment with the Clustal series of programs. Nucleic Acids Res 2003; 31: 3497.

[98]   Williams Jr CH, Zanetti G, Arscott LD, McAllister JK. Lipoamide Dehydrogenase, Glutathione Reductase, Thioredoxin Reductase, and Thioredoxin: A simultaneous purification and characterization of the four proteins from Escherichia Coli. J Biol Chem 1967; 242: 5226.

[99]   Andricopulo AD, Akoachere MB, Krogh R, Nickel C, McLeish MJ, Kenyon GL, Arscott LD, Williams Jr. CH, Davioud-Charvet E, Becker K. Specific inhibitors of *Plasmodium falciparum* thioredoxin reductase as potential antimalarial agents. Bioorg Med Chem Lett 2006; 16: 2283.

[100]   Eswar NM. Marti-Renom A, Webb B, Madhusudhan MS, Eramian D, Shen M, Pieper U, Sali A. Comparative protein structure modeling with MODELLER. Cur Protocols Bioinform Suppl 15 2006; 5.6.1.

[101]   Deleage G, Roux B. An algorithm for protein secondary structure prediction based on class prediction. Protein Eng 1987; 1: 289.

[102]   King RD, Sternberg MJ. Identification and application of the concepts important for accurate and reliable protein secondary structure prediction. Protein Sci 1996; 5: 2298.

[103]   Garnier J, Gibrat JF, Robson B. GOR method for predicting protein secondary structure from amino acid sequence. Meth Enzymol 1996; 266: 540.

[104]   Guermeur Y, Geourjon C, Gallinari P, Deleage G. Improved performance in protein secondary structure prediction by inhomogeneous score combination. Bioinformatics 1999; 15: 413.

[105]   Rost B, Sander C. Prediction of protein secondary structure at better than 70% accuracy. J Mol Biol 1993; 232: 584.

[106]   Frishman D, Argos P. Incorporation of non-local interactions in protein secondary structure prediction from the amino acid sequence. Protein Eng 1996; 9: 133.

[107]   Levin JM, Robson B, Garnier J. An Algorithm for secondary structure determination in proteins based on sequence similarity. FEBS Lett 1986; 205: 303.

[108]   Geourjon C, Deleage G. SOPMA: significant improvements in protein secondary structure prediction by consensus prediction from multiple alignments. Comput Appl Biosci 1995; 11: 681.

[109] Deleage G, Blanchet C, Geourjon C. Protein structure prediction. Implications for the biologist. Biochimie 1997; 79: 681.

[110] Linding R, Jensen LJ, Diella F, Bork P, Gibson TJ, Russell RB. Protein disorder prediction: implications for structural proteomics. Structure 2003; 11: 1453.

[111] Linding R, Russell RB, Neduva V, Gibson TJ. GlobPlot: exploring protein sequences for globularity and disorder. Nucleic Acids Res 2003; 31: 3701.

[112] Yang ZR, Thomson R, McNeil P, Esnouf RM. RONN: the bio-basis function neural network technique applied to the detection of natively disordered regions in proteins. Bioinformatics 2005; 21: 3369.

[113] Brooks BR, Bruccoleri RE, Olafson BD, States DJ, Swaminathan S, Karplus M. CHARMM: A program for macromolecular energy, minimization, and dynamics calculations. J Comp Chem 1983; 4: 187.

[114] Eisenberg D, Lüthy R, Bowie JU. VERIFY3D: Assessment of protein models with three-dimensional profiles. Methods enzymol 1997; 277: 396.

[115] Colovos C, Yeates TO. Verification of protein structures: Patterns of nonbonded atomic interactions. Protein Sci 1993; 2: 1511.

[116] Boutin JA. Myristoylation. Cell Signal 1997; 9: 15.

[117] Farazi TA, Waksman G, Gordon JI. The biology and enzymology of protein n-myristoylation. J Biol Chem 2001; 276: 39501.

[118] Knoll LJ, Johnson DR, Bryant ML, Gordon JI. Functional significance of myristoyl moiety in N-myristoyl proteins. Methods Enzymol 1995; 250: 405.

[119] Olson EN, Towler DA, Glaser L. Specificity of fatty acid acylation of cellular proteins. J Biol Chem 1985; 260: 3784.

[120] Towler DA, Adams SP, Eubanks SR, Towery DS, Jackson-Machelski E, Glaser L, Gordon JI. Purification and characterization of yeast myristoyl CoA:protein N-myristoyltransferase. Proc Natl Acad Sci USA 1987; 84: 2708.

[121] Gordon JI, Duronio RJ, Rudnick DA, Adams SP, Gokel GW. Protein N-Myristoylation. J Biol Chem 1991; 266: 8647.

[122] Selvakumar P, Lakshmikuttyamma A, Shrivastav A, Das SB, Dimmock JR, Sharma RK. Potential role of N-myristoyltransferase in cancer. Prog Lipid Res 2007; 46: 1.

[123] Hill BT, Skowronski J. Human n-Myristoyltransferases form stable complexes with lentiviral nef and other viral and cellular substrate proteins. J Virol 2005; 79: 1133.

[124] Georgopapadakou NH. Fungal N-myristoyltransferase inhibitors. Expert Opin Investig Drugs 2002; 11: 1117.

[125] Gunaratne RS, Sajid M, Ling IT, Tripathi R, Pachebat JA, Holder AA. Characterization of N-myristoyltransferase from *Plasmodium falciparum*. Biochem J 2000; 348: 459.

[126] Panethymitaki C, Bowyer PW, Price HP, Leatherbarrow RJ, Brown KA, Smith DF. characterization and selective inhibition of myristoyl-CoA:protein N-myristoyltransferase from *Trypanosoma brucei* and *Leishmania major*. Biochem J 2006; 396: 277.

[127] Bowyer PW, Gunaratne RS, Grainger M, Withers-Martinez C, Wickramsinghe SR, Tate EW, Leatherbarrow RJ, Brown KA, Holder AA, Smith DF. Molecules incorporating a benzothiazole core scaffold inhibit the N-myristoyltransferase of Plasmodium falciparum. Biochem J 2007; 408: 173.

[128] Luthy R, Bowie JU, Eisenberg D. Assessment of protein models with three-dimensional profiles. Nature 1992; 356: 83.

CHAPTER 4

Antitubercular Agents: Quantitative Structure Activity Relationships and Drug Design

Mushtaque S. Shaikh, Vijay M. Khedkar and Evans C. Coutinho[*]

Deptartment of Pharmaceutical Chemistry, Bombay College of Pharmacy, Kalina, Santacruz (E), Mumbai, 400 098. India

Abstract: Since its discovery in 1964 by Hansch, the Quantitative Structure Activity Relationship (QSAR) has remained an important tool in drug design. The work of a huge number of scientists has improved the strength, utility and efficiency of this vital technique in molecular modeling. The original formulation of the method was in two dimensions, the molecular descriptors *i.e.,* the physico-chemical constants were correlated with the biological activity, however, advances in technology, computational efficiency and the brilliant ideas of researchers have added many descriptors/dimensions leading to the 3D, 4D, 5D and 6D-QSAR techniques. The different forms of QSAR have not only contributed to understanding the pharmacophoric features required for improvement in the activity but has also helped to improve the pharmacokinetic and pharmacodynamic characteristics of drug candidates. The beauty of the QSAR technique is that it does not require information about the receptor (though well and good if known) and hence is helpful in the design and improvement of probable drug molecules not only against vital diseases/disorders but even against those which were long neglected. The diseases of tropical countries have been neglected for two reasons, poverty in these regions and remoteness to the developed parts of the world. The diseases which are top on this list are malaria, tuberculosis, leshmaniasis *etc.* However, in recent times the scenario has changed with many organizations, governments and research institutions showing an interest in eradicating such diseases mainly due to the serious problems of drug resistance. Under these circumstances, QSAR provides a good weapon for the design of novel candidates. Many QSAR studies have been reported in the literature both on the molecules synthesized and tested against the whole micro organism and also on molecules directed against specific targets of the micro organisms. This chapter will briefly cover the basics of the QSAR technique and will be followed by examples of the discovery of antitubercular agents through the QSAR methodology.

Keywords: SAR, QSAR-3D, Antitubercular Agents.

1. INTRODUCTION

When you can measure what you are speaking about, and express it in numbers, you know something about it; but when you cannot measure it, when you cannot express it in numbers, your knowledge is of a meagre and unsatisfactory kind. It may be the beginning of knowledge, but you have scarcely, in your thoughts, advanced to the stage of science. William Thomson (Lord Kelvin).

Molecular modeling and its tools are an important component of the drug discovery process. Through its tools *viz.* energy minimization, conformational search, molecular dynamics and Monte Carlo simulations, Free Energy Perturbation (FEP), homology modeling, molecular docking, pharmacophore mapping, and Quantitative Structure Activity Relationship (QSAR), molecular modeling has become the right hand of drug discovery scientists. Though many scientists believe that molecular modeling is a qualitative tool, it is very well justified that the associated computer graphics capabilities gives a realistic view of the scenario at the molecular level through pictures which are worth thousands of words.

A long cherished goal of medicinal chemists has been to design molecules with specific physico-chemical and biological properties. However, because of a combination of serendipity and empiricism as the foundation

*Address correspondence to Evans C. Coutinho: Department of Pharmaceutical Chemistry, Bombay College of Pharmacy, Kalina, Santacruz (E), Mumbai, 400 098. India; Tel: +91 022 26670905, Tele-fax: +91 022 26670816; E-mail: evans@bcpindia.org

of new drug discovery, finding new drug molecules has been an extremely challenging, time consuming and uneconomical process. In order to address these issues, molecular modelling tool *viz.* Quantitative Structure–Activity Relationships (QSAR) has been successfully implemented for decades in the development of new drug candidates. Although it does not completely eliminate the trial and error factor involved in the development of a new drug, it certainly decreases the number of compounds synthesized by facilitating the selection of the most promising candidates. The success of QSAR has attracted scientific community in the pharmaceutical arena to investigate relationships of molecular parameters with properties other than activity.

Originally based on the idea that compounds with similar physico-chemical properties trigger similar biological effects, Quantitative Structure–Activity Relationships (QSAR) builds virtual statistical models to establish a correlation between structural and electronic properties of potential drug candidates and their binding affinity towards a common macromolecular target. It includes all statistical methods, by which biological activities are related with structural elements (Free Wilson analysis), physico-chemical properties (Hansch analysis), or fields (3D-QSAR). QSAR studies have meant that, once a correlation between structure and property is found it is possible to predict quantities such as the binding affinity or the toxic potential of existing or hypothetical molecules.

The origin of QSAR dates back to 19th Century, when Crum-Brown and Fraser [1,2] published the first quantitative relationship between 'pharmacological activity' and 'chemical structure' in 1868 followed by the seminal contribution of Meyer and Overton in 1890s who noted that the toxicity of organic components depended on their lipophilicity [3,4]. True progress in the 'calculus' of Structure–Activity Relationships (SAR) was made in the 1960s, when Hansch and Fujita published a LFER related model considered to be the formal beginning of QSAR [5].

The first approach to building quantitative relationships which described activity as a function of chemical structure was based on the principles of thermodynamics, wherein the free-energy terms ΔE, ΔH and ΔS were represented by a set of parameters which could be derived for a molecule. Electronic effects *viz.* electron donating and withdrawing abilities, partial atomic charges and electrostatic field densities were represented by Hammett sigma (σ) values, resonance parameters (R values), inductive parameters (F values) and Taft substituent values (ρ^*, σ^*,) while steric effects (E_s) and molecular volume and surface area were represented by values calculated for Molar Refractivity and the Taft steric parameter. The steric effects were calculated using partition coefficient (log P) or the hydrophobic parameter (π), derived from the partition coefficient. The linear mathematical models (equations 1 and 2) which described the relationship between this physico-chemical properties and the observed Biological Response (BR) was the typical Hansch equation as shown below (see equations 1 and 2).

$$\Delta BR = f\left(\Delta electronic + \Delta steric + \Delta hydrophobic\right) \qquad \text{... (equation 1)}$$

$$\log 1/[C] = A(logP) - B(logP)^2 + C(E_s) + D(\sigma\rho) + E + ... \qquad \text{... (equation 2)}$$

For developing QSAR models the BR (biological response) is usually written as log $1/C$ (where C is the molar concentration of drug producing a standard response) since it only defines the quantitative measure of potency in our system (degree of perturbation). With the developments in biological assay techniques, Minimum Inhibitory Concentrations (MICs), IC_{50} values, K_i values and their ratios are used as values for C. The use and development of QSAR methodology have then never looked back, Hansch and his group alone have performed almost 2000 biomedical QSAR studies.

Though the Hansch approach was not free from limitations, it permitted complex biological systems to be modelled successfully using simple parameters. An alternative approach to compound design was suggested by Free and Wilson [6] where they used a series of substituent constants which related biological activity to the presence of a specific functional group at a specific location on the parent molecule. The mathematical equation which explained the relationship between biological activity and the presence or absence of a substituent was expressed as described in equation 3:

$$\text{Activity} = A + \sum_i \sum_j G_{ij} X_{ij} \qquad\qquad \dots \text{(equation 3)}$$

where A is the average biological response for the series; G_{ij} the contribution to activity of a functional group i in the jth position and X_{ij} the presence (1) or absence (0) of the functional group i in the jth position.

This approach involved use of the above equation to build a matrix for the series and represented this matrix as a series of equations. Subsequently, substituent constants were derived for every functional group at every position. The importance of the constants was estimated using the statistical tests. As a standard practice the models were tested for their validity and used to predict activity values for yet to be synthesized molecules.

The early QSAR studies typically relied on a single physico-chemical property, such as the solubility or the pK_a value, to explain the variance in biological effect of a molecule (1D QSAR). Later on, Hansch, Fujita, Free and Wilson extended the technique to include the connectivity of a compound by considering physico-chemical properties of single atoms and functional groups and their contribution to biological activity and this method was then referred to as 2D-QSAR [7-18]. Subsequently, realizing the limitation in the Hansch formulation, QSAR was further taken in for development by a number of scientists who added new dimensions leading to the nD-QSAR methods. The word dimensions referred to here are the properties (independent variables) added to develop the QSAR models. With the introduction of added dimensions, the original Hansch QSAR was referred to as 2D-QSAR. The third dimension came with the consideration of the 3D molecular properties of molecules. A number of 3D-QSAR techniques are now known. CoMFA [19,20], AFMoC [21,22], COMMA [23-25], CoMSIA [26,27], CoMSA [28-35], CoMPASS [36,37], CoRIA [38-40], CoRSA, SoMFA [41] and many more make the list of 3D-QSAR techniques. 3D-QSAR can be further divided into grid based and non-grid based methods.

The introduction of Comparative Molecular Field Analysis (CoMFA) [19,20] in 1988 by Cramer marked another milestone in the arena of QSAR paradigm as, for the first time, structure–activity relationship models were developed on the three-dimensional structural properties of the molecules (3D-QSAR). The widely used CoMFA method is based on molecular field analysis and represents real 3D-QSAR methods wherein the ligands' interaction with chemical probes is mapped onto a surface or grid surrounding a series of compounds (aligned in 3D space). This grid mimics a surrogate of the active site of the true biological receptor. 3D-QSAR methods are an extension of the traditional 2D-QSAR approach, wherein the physico-chemical descriptors like molecular volume, molecular shape, HOMO and LUMO energies, and ionization potential calculated from the 3D structures of the chemicals are correlated with the observed biological activity. A similar approach was adopted in developing other known 3D-QSAR formalisms like CoMSIA (Comparative Molecular Similarity Index Analysis) [26,27], SOMFA (self - organizing molecular field analysis) [41], and COMMA (Comparative Molecular Moment Analysis) [23-25]. The main problems encountered in building 3D-QSAR models are related to improper superposition of the ligands, greater flexibility of the ligands, identification of the bioactive conformation, and more than one binding mode of ligands.

While these problems have long been recognized, only recently, 4D-QSAR and 5D-QSAR have been developed allowing for the representation of multiple conformations, orientation and protonation states. As an example, Vedani *et al.* have developed QSAR techniques beyond the third dimension by accounting for the effect of multiple conformations as the 4th dimension, the induced-fit mechanism as the 5th dimension, and evaluation of solvation models as the 6th dimension [42]. A latest extension of the QSAR concept to 6th dimension (6D QSAR) allows for the simultaneous consideration of different solvation models [43-45].

In a nutshell, QSAR tries to correlate the physicochemical properties to the biological response, thus indirectly these physicochemical properties represent not only the action of the drug at the receptor site but its Absorption, Distribution, Metabolism Excretion and Toxicity (ADMET). ADMET properties of drug candidates were most times not considered implicitly. More recently, the technique has been extended to predict Absorption, Distribution, Metabolism, Elimination, Toxicity (ADMET) properties of compounds. Interestingly, prediction of the toxic potential of a drug or environmental chemical using QSAR has attracted much attention and various readymade tools (software or web based tools) are even available which work on the same principles. This outgrowth of QSAR which deals with pharmacokinetic issues is referred as Quantitative Structure Property Relationship (QSPR).

2. STATISTICAL TECHNIQUES FOR THE DEVELOPMENT OF A QSAR MODEL

The QSAR equation is a statistical model which correlates variations in biological activity to computed (or measured) physico-chemical properties for a series of molecules. Among the myriad of multivariate statistical techniques, Multiple Linear Regression Analysis (MLR) has been the most popular chemometric technique to be employed in QSAR studies. It quantitatively identifies the dependence of a drug property on any or all of the physico-chemical descriptors under investigation. These regression techniques generate a statistical model to represent the correlation of several independent X-variables (structural parameters) with dependent explicative variables Y (biological activity, binding affinity, toxicity). The model can then be used to predict the Y-variable from the knowledge of the X-variables, which can be either quantitative or qualitative. The validity of the MLR equation can be judged from the values of the multiple correlation coefficient (R); Fisher's F ratio; Student's t-value and standard deviation(s), the details of which are discussed in the later section.

In recent times, MLR has been substituted by multivariate projection methods, such as Principle Component Analysis (PCA), Principle Component Regression (PCR) and Partial Least Squares (PLS). These techniques project multivariate data into a space of lower dimensions, reducing the information content of data matrices and provide insight to visualise, classify, and model large sets of data.

Principle Component Analysis (PCA) is based on the assumption that the variations of different observable parameters may be approximated most effectively by the leading principle components of the covariance matrix that influence the observable effects. It does not create a model but is used to reduce the dimensionality of descriptors and provide orthogonal descriptors that retain a large portion of the information provided by the original descriptors. It seeks to find simplified relationships in data by transforming the original descriptors into a new set of uncorrelated variables termed as Principle components [46,47].

Partial Least Squares (PLS) analysis iteratively extracts components, each one of which satisfies the form of the least-squares equation. Each of the components is linearly independent of the rest and the components are extracted until no more components can be constructed. It carries out regression using latent variables from the independent and dependent data that are along their axes of greatest variation and are maximally correlated. PLS offers better opportunities than MLR when the independent variables are highly correlated or the number of independent variables exceeds the number of observations. It is a popular approach for the development of predictive models. Under such circumstances, it gives a more robust QSAR equation than multiple linear regression analyses. Another advantage of PLS is that it can be used with more than one dependent variable [48].

Principle Components Regression (PCR) is another statistical approach that applies the scores from PCA decomposition as regressors in building a QSAR model. The method uses multiple linear regression analysis to create a multiple-term linear equation using the Principle components as independent variables. PCR differs from PLS regression in the method adopted for extracting factor scores. While PCR generates the weight matrix reflecting only the covariance of the predictor variables, PLS regression includes the dependent variables Y in the process of data reduction, and hence the covariance is between the independent and dependent variables [46,47].

Genetic Function Approximation (GFA) is a novel evolutionary algorithm that has been developed for constructing QSAR models. It allows for the construction of QSAR models that are competitive with standard techniques, and provides additional information not obtained by other statistical techniques. In addition to linear polynomials, GFA can build higher-order polynomials, splines, and other nonlinear equations. It has a specific advantage of being applicable to data sets containing many more variables than samples, or data sets which contain nonlinear relationships between the independent variables and the biological activity.

GFA applies the natural principles of evolution of species, wherein improvement is achieved by recombination of independent variables, that is, reproduction, mutation, and crossover. This method begins with randomly constructed multiple QSAR models that are further evolved by repeatedly selecting two better-rated models to serve as parents, then creates a new child model by using terms from each of the parent models. As the evolution progresses, the population becomes enriched with better quality QSAR models. GFA also belongs to the class of variable reduction algorithms and is most useful if the critical information in the data set is concentrated in a few variables [49].

Artificial Neural Network (ANN) has its genesis from the neural networks present in the human brain, which is composed of many simple processing units operating in parallel, the so-called neurons. The objective is to simulate the multiple shells of the simple neurons, where each neuron is connected to a number of neighbouring neurons with variable coefficients of connectivity each representing the force of these connections. The learning process involves adjusting the coefficient so that the network provides the appropriate results as an output. The technique has been widely accepted over the last few decades for QSAR and QSPR modelling, especially in situations marked by ill-defined cause–effect relationships, and non-normal distributions [50].

Genetic Partial Least Squares (G/PLS) is a chemometric technique evolved by combining the best features of both GFA and PLS. Both of these techniques serve as a valuable analytical tool in situations where datasets have more descriptors than observations [51,52].

3. VALIDATION OF QSAR MODEL

The reliability a QSAR model depends on how well the model can predict the activity of compounds not used to build the model, rather than how well the model reproduces the biological activity of compounds used to build the model [53]. Various approaches used for validation of any Quantitative Structure–Activity Relationship (QSAR) study are discussed below:

Cross-Validation (CV) Technique is an internal validation method for statistical models wherein the predictive ability of model is estimated using a reduced set of structural data. Usually, one molecule of the dataset is extracted each time, and the model recalculated using the remaining $n-1$ objects as a training set, so that the property value for the extracted element is predicted once for all compounds. This is the called leave-one-out (LOO) method. Analogously, when more than one molecule is extracted from the dataset of the system each time, then the procedures are generally referred to as leave-n-out, leave-group-out or leave-many-out (LMO) CV method. The predictive ability of the model is defined by the cross-validated correlation coefficient r^2_{cv} (q^2) which is also a criterion of robustness of the derived statistical model (equation 4).

$$r^2_{cv} = (PRESS_o - PRESS)/ (PRESS_o) \qquad \qquad \text{...(equation 4)}$$

where $PRESS_o$ is the standard deviation of the observed biological activity while PRESS is the sum of the squares of the differences between the predicted and the observed biological activity values [54,55].

The correlation coefficient (r) constitutes the variance in the data and is a measure of the degree of linearity of the relationship. Ideally it should be equal to or approach unity, but in reality due the complexity of biological data, any value above 0.90 is considered to be satisfactory. The **Pearson's correlation coefficient (r^2)** is the squared correlation coefficient which signifies how well the model reproduces experimental data. It is noteworthy that, since it is easier to fit the experimentally observed biological data than to predict them from the QSAR model, r^2 of the model is always higher than q^2 [56].

Bootstrapping is another technique used concomitantly with cross-validation to evaluate the robustness and the statistical confidence of the QSAR model. The technique simulates a large number of datasets which are of the same size as the original and are produced by randomly selecting samples from the original dataset. The PLS statistical calculation is performed on each of these bootstrap samplings. The difference between the parameters calculated from the original dataset and the mean of the parameters calculated from the many bootstrap samplings signifies the bias of the original calculations [57-66].

External validation of the QSAR model involves predicting the activity of an external test set of molecules using the models derived from the training set. **Predictive r^2 (r^2_{pred})**, is a measure of the predictive ability of the derived QSAR model and is calculated by the formula (equation 5).

$$r_{pred}^2 = SD - PRESS/SD \qquad \qquad \ldots \text{(equation 5)}$$

where SD is the sum of squared deviation of the biological activities of the test set and the mean activity of the training set molecules; while PRESS is the sum of squared deviations of the experimental and the predicted activities of the test set compounds [53].

The **Fischer statistic** parameter is another variance-related parameter which is used as a measure of the level of statistical significance of the regression model; a higher **F** value implies that a more significant correlation has been reached [67].

Randomization or **y-scrambling** is a rigorous alternative to cross-validation and bootstrapping wherein activity values are re-assigned arbitrarily to different molecules, and a new regression is performed. If the results of a PLS model derived using the original sequence of the biological data is significantly better than the results from the 'scrambled' models, one can then be sure that significant correlation indeed exists between the biological data and the independent variables. Randomization analyzes the ability of the statistical model to derive real structure-activity relationships [68].

4. QSAR STUDIES ON ANTITUBERCULAR DRUGS

Over a period of time QSAR techniques have matured into a widely used predictive tool, making substantial contribution to the drug-discovery process. By now almost all the known classes of compounds must have been subjected to QSAR for one activity or aother. Here a special mention of antitubercular drugs needs to be made, where the QSAR studies have been done for both, the antitubercular activities obtained from whole cell assays and activities against specific targets. In this chapter we have also discussed some QSPR studies which were performed on antitubercular drugs.

5. QSAR STUDIES ON UNTARGETED ANTITUBERCULAR DRUGS

A number of classes of compounds have been studied using QSAR methodologies for antitubercular activity. Some drugs and chemical classes which have been extensively studied by QSAR are quinolones, rifampicin, isoniazid, pyridines, thiadiazoles, oxadiazoles, coumarins, chalcones and adamantanyl-quinoline derivatives. Besides these classes, some other novel compounds which have been analyzed by QSAR are, cinnamates, hydrazides, phenylcarbamates, indolediones, benzamides, benzoxazines, alkenols, myristates, nitrofurans and nitroimidazoles *etc.*

Quinoline Derivatives

Quinolines are the earliest known and the most widely studied antitubercular agents with an extensive literature on QSAR. Several derivatives of quinoline have been made and tested and found to exhibit antitubercular activities *viz.* fluoroquinolones, admantanyl quinolines, quinolinecarbaldehyde, quinoxaline *etc.* We shall start our discussion on the quinolone scaffold represented in **Scheme 1**.

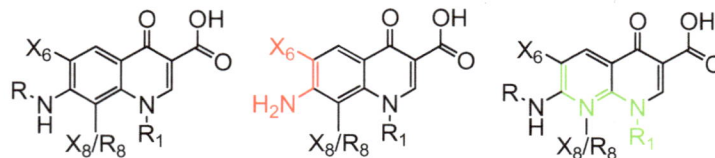

Scheme 1: R= alkyl, cycloalkyl, X= halides, specially –F, The part of the structure colored red corresponds to the improvisation of the biophores while the part structure highlighted as green is a generalized biophobe.

In the year 1993, Klopman *et al.* discovered and applied a new technique called MULTICASE methodology [69-72] which is a hierarchical computer automated structure evaluation program, to datasets of quinolone derivatives tested against 14 strains of *M. avium* [73]. In this approach, the significant substructures, the log P, the quantum mechanics parameters, and graph indices were used as potential parameters for a regression analysis to generate a linear QSAR. With MULTICASE methodology the fragments having relationship with the activity were identified and called **biophores**. This study not only helped to identify a number of key structural features (biophores and biophobes as shown in **Scheme 1**) relevant to the activity of quinolones against *M. avium* but also generated predictive QSAR models. This helped to identify new potent quinolones [74].

To derive a deeper insight Klopman and co-workers then studied a larger set of 116 fluoroquinolones (**Scheme 2a**).

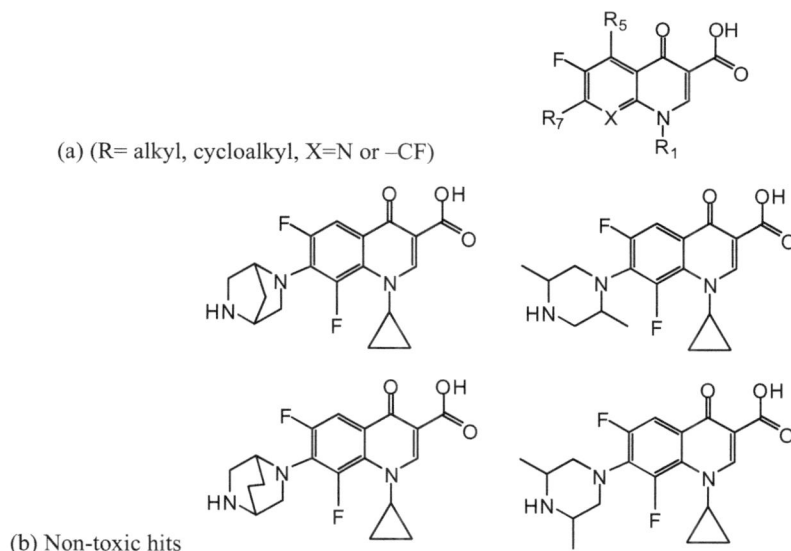

(a) (R= alkyl, cycloalkyl, X=N or –CF)

(b) Non-toxic hits

Scheme 2: Fluoroquinolones potent against *M. avium.*

The methodology and tools as described above were applied to understand the QSAR pattern for this dataset, but this time the cytotoxicity of these compounds in humans was also considered as a dependent variable. Their efforts resulted in the identification of biophores responsible for the toxicity. Though some similarity was observed between the biophores relevant to cytotoxicity and antimycobacterial activity, nevertheless it gave to them a handle on the problem by identifying associated modulators for both the activities. Having these models in hand Klopman *et al.* wisely applied them to the set of molecules designed in their work cited above. This resulted in scrutinizing four molecules (**Scheme 2b**) which were predicted to be non-toxic and potent and were later confirmed by experimental evidences [75].

Encouraged with the fruitful results from their previous work, Klopman *et al.* determined the MICs of N^1-^{t}butyl-substituted quinolones from the quinolone library at Parke-Davis, Ann Arbor, Michigan, against 14 selected reference and clinical strains of the *M. avium-M. Intracellular complex*. This resulted in the identification of 45 quinolones which were active against the *M. avium-M. Intracellular complex*, the N^1-^{t}butyl-substituted quinolones being the most active of the set. Along with the aforementioned series, some N^1-cyclopropyl-substituted quinolones, a few other quinolone derivatives and quinolones previously studied by the MULTICASE approach, a structure-activity relationship analysis was performed, which led to the identification of two major biophores and two biophobes [76].

In continuation, a slightly different approach was followed by Gozalbes and his group to a quinolone dataset containing 158 molecules (**Scheme 3**) [73,76,77]. Using topological indices as structural descriptors, a relationship to the anti-*M. avium-M. intracellular* complex activity was established by the Linear Discriminant Analysis (LDA) technique (equation 6)*.

$$M_{Ma} = -2.6 + 20.1[^3\chi_{ch}] - 12.9^4{}_c + 42.5[^4\chi_c{}'] + 25.6[^6\chi_{ch}] - 2.2[G_3{}'] + 2.4[G_4{}'] \qquad \dots \text{(equation 6)}$$

---

*Subscripts to the activity terms in the QSAR equations indicate the species of mycobacterium against which the activity was reported.

In the best LDA model (equation 6) the value of M indicates the actives (if M>0) or inactives (if M<0) and the indices $^4\chi_c$ and $^4\chi_c^v$ represent the quaternary ramifications; $^3\chi_{ch}$ and $^6\chi_{ch}$ reflect the presence of cycles of three and six atoms.

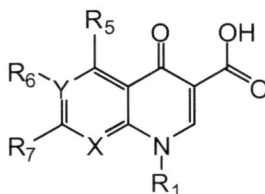

Scheme 3: X,Y = C or N

Further, in this study the correlations between the variables and activity were also found with the help of multilinear regression (MLR), equation 7.

$$\log(1/MIC_{Ma}) = -7.6 + 1.0[5\chi p] - 2.0[9\chi p] + 0.3[0\chi v] \qquad \text{... (equation 7)}$$

According to their models (*e.g.* equation 7) the presence of rings of three and six atoms at N_1 make a marked contribution to the improvement of activity, reflecting the role of the cyclopropyl substituent at N^1 against the *M. avium- M. intracellular complex* activity. Like the work by Klopman *et al.*, the predictive power of the QSAR models was also evaluated and models were further utilized to design derivatives of moxifloxacin, sparfloxacin, and gatifloxacin [78].

The advent of a new descriptor the Szeged index (Sz) provoked Agarwal *et al.* to carry out a 2D-QSPR study on N-2,4-difluorophenyl quinolones (**Scheme 4**).

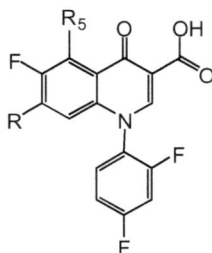

Scheme 4: *N*-2,4-difluorophenyl quinolones used to derive 2D-QSPR modeling of antitubercular activities.

In this approach a number of descriptors like Sz, the Weiner index (W) and molecular orbital energies (E_{HOMO}, E_{LUMO} and E_{total}) were calculated and related to clogP by univariate as well as multivariate regression analyses. The predicted clogP values (equations 8-10) were found to be within the range of experimental errors [79].

$$\text{clogP} = (0.4692)E_{total} - 24.9894 \qquad \text{... (equation 8)}$$

$$\text{clogP} = (6.9445 \times 10^{-4})Sz + (0.6669)E_{total} - 34.1581 \qquad \text{... (equation 9)}$$

$$\text{clogP} = (-6.6051 \times 10^{-4})Sz - (0.4718)E_{LUMO} - (0.6650)E_{total} - 34.3017 \qquad \text{... (equation 10)}$$

By this time a good dataset of N^1, C^7 and C^8 substituted quinolones was available and taking this as an opportunity Ghosh *et al.* undertook efforts to develop QSAR models using a variety of physico-chemical, constitutional, geometrical, electrostatic and topological indices. The descriptors calculated were studied through Ridge Regression (RR) models to better explain structure-activity relationship. This study used a novel method of dividing the activity data by means of structural similarity and thus attempts were made to investigate and compare the pattern of QSAR models. Using an in-house computer program Ghosh *et al.* classified the data set of 69 compounds into three classes, based on the degree of similarity among atom pairs with sparfloxacin and the mathematical derived descriptor based models. Of the three methods of analysis - ridge regression, principle component analysis and partial least squares, ridge regression produced the best model. It was found that the RR model was the best amongst the three and in the complete set of 69 compounds, 51 compounds had 50% similarity with sparfloxacin and 22 compounds were 60% similar to sparfloxacin in terms of descriptors. The models built from 51 and 22 molecules for inhibitory activity against *M. fortuitum* and *M. smegmatis* respectively, showed a good fit only with the topological descriptor, however, consideration of other theoretical descriptors like constitutional, geometrical and electrostatic indices improved the quality of the model [80].

Bagchi *et al.*, have also investigated quinolone derivatives assayed against *M. fortuitum* and *M. smegmatis*. They used descriptors like the triplet index from adjacency matrix, atomic number, and graph order, distance sum, sum of cross-product of degrees over all connected vertices, number of paths of length, flexibility index, sum of square of degrees over all vertices, chain connectivity index of order n *etc*. The triplet index from adjacency matrix (AZN_4, AZS_4), graph order and vertex degree (ANS_4, ASV_3, ANV_2) were found to correlate well with the activity against *M. fortuitum* (equations 11 and 12).

$$\log MIC_{Mf} = -(8.41)AZN_4 - (7.44)AZS_4 - (7.37)M_2 - (6.88)P_2 + (6.82)phia - (6.50)M_1 - (6.332)^3\chi_{Ch} - (5.93)ANS_4 - (5.32)ASV_3 - (4.94)ANV_2$$

       ... (equation 11)

$$\log MIC_{Ms} = -(5.47)SssO + (5.14)NHB_{int5} - (5.08)^3\chi_{Ch} - (4.89)^6\chi_{Ch} + (4.81)SsF + (4.77)phia - (4.56)P_5 + (4.26)SsssssC - 4.03(AZN_4) - (4.03)SHB_{int2}$$

       ... (equation 12)

For the activity against *M. smegmatis*, sum of the E-states for –O–, number of potential internal hydrogen bonds separated by 5 edges, chain connectivity index of order 3, chain connectivity index of order 6, sum of the E-states for –F, sum of square of degrees over all vertices, number of paths of length 5, sum of the E-states carbon with four single bonds, triplet from adjacency matrix, atomic number, and graph order, sum of E-state products for potential internal hydrogen bonds separated by 2 edges were found essential [81].

Meanwhile Anquetin *et al.* reported the rational design of a series of new 6-fluoroquinolones (substituted with groups shown in **Scheme 5**) from a QSAR analysis using molecular connectivity of molecules active against *T. gondii* [82] which were later also tested against the *Plasmodium* spp. and *M. tuberculosis*.

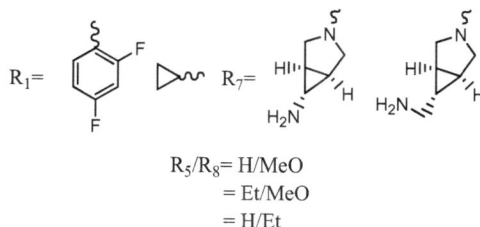

Scheme 5: R_1, R_5, R_7, R_8 = are the substituents at the 1, 5, 7 and 8 positions on the quinolone rings.

The molecules were found to inhibit the DNA supercoiling caused by *M. tuberculosis gyrase* like the most active quinolones but are poor inhibitors of *M. tuberculosis* growth. This is an example where QSAR carried out for other species of microorganisms was used to direct the design of antitubercular drugs. Through prediction of activity with models based on parameters like Topological Indices (TIs) and the E-state indices, some structures were identified as hits. The success of this project revealed some similarity between the SARs of 6-fluoroquinolones for different species and the discovery of the wide spectrum of activity of 6-fluoro-8-methoxy-quinolones [83] .

Recently in 2003-04, Nayyar *et al.* took up the design and synthesis of some novel derivatives of quinoline and they ended up identifying eleven series of quinolines (**Series 1-11** in **Scheme 6**) [84-87]. Later in 2006, they reported an extensive 3D-QSAR study on 108 molecules belonging to these eleven series.

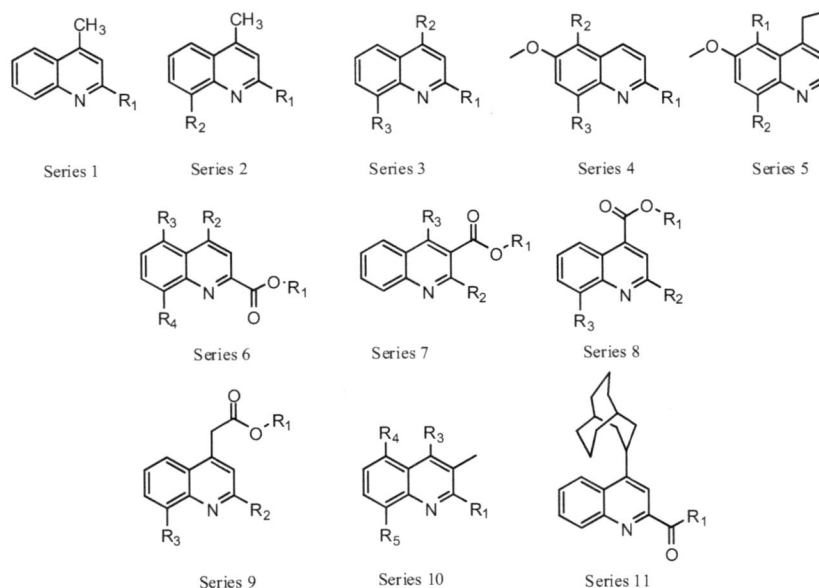

Scheme 6: Series of quinoline derivatives with antitubercular activity.

The 3D-QSAR studies required the energetically and conformationally stable molecules to be aligned in 3D-space based on common features in the molecules. Two types of alignments *viz. atom fit* alignment and *field fit* alignment were employed. In *atom fit* alignment the common atoms of molecules in the dataset are matched thus bringing them in one co-ordinate frame, while in case of field-fit alignment the fields (steric/electrostatic) calculated around a grid surrounding the molecules are matched. On these aligned sets the 3D-QSAR techniques CoMFA and CoMSIA were run. With Partial Least Square (PLS) analysis [48] the regression coefficient between the descriptors and the *logit* activity was determined. According to this study the hydropathic values calculated with HINT [88] and included in CoMFA models did not seem to be essential for antitubercular activities. The study helped to identify the positions where electropositive, electronegative and sterically bulky groups could be placed to improve the activity [89].

Based on the results of the above study Nayyar *et al.* modified the structure of quinolines to give the 2/4-quinolinecarbaldehyde (hydrazine/hydrazide) derivatives (**Series 12-15** in **Scheme 7**).

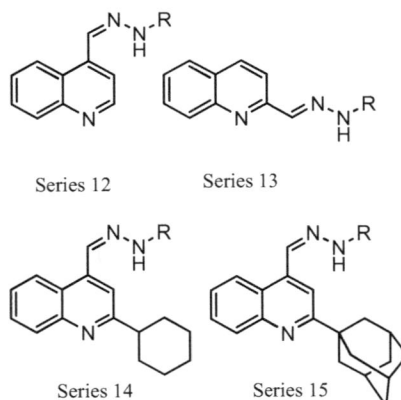

Scheme 7: Series of 2/4-quinolinecarbaldehyde (hydrazine/hydrazide) derivatives with antitubercular activity.

This time because of the way the biological data was generated, a different methodology was utilized to gain a better insight into the QSAR of quinolines derivatives. The binary classification as active or inactive (1 or 0) was used for the QSAR study. This data was correlated with similarity indices which were utilized as 3D-descriptors. The similarity indices were calculated based on atomic charges, shape, molar refractivity (MR) and lipophilicity (logP). The 'Carbo molecular similarity index' in terms of both 'individual similarity indices' and the 'combined similarity index', were calculated for 24 molecules belonging to the four series mentioned above. Principle Component Analysis (PCA) was used to find the appropriate numbers of components (independent variables). Stepping Discriminant Analysis (SDA) was carried out on the binary activity data to derive statistically sound models. The results obtained were in the form of "degree of confidence" *i.e.,* confidence for active, inactive and the overall confidence. This enabled them to classify newly designed molecules as active or inactive within the limits of statistical confidence and assisted in developing new ring-substituted quinolines compounds effective for the treatment of TB [90].

Motivated by these interesting results the work was taken to one more step of development which yielded the 4-adamantan-1-yl-quinoline-2-carboxylic acid *N*-mono and di-alkylhydrazides (**Scheme 8**) and 4-adamantan-1-yl-quinoline-2-carboxylic acid alkylidene hydrazides (**Scheme 8**) as promising antitubercular agents.

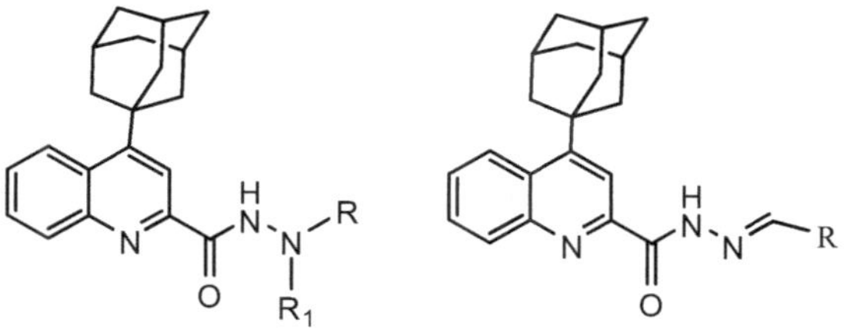

Series 16   Series 17

Scheme 8: Promising 4-adamantanyl quinoline derivatives as antitubercular agents.

A 3D-QSAR study on the lines discussed above was carried out and the study revealed that there is even more scope for modification of the 2-substituent on the quinoline framework in that there is enough space in the receptor (yet unknown) to accommodate these bulky substituents. In addition if this side chain at position-2 contains an electrostatic group it would further enhance the antitubercular activity [91].

The interesting outcomes for 4-adamantanyl quinolines stimulated Nayyar *et al.* to move ahead one step further, and embarked on making hybrids of 4-(adamantan-1-yl)-2-quinoline carboxylic acid and 4-(adamantan-1-yl)-2-quinoline carboxylic amides with hydrazides (**Series** 18 in **Scheme 9**) and with amino acids (Series 19) since such hybrids satisfy both the requirements as discussed above.

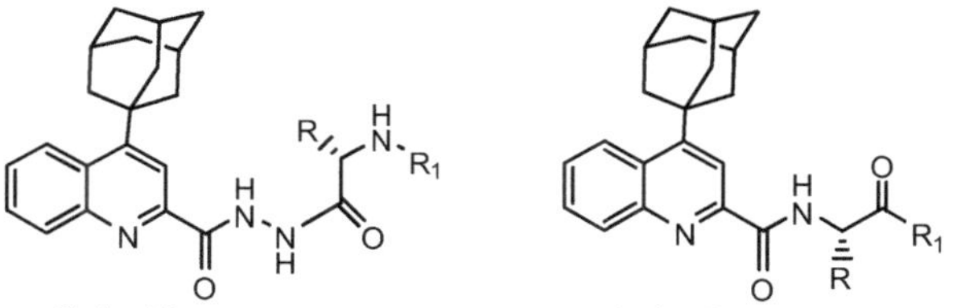

Series 18   Series 19

Scheme 9: Derivatives of 4-adamantanyl quinoline of Scheme 8.

The amino acids used for conjugation were of the type basic, aromatic, aliphatic and some unnatural. In the two series some conjugated amino acids still retained the ᵗBoc and Bzl protecting groups, so as to validate the effect of bulk at position-2 as gathered from the QSAR study. The 3D-QSAR studies carried out on the same lines as mentioned above revealed that more potent analogs could be obtained if the derivatives with aromatic amino acids carry an electropositive group on the aromatic ring while aliphatic side chains should bear electronegative groups. In contrast to earlier studies it was found that any increase in steric bulk in certain positions would affect the potency, thus indicating a ceiling for addition of bulky groups [92].

While the quinoline derivatives were under development and optimization, a group of Spanish scientists announced the discovery of quinoxaline-2-carboxylate-1,4-di-N-oxides (**Scheme 10**) as potential antitubercular agents [93,94]. The group also looked at the relationship between structure and activity.

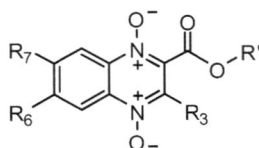

Scheme 10: Quinoxaline-2-carboxylate-1,4-di-*N*-oxides as potential antitubercular agents.

It was an extensive piece of work where 1497 theoretical descriptors calculated with the software DRAGON were utilized to build linear regression models. The models were prepared for five different sets of activities determined from earlier experiments. The descriptors which were seen in some of best models were topological descriptors that characterize shape (describing different structural elements such as degree of branching, size, symmetry, flexibility, cyclicity, centricity, *etc.*) atomic property (with different weighting schemes to differentiate the nature of atoms: mass (*m*), polarizability (*p*), electronegativity (*e*) or volume (*v*). 2D-autocorrelations), BCUT descriptors (eigen values of the Burden matrices), geometrical variables (3D-descriptors obtained from moment expansions), GETAWAY (GEometry, Topology, and Atom-Weights AssemblY) descriptors, 3D-MoRSE descriptors, weighted holistic invariant molecular (WHIM) descriptors *etc.* These QSAR models helped in the proposal of 12 new quinoxaline derivatives. The best model for antitubercular activity is equation 13. With the same strategy Vicente *et al.* also focused on cytotoxicity issues with the best model represented by equation 14. Additionally they also tried to tackle the selectivity and efficacy issues through QSAR modeling and the results can be seen as equations 13 to 16 [95].

$$\log_{10}(MIC_{Mt}) = -0.144\,(\pm 0.04) - 3.492\,(\pm 0.3)\Omega_{MTS7p} + 8.465\,(\pm 1.0)\Omega_{R2u} + 3.109\,(\pm 0.5)\Omega_{BElm6} - 1.987$$
$$(\pm 0.4)\Omega_{R3p} + 67.114\,(\pm 21.0)\Omega_{R7p} \qquad \dots \text{(equation 13)}$$

$$\log_{10}(IC_{50}) = 43.160\,(\pm 3.0) - 43.093\,(\pm 0.04)\Omega_{MATS2m} - 3.340\,(\pm 0.3)\Omega_{GCI8} - 0.285\,(\pm 0.03)\Omega_{T(F\dots F)} + 0.070$$
$$(\pm 0.009)\,\Omega_{Mor02e} \dots \qquad \text{(equation 14)}$$

$$\log_{10}(IC_{50}/MIC_{Mt}) = 13.910\,(\pm 2) + 221.107\,(\pm 23)\Omega_{X5A} - 0.915\,(\pm 0.2)\Omega_{DISPe} - 8.298\,(\pm 2)\Omega_{RTe+} -$$
$$4.259(\pm 1)\Omega_{IVDE} - 1.373\,(\pm 0.4)\Omega_{Mor19u} \qquad \dots \text{(equation 15)}$$

$$\log_{10}(EC_{90}/MIC_{Mt}) = 15.192\,(\pm 2) - 0.0009\,(\pm 0.0007)\Omega_{TIE} + 1.908\,(\pm 0.2)\Omega_{S3K} + 79.972\,(\pm 10)\Omega_{G2v}$$
$$\dots \text{(equation 16)}$$

Further the quinoxaline derivatives were studied by combining two different quinoxaline datasets with a wide range of substituents in the 2, 3, 7, and 8 positions, on the basis of constitutional descriptors. Principle component analysis was run on the unified quinoxaline dataset, identified from score plots. These score plots were based on constitutional descriptors. A comparison based on leave-one-out cross-validation technique, between the models built with the constitutional descriptors and the calculated molecular descriptors, suggested that the former were more robust, stable and reliable [96].

Isonicotinyl Hydrazide (INH) Derivatives

Isoniazid is still an effective drug in the current therapy of tuberculosis. The first instance of a QSAR study was on a series of 2-substituted isonicotinic acid hydrazides (**Scheme 11**) where descriptors representing electronic, steric, and lipophilic properties of the substituents were correlated with the biological activity data (MIC) against *M. tuberculosis* (strain *H37Rv*). A remarkable aspect of this study was the attempt to correlate the reactivity of INH derivatives with the cofactor NAD in the receptor cavity.

Seydel *et al.* derived different QSAR models (equation 17) equating -logMIC with π (partition coefficient), pK$_a$ (dissociation constant) and V$_w$ (van der Waals volume) through multiparameter regression analysis (MLR). In order to compare the biological and the chemical reactivities (rates of quarternization) of the INH analogs the logarithm of the reciprocal MIC was plotted against the logarithm of the relative rate constants. A linear correlation was obtained for those pairs of compounds where the relative rate of quarternization is at least ≈2% of the rate of the unsubstituted compound.

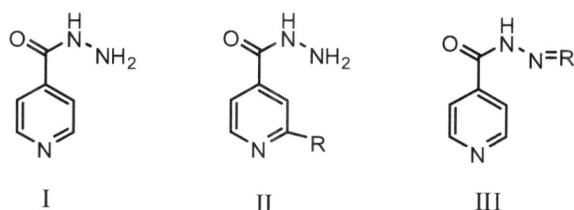

Scheme 11: Series of 2-substituted isonicotinic acid hydrazides as antitubercular agents.

$$-\log MIC_{Mt} = 0.533(5.79)\, pK_a - 2.469(6.57)\pi - 3.347 \qquad \text{... (equation 17)}$$

This model validated their hypothesis that there is a similar dependence upon steric and electronic influences of the substituents on the antibacterial activity of 2-substituted INH derivatives as for the reactivity of 2-substituted pyridines toward quarternizing agents [97].

Inspired by these results Bagchi *et al.* applied graph theory on a series of 2-substituted isonicotinic acid hydrazide (INH) molecules to establish a relationship between activity and some topological indices often used in QSAR studies. They calculated the Wiener index, molecular complexity indices and inter-molecular similarity of INH and its derivatives along with the pK_a, van der Waals volume and the estimated π values. The calculated topological parameters along with pK_a, van der Waals volume and π values were correlated with the antibacterial action by means of multiparameter regression analyses. It was found that three graphical invariants - atom pair based inter-molecular similarity (SIM), Structural Information Content (SIC) and Wiener index (W) alone could provide a good quality predictive QSAR model (equation 18).

$$\log(1/MIC_{Mt}) = 7.130(6.927)\, SIM + 0.0004(0.0148)W - 8.138(10.633)\, SIC + 0.407\,(0.226)pK_a -$$
$$0.095(0.072)VdW - 3.058 \qquad \text{... (equation 18)}$$

The best models generated by Bagchi and co-workers also show that pK_a and the estimated π values are important for the antitubercular activity in agreement with Seydel's observations [98].

Bagchi *et al.* took their work to the next level and published QSARs of INH derivatives using graph invariants. In this attempt they again explored the 2-substituted INH derivatives. In this study the theoretical molecular descriptors like Wiener index, molecular complexity indices and inter molecular similarity were introduced in the physico-chemical property based models and the superiority of the models over property–property correlation was shown. Here Ridge Regression (RR), Principle Component Regression (PCR) and Partial Least Squares (PLS) analyses were performed. A set of 230 molecular descriptors were calculated and partitioned into Topostructural (TS), Topochemical (TC), and geometrical (3D) classes and the RR models were developed utilizing these classes in a hierarchical fashion. According to this model the topochemical indices alone provided good quality predictive models for 2-substituted INH derivatives. Comparatively, the QSPR studies that used the small set of physico-chemical properties as molecular descriptors delivered inferior models [99].

As for the quinolone derivatives, Klopman *et al.* also applied the MULTICASE methodology to analyze the QSAR of 136 INH derivatives. Based on these models and the previous biophores generated, Klopman *et al.* predicted that Schiff base chemistry similar to vitamin B operation might be involved in the mode of action of isoniazid and its related compounds. Accordingly, in the very early stages in the mycolic acid biosynthetic pathway, INH binds to the cofactor ACP and then interacts covalently to form a α-ketoacid. This interaction slows down the mycolic acid synthesis, resulting in a weak mycobacterial cell wall. In addition it was also remarked that probably INH molecule as a whole is involved in this interaction and this mechanism gets a justification from the structural features of the less active hydrazides [100]

Lately, in 2004 Pasqualoto *et al.* reported a receptor-independent 4D-QSAR study on hydrazide derivatives (**Scheme 12**) in which the conformational *ensemble* of the 3D structures of the molecules combined with NAD (the cofactor for the enzyme *InhA* to which INH bind) were constructed with molecular dynamic simulations.

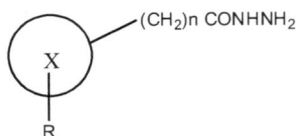

Scheme 12: Hydrazide derivatives used by Pasqualoto *et al.* for their 4D-QSAR study. The ring X represents substituted aromatic rings like pyridine, pyrimidine, pyrazine, imidazole, benzofuran, quinoline, thiazole, thiadiazole, piperidine, phenyl and benzyl.

Different sets of the conformers were aligned using interaction pharmacophore elements and then G/PLS analyses (genetic algorithm with PLS) were run to obtained a correlation between pMIC's and 3D-descriptors *i.e.,* Grid Cell Occupancy Descriptor (GCOD). This methodology not only resulted in good models (*e.g.* equation 19) but also in finding the possible bio-conformations which were comparable to those generated by docking the same derivatives in the enzyme *InhA*.

$$pMIC_{Mt} = -11.94GC1(np) + 11.37GC2(any) + 23.07GC3(any) + 1.57GC4(any) + 6.43GC5(any) + 30.7CG(np) - 0.23 \qquad \text{... (equation 19)}$$

According to this analysis (equation 19) the nonpolar groups in the acyl moiety of hydrazines markedly decrease biological activity [101].

Very recently in the year 2008, Ventura *et al.* used QSAR/QSPR methodology to analyze a huge set of 173 hydrazides, a great part of which were isoniazid (INH) derivatives. Various molecular descriptors belonging to physico-chemical, steric, geometrical and electronic categories were studied through MLR analyses. A very interesting outcome of this research was that biological activity of these compounds does not depend on lipophilicity (log*P*) (equation 20).

$$\log(1/MIC_{Mt}) = (1.892\pm0.142) + (-0.436\pm0.091)B + (1.201\pm0.126)V \qquad \text{... (equation 20)}$$

where B stands for Verloop's parameters while *V* is the McGowan characteristic molar volume.

Here again some light was shed on the mechanism of INH derivatives, which probably involves the formation of radical species. This assertion was made based on presence of electronic factors (*c* and *μ*) that contribute inversely to the activity as seen in equation 20. Interestingly this correlation shows a probability that these compounds go through an initial process of hydrolysis, producing INH and the respective ketone. It was also hypothesized that the activation of the prodrugs of INH and its analogs (hydrazines and hydrazones) should occur near the cellular surface of *M. tuberculosis* [102].

Around the same time, Andrade *et al.* were busy working with classical 2D-QSAR and hologram QSAR (H-QSAR) studies on isoniazid derivatives shown in **Scheme 13** below.

Scheme 13: R = pyridine, methylpyridine, pyrimidine, pyrazine, imidazole, benzofuran, quinoline, thiophene, thiazole, thiadiazole, benzothiazole, piperidine, phenyl and benzyl; R'= various alkyl, cycloalkyl, aromatic, saturated and unsaturated heterocycles.

For the traditional 2D-QSAR around 900 descriptors were calculated and only those which were found correlated with the activity were considered. While for H-QSAR, holograms were generated using 6 distinct fragment sizes over the 12 default series of hologram lengths. These molecular holograms were used as independent variables during the PLS regression analyses to derive the H-QSAR models. Cross-validation using the leave one out and leave group out helped to find robust models. It was observed that despite the strong influence of the topological descriptor Lopping centric index (Lop) in the final QSAR model, its raw correlation to potency (pMIC) was insignificant ($r^2 = 0.10$) indicating that steric constraints alone are not sufficient to describe inhibitory potency in this series. With this insight, the effect of the topological charge indexes that account for the mean value of charge transfer for each bond between atoms placed at a defined topological distance (three or nine for JGI3 and JGI9, respectively) was investigated and it was found that these descriptors were the most significant in the final QSAR model and a stringent charge balance is essential for biological potency. Similarly nN–N and nC=N–N< indices, which simply represent the number of aliphatic hydrazines and the number of hydrazones in the data set were found to have weak negative correlation and a similar positive correlation respectively with the potency. It affirmed that the conversion of hydrazine to the corresponding hydrazone derivative should improve the inhibitory potency. H-QSAR was performed to investigate the effect of different fragment sizes on the activity. Eventually it was verified that the molecular fragment corresponding to the hydrazide group positively correlates with the biological activity, and also the substitution of the hydrogen atom in the hydrazide group would be favorable towards activity [103].

Rifamycin Derivatives

Like isoniazid, rifamycin and its analogs have been used for decades against tuberculosis as reliable and effective drugs. The pioneering work in the field of QSAR on rifamycin analogs (hydrazides and amides) can be traced back to Corwin Hansch and his associates.

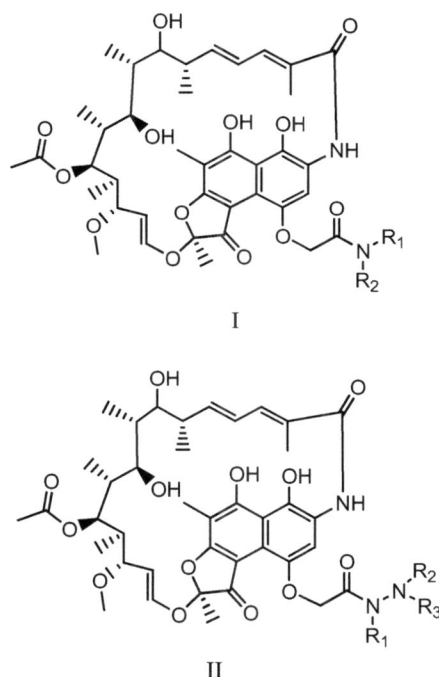

Scheme 14: Series of amides (I) and hydrazides (II) with antitubercular activity.

Quinn *et al.* and Hansch developed QSAR models for a series of 44 amides (I in **Scheme 14**) and 25 hydrazides (II in **Scheme 14**) of rifamycin B in five bacterial systems including *M. tuberculosis*. The best equations for the amides showed that activity was a parabolic function of log P (equations 21 and 22). A wide variation was found in log P for the various bacterial systems.

For rifamycin amides tested against *M. aureus*, the predicted activity is given by

$$\log(1/C_{Ma}) = 7.68 \ (\pm 0.15) - 0.41 \ (\pm 0.14)E_S - 1.27 \ (\pm 0.28)\sigma^* \qquad \dots \text{(equation 21)}$$

$$\log(1/C_{Ma}) = 7.41 \ (\pm 0.18) - 0.25 \ (\pm 0.15)E_S + 0.22 \ (\pm 0.11)(\log P)^2 - 1.20(\pm 0.24)\sigma^* \qquad \dots \text{(equation 22)}$$

The most important correlation parameter in the hydrazide equations was Taft's aliphatic substituent constant σ^*. Unfortunately, with these descriptors no adequate quantitative correlation could be obtained for the rifamycin amides in the *M. tuberculosis* system. Hansch *et al.* then introduced some dummy parameters which then were able to produce good correlation for the antitubercular activities of rifamycin hydrazide derivatives (equations 23 & 24) [104].

For rifamycin B hydrazides

$$\log(1/C_{Ma}) = 6.20\ (\pm0.26) - 0.51\ (\pm0.23)\log P - 0.06\ (\pm0.06)(\log P)^2 + 1.30\ (\pm0.24)D \quad \dots \text{(equation 23)}$$

$$\log(1/C_{Mt}) = 6.58\ (\pm0.09) - 0.40\ (\pm0.16)\sigma^* + 0.17\ (\pm0.06)\log P - 0.44\ (\pm0.17)A \qquad \dots \text{(equation 24)}$$

The work on rifamycin hydrazide derivatives continued for some time and this dataset was used as reference for the development of QSAR methods by researchers in the field. Kubinyi, another renowned computational chemist and QSAR expert, in 1976 applied a mixed approach based on Hansch and Free-Wilson analysis to a series of rifamycin hydrazides which were tested for antibacterial and mycobacterial activity. In this work, he intended to correlate the activities with both the structural as well as chemical features (descriptors). Along with other antibacterial activity, antitubercular activity was also used to build QSAR models (equations 25 and 26).

$$\log(1/C_{Mt}) = 0.007[\text{Et}] + 0.119[\text{Pr}] + 0.160[\text{Bu}] + 0.472[\text{Pip}] + 6.594 \qquad \dots \text{(equation 25)}$$

$$\log(1/C_{Ma}) = -0.137\ (\pm0.072)(\log P)^2 + 0.255[\text{Et}] + 0.444[\text{Pr}] + 0.777[\text{Bu}] + 0.362[\text{Pip}] + 8.231$$
$$\dots \text{(equation 26)}$$

The parabolic relationships were not observed for *M. tuberculosis* nor *M. aureus*, between the functional groups and the lipophilicities with the Free Wilson analysis. Surprisingly it was also projected that $(\log P)^2$ term was found less significant for antitubercular activity (equations 25 and 26).

Hansch analysis resulted in equation 27 where a good correlation was observed with σ^* and anti-*M. aureus* activity. Unfortunately, Hansch analysis was unsuccessful for antitubercular activity against *M. tuberculosis* because of the collinearity problem [105].

$$\log(1/C_{Ma}) = -0.96(\pm0.16)\sigma^* + 8.43 \qquad \dots \text{(equation 27)}$$

Meanwhile, a series of 3-formylrifamycin SV piperatinoacethydrazones (**Scheme 15**) was reported by Kiritsy *et al.*

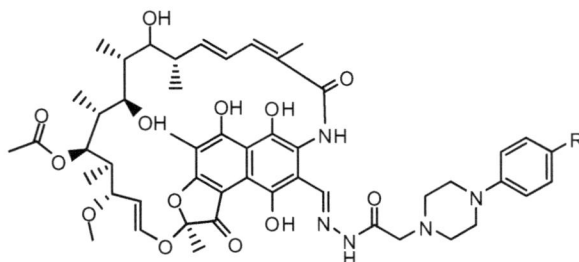

Scheme 15: 3-Formylrifamycin SV piperatinoacethydrazones with antitubercular activity.

Though these molecules were not as active as other rifamycins, they are important from a historical and medicinal chemistry point of view. Regression equations were generated for activity against *Bacillus subtilis, Staphylococcus aureus, C. perfringens, M. phlei,* and *M. tuberculosis* bacterial systems for all combinations of $\pi, \pi^2, \mathcal{F}, \mathcal{R}$, MR, and σ and without $\mathcal{F}, \sigma$ and $\mathcal{R}$. In this study the sensitivity of *M. phlei* to the compounds was found to correlate well with a linear combination of hydrophobic, electronic, and steric parameters (equations 28 and 29).

$$\log(1/C_{Mp}) = 0.27\ (\pm0.10)\pi - 0.23\ (\pm0.22)\sigma - 0.02\ (\pm0.01)MR + 5.16(\pm0.13) \qquad \text{... (equation 28)}$$

$$\log(1/C_{Mp}) = 0.30\ (\pm0.09)\pi - 0.42\ (\pm0.32)\Re - 0.02\ (\pm0.01)MR + 5.09(\pm0.10) \qquad \text{... (equation 29)}$$

Unfortunately, again no statistically significant correlation could be determined between the physico-chemical parameters studied and the activity of the compounds against *M. tuberculosis* [106].

Dimov *et al.* undertook the study of the 3-formylrifamycin SV derivatives (**Scheme 16**) synthesized by Maggi *et al.* [107] for QSAR analysis.

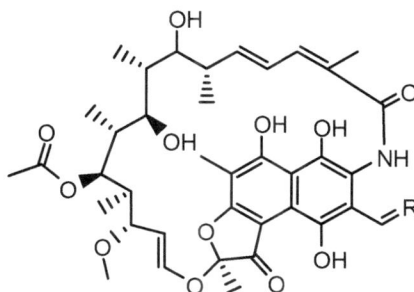

Scheme 16: 3-Formylrifamycin SV derivatives with antitubercular activity.

This was really a beautiful piece of work where Dimov *et al.* not only introduced new descriptors but the outcome was a new method and novel applicability of models/results. In their pioneering research, for the first time the AM1-based quantum chemical descriptors were utilized in their dynamic QSAR method that allowed explicit consideration of the conformational space of properly generated 3D structures. In this approach a diverse set of conformation of the molecules were generated using a genetic algorithm and then these sets were subjected to AM1 calculations, by which the descriptors like heat of formation, LUMO energy, HOMO energy, HOMO–LUMO energy gap, electronegativity, dipole moment, volume polarizability, degree of stretching or compactness (quantified as sum of interatomic steric distances), greatest interatomic distance, steric distance between atoms *i* and *j*, and planarity (normalized sum of torsion angles in a molecule) were calculated. Local electronic descriptors included atomic charges, frontier atomic charges, donor and acceptor super-delocalizabilities and self-atomic polarizability. The assumption supporting the selection of these descriptors was that hydrophobicity, steric bulk and size constraints were expected to play an important role in antibacterial effects (equation 30).

$$\log(MIC^{std}{}_{Mt}/MIC^{test}{}_{Mt}) = 1.907(\pm0.264)\ E_{HOMO} + 0.074\ (\pm0.028)\log K_{ow} - 150.116\ (\pm25.114)S^N{}_{16} + 32.844\ (\pm4.495)a_o \qquad \text{... (equation 30)}$$

It was observed in the models that activity [here log (MIC^{std}/MIC^{test}) in equation 30] is directly related to HOMO energy. This is clearly reflected in the mechanism of action of these compounds. Through increased electron donor capability the molecules were suggested to possess a tendency to undergo the hydroquinone-semiquinone-quinone oxidation reaction thus pointing that for the biological activity of the rifamycins, their readiness for an oxidation of the naphthohydroquinone moiety to the corresponding quinone is essential [108].

Taking the legacy one step ahead, Deeb *et al.* performed QSAR modeling of antimycobacterial activity of 3-formylrifamycin SV derivatives using topological descriptors and multivariate analysis (equation 31).

$$\log A_{Mt} = 1445.987\ (\pm663.583) - 27.100\ (\pm10.586)BIC4 - 717.297\ (\pm274.463)LP1 - 0.001\ (\pm0.000)PIPC10 + 25.844\ (\pm17.866)VEA1 - 298.788\ (\pm88.700)X5AV + 0.083\ (\pm0.026)MPC10 - 0.063\ (\pm0.023)EIG1M - 1.734\ (\pm0.557)SEIGE + 0.106\ (\pm0.034)MPC09 - 0.047\ (\pm0.012)BAC + 0.109\ (\pm0.024)VRP1 + 468.834\ (\pm124.995)XDA \qquad \text{... (equation 31)}$$

As seen in equation 31, it was concluded that antimycobacterial activity of 3-formylrifamycin SV derivatives could be explained with a twelve-parametric model containing a variety of molecular descriptors including distance-based and connectivity indices [109].

Pyridine Derivatives

Beside INH analogs other pyridine derivatives have also been a subject of interest as antitubercular drugs. In a study it was found that pyridine derivatives substituted with alkylated tetrazoles at the 3 and 5 positions showed *in vitro* activity that was better than pyrazinamide against *H37Rv M. tuberculosis* strain [110]. Desai *et al.* have synthesized and performed QSAR studies on the 1,4-dihydropyridine derivatives (**Scheme 17**) and have identified that structural and physico-chemical parameters (MR, σ^{ortho}, σ^{meta}, σ^{para}) contribute significantly to antitubercular activity of this series of compounds (equation 32).

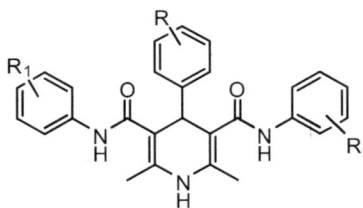

Scheme 17: 1,4-dihydropyridine derivatives with antitubercular activity.

$$\log C_{Mt} = 0.733 \ (\pm 0.177)\sigma^p{}_{R1} + 1.153 \ (\pm 0.131)\sigma^m{}_{R1} - 0.974 \ (\pm 0.204)\sigma^o{}_{R1} + 0.078 \ (\pm 0.014)MR - 1.070$$
$$(\pm 0.199) \qquad\qquad\qquad\qquad\qquad\qquad\qquad\qquad\qquad\qquad\qquad\qquad\qquad \dots \text{(equation 32)}$$

The QSAR models (*e.g.* equation 32) were able to illustrate the structural requirement for betterment of activity like the electronic influence of the substituents at the carbamoyl phenyl ring at the 2, 3 and 5 positions of DHP follows the order $\sigma^m > \sigma^p > \sigma^o$ [111].

To elaborate more on the QSAR, Desai *et al.* also reported CoMFA and CoMSIA studies on the same dataset, where it was once again shown that steric and electrostatic fields of the inhibitors were relevant descriptors for SAR. Some additional models built using genetic function approximation revealed the importance of spatial properties and conformational flexibility of the side chains for antitubercular activity. In addition, inclusion of fractional polar solvent accessible surface area showed a direct correlation and improved the correlation. Interestingly, a close observation of the models also revealed that placing –Cl and -CH₃ groups in certain positions would hamper metabolic degradation and improve the residence time of these compounds in the patient [112].

Motivated by the exciting results, Manvar *et al.* from the same laboratory synthesized 97 different symmetrical, asymmetrical, and N-substituted 1,4-dihydropyridines (**Scheme 18**).

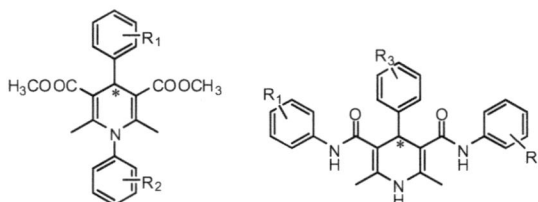

Scheme 18: *N*-substituted 1,4-dihydropyridines with antitubercular activity.

For the detailed 3D-QSAR studies the percentage inhibition values at different concentrations (µg/ml) were converted to pA$_{logit}$ values. With the diverse molecules aligned by the atom fit method, CoMFA and CoMSIA models were built and with PLS analysis, models with statistically sound parameters chosen. The models were validated by internal validation using bootstrap analysis, leave group out and y-scrambling, and the predictive power of the models were tested on an external set of molecules not included in the training set. The novel feature in this work was that the authors detailed on the ADMET properties through *in silico* and intuitional interventions. Thus new molecules were built based on the QSAR results and representative molecules were selected based on their predicted solubilities, log D, log *P*, pK_a values, oral bioavailability, protein binding, volume of distribution, genotoxicity, oral toxicity *etc.* [113].

Klimešová *et al.* analyzed 4-benzylsulfanyl derivatives of pyridine-2-carbonitriles and pyridine-2-carbothioamides (**Scheme 19**) through QSAR, using some physico-chemical and quantum–chemical parameters (equations 33 -36).

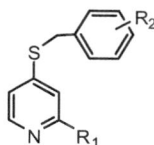

Scheme 19: (R_1 = CN, CSNH$_2$; R_2 = H, Cl, Br, F, CH$_3$, NO$_2$, OCH$_3$).

$$\log(1/C_{Mt}^{26\,mol}) = 0.367\ (\pm0.101)\sigma - 0.082\ (\pm0.118)\log P - 0.761\ (\pm0.093)I + 5.184\ (\pm0.370)$$
$$\dots \text{(equation 33)}$$

$$\log(1/C_{Mt}^{31\,mol}) = -0.797\ (\pm0.291)HOMO - 1.206\ (\pm0.145) \qquad \dots \text{(equation 34)}$$

$$\log(1/C_{Ma}^{33\,mol}) = -0.723\ (\pm0.348)HOMO + 0.498\ (\pm0.139)\log P - 1.166\ (\pm0.199)I - 2.660\ (\pm3.168)$$
$$\dots \text{(equation 35)}$$

$$\log(1/CMa27\,mol) = 0.186\ (\pm0.117)\sigma + 0.401\ (\pm0.138)\log P - 0.780\ (\pm0.113) \qquad \dots \text{(equation 36)}$$

A set of descriptor was calculated for this class of molecules which mainly included AM1 descriptors (HOMO orbitals, LUMO orbitals), Hammett constants σ, log *P*, MR and indicator variables for the different derivatives (for –CN =1 and –CSNH$_2$ =0). The molecular orbital calculation indicated a difference in the HOMO orbital for thioamide and nitrile derivatives and it was well justified through the QSAR models (equations 33 and 34). In addition, the HOMO orbitals helped in describing possible points of interaction (predominantly the -thioamide group) with the as yet unknown receptor. The σ descriptor (representing electronic effects) was seen to have a greater correlation than MR (representing steric effects) with log *1/C* for antitubercular activity (equations 33-36) and this has led to the interpretation that electron withdrawing substituents in the benzyl moiety would enhance activity. Essentially the models were built, through multiple linear regressions, to delineate between antitubercular activity against *M. tuberculosis* and *M. avium*. The result is that hydrophobicity is more important for activity against *M. avium* [114].

While others were engaged in performing QSAR for improving the activity of antitubercular agents, Coleman *et al.* used QSAR to evaluate the toxicity of the 2-arylcarbohydrozonamide series of compounds synthesized as antitubercular agents.

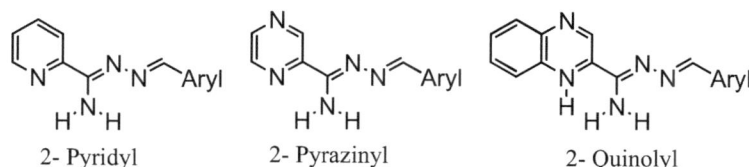

2- Pyridyl   2- Pyrazinyl   2- Quinolyl

Scheme 20: *N*-benzylidene-pyridine/pyrazine/quinoline-2-carbohydrozonamide with antitubercular activity.

In reality the compounds are analogs of N^1-benzylidene-pyridine/pyrazine/quinoline-2-carbohydrozonamide (**Scheme 20**). The cytotoxic activities (relative to INH) of this series of molecules were estimated against human mononuclear leucocytes. A range of QSAR descriptors were calculate on geometry-optimized structures which included mean polarizability, LUMO, HOMO, LUMO-HOMO gap, total dipole moment (from the AM1 semi-empirical calculations), ellipsoidal volume, log P, total dipole moment and molecular refractivity, see equation 37.

$$Activity_{Mt} = -0.059(\log P)^3 - 0.088(dipole\ moment)^3 + 0.002(lipole)^3 + 7.61 \qquad \text{... (equation 37)}$$

The 25 compounds were divided into two groups, for Group1 (N^1-benzylidene-pyridinyl-2-carboxamidrazone) compounds, toxicity was found to be related to hydrophobicity represented by log P (equation 37), while for Group2 (N^1-benzylidene-pyrazine/quinine-2-carboxamidrazone) compounds, the toxicity was found to be strongly related to the ionization potential (equation 38).

$$Toxicity = 14.35(ionization\ potential) - 119.11 \qquad \text{... (equation 38)}$$

From the QSAR studies, the dependency of toxicity on different parameters indicated a difference of mode of toxicities for the two groups of compounds. The models contradicted the previously assumed "detergent like cytotoxic action" and indicated that for Group1 compounds, by virtue of their lipophilicity (represented by log P) they are capable of entering into the cytoplasm, following which they exert their cytotoxic effects; while for Group2 compounds, the importance of ionization potential in the model, points towards intracellular redox reactions being responsible for their toxicity [115].

Pyrimidine Derivatives

Like pyridines, molecules with a pyrimidine nucleus had also been focus of research. Though most of the molecules of this class were tested and found active against the enzyme *dihydrofolate reductase*, a set of generalized data was available on which QSAR studies have been reported.

Virsodia *et al.* reported the synthesis, screening and 3D-QSAR studies of substituted tetrahydropyrimidine-5-carboxamides derivatives as antitubercular agents (**Scheme 21**).

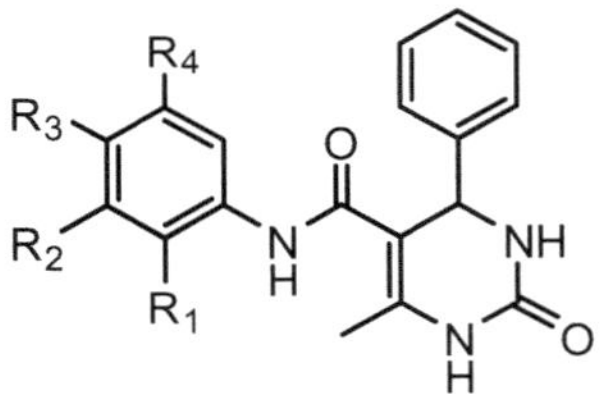

Scheme 21: Substituted tetrahydropyrimidine-5-carboxamides derivatives as antitubercular agents.

The CoMFA and CoMSIA models generated on the molecules aligned by the atom-fit method showed a good correlation with activity and the models were used to design some new molecules with improved activity. The contour maps revealed that there is a good scope for steric substitution at the 5-position. This indicated that the sizes of these molecules must match with the volume of receptor active site; one other aspect of the receptor active pocket is the presence of a negatively charged or H-bonding region that will strengthen binding with electropositive donor groups at the 4-position and acceptor groups in 5-carboxamide side chain [116].

Piperazines and Pyrazines

The piperazine derivatives are known to possess activity against a variety of infectious diseases including tuberculosis. It was astonishing to note that piperazines have not been studied as a separate class of antitubercular agents but have been mainly exploited as a side chain substituent to improve on the activities of various different core nuclei *viz.* fluoroquinolones, rifamycins, and pyrroles to name a few. Some of these have already been discussed in this context and the remaining would be discussed later. The unsaturated form of piperazine *i.e.*, pyrazine has been studied as a structural class of antitubercular agents. One of the oldest illustrations can be located in 1978, where pyrazine carbothiomides were prepared and QSAR studies showed that the activity was related both to the size of the substituent and the electronic properties of the molecule. In this venture, Kaliszan *et al.* used spectroscopic data to develop the QSAR models [117].

After a long break, spanning a few decades, another QSAR study appeared on the antimycobacterial activity of pyrazinoic acid esters (**Scheme 22**) by Bergmann *et al.* This class of compounds was earlier reported in 1950s with activity against *M. avium*, *M. kansasii* as well as *M. tuberculosis* strains.

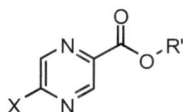

Scheme 22: Pyrazinoic acid esters with antimycobacterial activity.

The QSAR study was carried out on 51 diverse pyrazinoic acid esters collected from various literature sources having known minimum inhibitory concentration against multiple strains of *mycobacteria*. As the descriptors, Bergmann *et al.* calculated the Taft constant E_s, Charton's steric substitution constant v^{13}, the steric parameter $v_{OR'}$ (related to the rate of hydrolysis of ester), to represent the steric demand. Similarly for the depiction of lipophilicity log P and π values were used, while electronic character was described through the substitution constant σ. One combined parameter for steric and electrostatic features was also calculated using the Taft-Ingold equation.

$$\log(1/MMIC_{Mt,Ma,Mk}) = a\sigma + bv_{OR'} + c(\log P) + d\log t + e \qquad \qquad \text{... (equation 39)}$$

The final model (equation 39) generated through linear regression analysis demonstrated that E_s, log P and σ were directly correlated with the antitubercular activity. The innovative aspect of this research was the incorporation of parameters representing the reactivity of this class of compounds, considering the fact that this series of molecules undergoe a process of hydrolysis during their course of biological action. It was assumed in this regard that the factors enhancing stability of esters of pyrazinoic acid would also contribute towards activity. Thus it was claimed that high steric demand and moderate lipophilicity for these compounds was the key for compounds with better activity. Moreover the electron releasing substitutions would reduce the stability of the pyrazinoic acid esters [118].

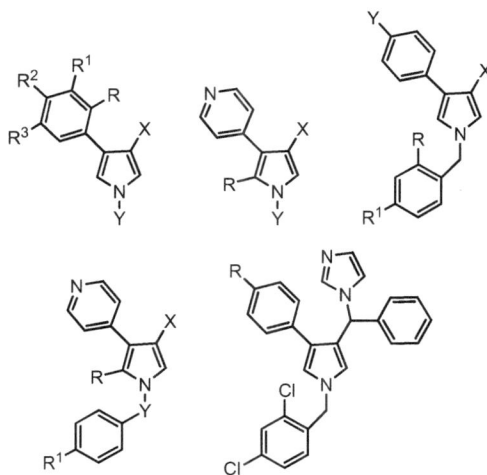

Scheme 23: Aryl-pyrroles used against *M. tuberculosis*.

Pyrroles

In the year 2000, Ragno *et al.* tested for antitubercular activity, compounds depicted in Scheme **23**, which were known to possess good antifungal activity. It was then discovered that, of this set, the 3-aryl-pyrroles had excellent activity against *M. tuberculosis*.

In order to design molecules with even better activities, a 3D-QSAR study was done in addition to the standard 2D-QSAR study. In all, three models were made with these two methodologies, with one of the models constituted from a mixture of 2D-QSAR and CoMFA descriptors. For the combined 2D-QSAR:CoMFA study, along with the CoMFA steric and electrostatic fields, the logarithm of the partition coefficient (HlogP), HOMO and LUMO energies, ionization potential, dipole moment, molecular volume (Vol) and the cavity-dispersion-solvent structure free energy were computed and used as descriptors. The PLS statistics led to QSAR models which revealed (through contours) that the presence of bulky substituents on the 2-position and a hydrogen bonding donor at the 4-position are vital for the activity. Similarly hydrophobicity was also found to correlate to the activity and high coefficients for both the dipole moment and the LUMO energy (coefficient with opposite sign) were also observed. The molecular volume had limited contribution and hence led to the conclusion that there was not much scope in increasing the volume through diverse substitution [119].

On the basis of a previous pharmacophore model (four points, represented by a hydrophobic region, a hydrogen-bond acceptor group and two aromatic rings) and structure-activity relationship of substituted pyrroles, Biava *et al.* reported the design and synthesis of new analogs of 1,5-(4-chlorophenyl)-2-methyl-3-(4-methylpiperazin-1-yl)methyl-1H-pyrrole (**Scheme 24**).

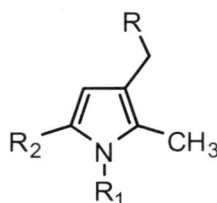

Scheme 24: 1,5-(4-chlorophenyl)-2-methyl-3-(4-methylpiperazin-1-yl)methyl-1H-pyrrole with antitubercular activity.

QSAR studies were performed to rationalize the activity of the new compounds in terms of their hydrophobic character. The log *P* values of the newly designed pyrrole derivatives were calculated with the aim of finding if any correlation between antitubercular activity and lipophilicity could be found. The results showed that the log *P* values of the 3-thiomorpholine derivatives were higher than the corresponding *N*-methylpiperazine counterparts. Thus it supported the hypothesis that a greater hydrophobic character is preferred for antitubercular potency of this class of compounds [120].

After a short break of a few years Biava *et al.* again tried to gain an insight into the activity of 1,5-diarylpyrroles (**Scheme 25**) [121-130] with a motto to combat MDR and XDR-TB.

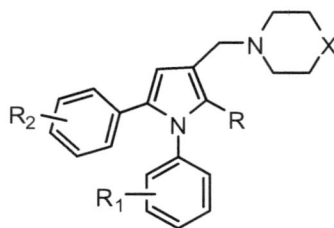

Scheme 25: 1,5-diarylpyrroles with antitubercular activity.

In this work it was shown that higher log *P* values were not always consistent with higher activity and this led them to revise their previous hypothesis that an increase in log *P* was in general associated with an improvement of activity. This was seen by the fact that the increase in log *P* due to replacement of the methyl at position 2 with an ethyl group was not profitable in improving activity. In this scenario, it was suggested that the substituent at position-2 was less effective in comparison to substituents at positions 1, 3, and 5. It was concluded that to enhance antimycobacterial activity, structural modifications should be made primarily at the phenyl rings at the N_1 and C_5 positions in the ring [131].

Indole Derivatives

Artificial intelligence and the neural network approaches are some of the most successful methods in many areas in science including molecular modeling. QSAR has also seen the adaption of these methods in its formalism.

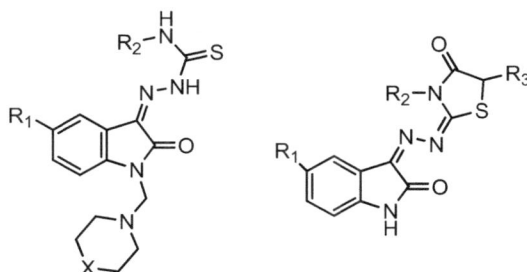

Scheme 26: Cyclized/non-cyclized thiosemicarbazones of 5-fluoro-1H-indole-2,3-dione derivatives with antitubercular activity.

One fine example is the QSAR studies of cyclized/non-cyclized thiosemicarbazones of 5-fluoro-1H-indole-2,3-dione derivatives (**Scheme 26**) where a combination of the electronic topological method with artificial neural network (ETM-ANN) was used. This method is actually a pharmacophore mapping method (through ETM) that generates models with a predictive capability (through ANN). This particular study started out with finding pharmacophores and "anti-pharmacophores" through ETM which is based on the system of "Electronic-Topological Matrices of Conjunction" (ETMC). The pharmacophore referred to here are the fragments which are essential/responsible for (and/or improving) activity while anti-pharmacophores are those fragments that diminish the activity. Through ETM, a continuous process of comparison among the ETMCs of molecules was carried out. In this manner, the pharmacophores were identified, and then the probability of finding common fragments among the pharmacophores was determined. All activity features detected by the ETM are characterized by their probabilities that lie in the limits 0.84-0.91. Comparison of the activity features with the 'breaks of activity' provided an explanation for the variation in the activity. The data obtained through ETM was weighted and represented in the form of ETSCs (sub-matrices of ETMCs) and then as descriptors (WDs) for the ASNNs training. To do this, the fragments were projected on the Kohonen's maps that correspond to their initial ETMCs. For the ANN training an *ensemble* of neural network was trained on a three-layered system with five neurons in one hidden layer and then the activity values were calculated. This data was subjected to cross validation process through the leave-out method. Lastly, pruning of the data led to the identification of the most relevant ETMC fragment. The ANN models were able to distinguish between the actives and inactives to quite a reliable extent. These models were efficient at this stage for not only prediction of activity but also in designing new antitubercular molecules [132].

Thiazole Derivatives

In order to have a glimpse whether the thiazole derivatives possess antitubercular activity, Vicini *et al.* synthesized three series - benzo[d]isothiazole, benzothiazole and thiazole Schiff bases (**Scheme 27**) and evaluated them against some atypical mycobacterial strains (*M. fortuitum* and *M. smegmatis*) [133]. From this initiative, many groups of scientists have synthesized derivatives of these classes, but it was only in the year 2008 that QSAR studies on thiazoles and thiazolidines have appeared.

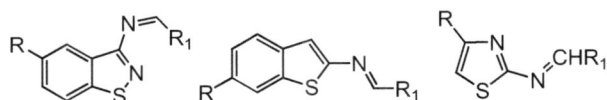

Scheme 27: Benzo[d]isothiazole, benzothiazole and thiazole Schiff bases used against atypical mycobacterial strains.

Sivakumar *et al.* carried out a QSAR study using Genetic Function Algorithm (GFA) to identify the best set of descriptors using a set of 55 pyrazine molecules containing thiazolines and thiazolidinones classes (**Scheme 28**). The equations 40 and 41 represent the best QSAR models for a 14 day and 21 day activity data.

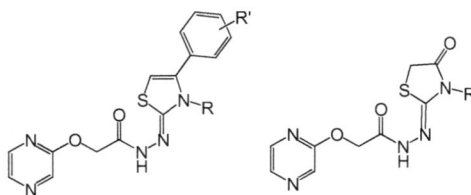

Scheme 28: Pyrazine molecules containing thiazolines and thiazolidinones as antitubercular agents.

$$\log(1/C_{Mt}^{14D}) = 1.793 - (0.253)Jurs_{-RPCS} - (1.014)Atype_{-C-40} + (0.053)Kappa_{-1-AM} \qquad ... \text{(equation 40)}$$

$$\log(1/C_{Mt}^{21D}) = 0.731 - (0.464)Jurs_{-RPCS} - (0.616)Atype_{-C-40} + (0.079)Kappa_{-1-AM} \qquad ... \text{(equation 41)}$$

This study revealed that the descriptors pertaining to electronic, topology, and hydrophobicity of the molecules are related to the activities. Ghosh and Crippen atom type descriptors were found to correlate negatively, indicating that there is a good scope for structural modification while maintaining the molecular shape, for improving the potency. The same descriptors also threw light on the hydrophobicity of the compounds, which was found to be important. It was realized that hydrophobic substitutions on the phenyl ring attached to the thiazoline ring contributes to the activity. The descriptors representing the topology (Kier's shape indices) affirmed that shape of the molecules was optimum for activity. Similarly, the presence of negative contributions from some spatial descriptors in the antitubercular models indicated that electropositive substitutions would decrease the activity [134].

Desai *et al.* have reported the synthesis and evaluation of several thiazolidine derivatives and their isosteric congeners as antibacterial agents and studied the QSAR. Their work started from the design and synthesis of thiosemicarbazides of 1,2,4-triazoles; 1,3,4-thiadiazole and 1,3,4-oxadiazole (**Scheme 29**) [135].

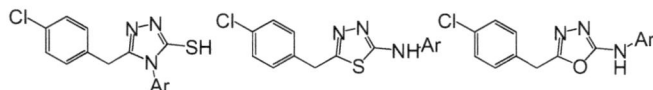

Scheme 29: Thiosemicarbazides of 1,2,4-triazoles; 1,3,4-thiadiazoles and 1,3,4-oxadiazoles with antibacterial activity.

Based on the outcome, they shifted their focus on 4-oxo-thiazolidines and 2-oxo-azetidines (**Scheme 30**). The inhibitory activities of these series of compounds were checked against only *E. coli* and *S. aureus* but unfortunately not against *Mycobacterium* [136].

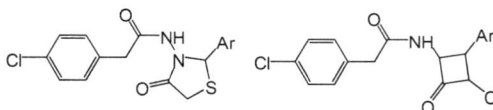

Scheme 30: 4-Oxo-thiazolidines and 2-oxo-azetidines with activity against *E. coli* and *S. aureus*.

This was taken further by Thaker *et al.* who replaced the N-phenylimidazolone present in previous molecules to the simple 3,5-*di*chlorobenzothiophene-2-carboxamide (**Scheme 31**) and tested them for antitubercular activity [137].

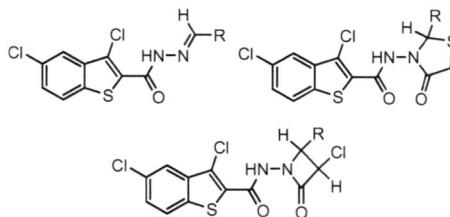

Scheme 31: 3,5-*Di*chlorobenzothiophene-2-carboxamide with antitubercular activity.

The same dataset was analyzed by Narute *et al.* by QSAR using sequential multivariate regression analysis to find some correlation between the biological activity and various 2D-descriptors. The descriptors included various physico-chemical, spatial, thermodynamic and electronic parameters (equation 42).

$$-\log BA = 4.453\ (\pm 2.671) + 0.0002\ (\pm 0.0001)PMX - 0.064\ (\pm 0.052)D + 0.458\ (\pm 0.301)HOMO - 0.023\ (\pm 0.014)E_t$$ ... (equation 42)

The final (validated) QSAR model (equation 42) for antitubercular activity revealed a dependence on the moment of inertia, dipole moment, HOMO and torsional energies. All this collectively pointed out that the antitubercular activity of this series of molecules was very much dependent on the partitioning coefficient and hence on lipophilicity. In addition, the positive contribution of HOMO indicates that the molecules need to be still more nucleophilic for better activity and adds another aspect on the mechanism of action. The torsional energy and dipole moment were seen to have negative coefficients indicating that there is still room for addition of bulky substitution on the molecules [138].

Thiadiazole Derivatives

Like thiazole derivatives, substituted thiadiazole analogs were also the focus of study of medicinal chemists in the past decades. The pioneering work on this series of molecules as antitubercular agents can be traced back to 1950s. Amazingly, the interest in this class of molecules has reemerged in recent times. In the above sections we have referred to the work of Desai *et al.* in 2008, however well before, in 2001, Foroumadi *et al.* at Kerman University of Medical Science, Iran had first investigated the potential of this series [139]. In association with Hemmateenejad and his colleagues at Shiraz University, Iran a QSAR study was undertaken.

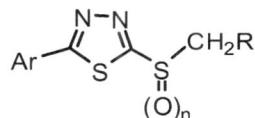

Scheme 32: 2-Alkylsulfide/sulfo(*i*)nyl-5-nitrofuryl/imidazolyl-substituted thiadiazones with antitubercular activity.

On a set of twenty seven 2-alkylsulfide/sulfo(*i*)nyl-5-nitrofuryl/imidazolyl-substituted thiadiazones (**Scheme 32**) Multiple Linear Regression (MLR), Free-Wilson Analysis (FWA) and Principle Component Analysis (PCA) were applied to derive QSAR models from a pool of chemical, topological, constitutional, quantum and descriptors related to functional groups. One interesting aspect of this study was the use of individual class of descriptors to build the models with MLR, so that the effect of each kind of descriptor could be traced, which might otherwise be subdued if all descriptors were used collectively. Nevertheless, one model was built with all the descriptors to identify the most prominent descriptors out of the mass. This model was built with contributions from the mean electro-topological state (*Ms*) as a constitutional descriptor, sum of the surface area of polar atoms (*PSA*) as the electronic parameter, number of sulfur atoms (*nRSR*) as a functional group parameter and a valence connectivity topological index (*X2Av*), see equation 43.

$$Activity_{Mt} = 14.01 - 2.37Ms - 0.002PSA - 0.76nRSR - 12.22X2Av$$ ... (equation 43)

The MLR equation (*e.g.* equation 43) describes the significant roles of hydrophobic and electronic interaction on the anti-tubercular activity; the effect of molecular topology and transient interactions on the anti-tuberculosis activity was also elaborated. As mentioned earlier, multiple models were generated based on one type of descriptor and it felt necessary to check if these equations are consistent in their correlation with the descriptors. For this, Hemmateenejad *et al.* formulated a correlation matrix by calculating the correlation coefficients among various descriptors observed in the equations. The outcome of this exercise was that several descriptors correlated well and hence the MLR equations were confirmed to be equivalent in terms of the results.

Further, PCA analysis on the descriptor matrices helped to classify the molecules as actives and inactives. The PCA scores were tested for their efficiency in discriminating the active from the inactive molecules. The patterns of distributions obtained by plotting the first three scores against each other helped in identifying related and outlier molecules in the dataset.

Since the chemical descriptors are functions of chemical structures, the FWA was also performed to obtain some information (equation 44) on the scope and effect of modifications. The FWA equations indicated that lower number of nitrogen atoms and sulfoxide group would result in compounds with higher activity [140].

$$Activity_{Mt} = 1.01 + 0.07 MN1 - 0.39n - 0.42CO_2Et - 1.04CONH_2 - 1.04CONHNH_2 + 0.31Me - 0.02Et - 1.08CO_2H \qquad \text{... (equation 44)}$$

In continuation, Foroumadi *et al.* synthesized and evaluated 2-nitroarylsulfo(*i*)nyl-5-nitrofuryl/imidazolyl-substituted thiadiazones (**Scheme 33**) to gain a better perception of the QSAR of this series as antitubercular agents.

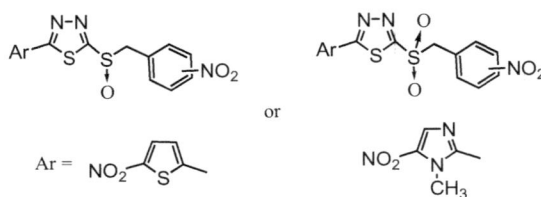

Scheme 33: 2-nitroarylsulfo(*i*)nyl-5-nitrofuryl/imidazolylsubstituted thiadiazones with antitubercular activity.

From the results of the QSAR study, Foroumadi *et al.* concluded that electronic distribution was one of the detrimental factors for activity (equation 45). Like their previous work, the QSAR analysis involved Multiple Linear Regression (MLR) to correlate between the activity (logarithm of reciprocal inhibition percentage, $(\log 1/K_i)$ and physico-chemical descriptors such as hydrophobicity, topological indices, electronic parameters and steric factors. To validate the models cross-validation and Y-scrambling analyses were run. This time, the sequence of events was modified a little by determining first the correlation matrix among the descriptors and including only those descriptors which were collinear in the statistical analysis. Eventually, models with significant correlations were found using constitutional and electronic descriptors.

$$\log(1/K_{iMt}) = 24.88 \ (\pm 3.71) + 32.46 \ (\pm 5.16)Mp - 1.49 \ (\pm 0.07)MPC \qquad \text{... (equation 45)}$$

The descriptors appearing in equation 45 are the mean atomic polarizability (Mp) and Maximum Positive Charge (MPC). Specifically, presence of the former indicated that nitroimidazole analogs are more effective than their nitrofuryl counterpart [141].

Oxazoles and Oxadiazole Derivatives

QSAR studies for oxadiazoles were reported prior to the oxazoles. Macaev *et al.* synthesized 5-aryl-2-thio-1,3,4-oxadiazoles derivatives (**Scheme 34**) and tested them for antitubercular activity.

With the interesting pattern of activities observed, the group was motivated to peep into the QSAR for this series of molecules and followed a similar method as discussed earlier for indole derivatives, *i.e.,* the Electronic-Topological Method-Artificial Neural Network (ETM-ANN) based method was followed. The Electronic-Topological Matrices of Contiguity (ETMCs) were formulated and then the pharmacophores and anti-pharmacophores were searched in the 3D-structures of the molecules in the series through comparisons of ETMCs. For ANN, three-layers with five neurons in one hidden layer were considered. The data generated was subjected to cross-validation process through the leave-one-out method. Further application of pruning techniques helped in the selection of only five of the most appropriate parameters responsible for the anti-tuberculosis activity [142].

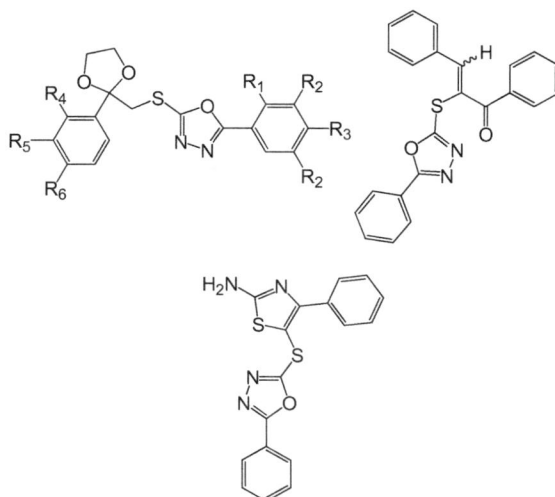

Scheme 34: 5-aryl-2-thio-1,3,4-oxadiazoles derivatives with antitubercular activity.

Sriram *et al.* synthesized and tested phenyl, benzyl and indolone derivatives of 5-cyclobutyl-oxazolylthiosemicarbazones (**Scheme 35**) against MDR-TB [143].

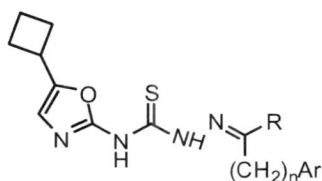

Scheme 35: (n=0 or 1).

The biological data was taken by Gupta *et al.* for QSAR studies wherein both the Fujita-Ban analysis [144,145] and Hansch QSAR were applied. Initially a *de novo* Fujita-Ban analysis indicated that many substituents have poor contribution to the inhibitory activity but further improvements revealed that electron withdrawing groups are favored at certain positions. Furthermore, the addition of one more aryl group on the terminal nitrogen of the thiosemicarbazone unit and retention of the nitroaryl substitution was suggested for improving the activity (equation 46).

$$p\mathit{MIC}_{Mt} = 0.416\ (\pm0.102)[R\text{-}CH_3] + 1.987\ (\pm0.205)[R\text{-}C_6H_5] + 0.069\ (\pm0.162)[R_1\text{-}2OH] + 1.059\ (\pm0.205)[R_1\text{-}2NO_2] + 1.059\ (\pm0.205)[R_1\text{-}3NO_2] - 0.166\ (\pm0.205)[R_1\text{-}3OCH_3] + 1.306\ (\pm0.162)[R_1\text{-}4NO_2] + 0.519\ (\pm0.162)[R_1\text{-}4CH_3] + 0.745\ (\pm0.205)[R_1\text{-}4Cl] + 0.455\ (\pm0.205)[R_1\text{-}4N(CH_3)_2] + 0.136\ (\pm0.205)[R_1\text{-}4OCH_3] + 0.022\ (\pm0.177)[R_1\text{-}4OH] + 0.622\ (\pm0.177)[R_1\text{-}4NH_2] + 0.384\ (\pm0.177)[R_1\text{-}4Br] + 4.587$$

… (equation 46)

For the Hansch analysis (equations 47 and 48), descriptors like substituent constants for hydrophobic (p), steric (Molar Refractivity or MR), Hydrogen Acceptor (HA), Hydrogen Donor (HD); the electronic descriptors (field effect $\mathscr{F}$, resonance effect $\mathscr{R}$, Hammet's constant σ) and Verloop parameters (shape of each substituent) were taken from various literature sources. The statistical method used to build the models was again MLR, while for validation of models cross-validation; bootstrapping and y-scrambling were used, while significance level of the equations was confirmed using sequential Fischer test. All the regression expressions were checked for the presence of outliers through Z-score method.

$$p\mathit{MIC}_{Mt} = 1.095\ (\pm0.287)Mor09u + 11.506\ (\pm1.475)ARR \qquad \text{… (equation 47)}$$

$$p\mathit{MIC}_{Mt} = 1.058\ (\pm0.330)Mor09m + 12.716\ (\pm1.649)ARR + 1.936 \qquad \text{… (equation 48)}$$

The best models showed that empirical aromatic index (ARR) contributed positively to the activity. Also, the 3D molecular representation of structure based on the electron diffraction (3D-MoRSE) code was found to be an important feature in the equations [146].

Benzoxazoles, Benzoxazines, Benzimidazoles and Benzmorpholines

A large series of 64 substituted N-phenylbenzoxazines and N-phenylquinazolines (**Scheme 36**) was synthesized by Waisser *et al.* during the decade 1992-2001 [147-150].

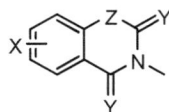

Scheme 36: (X=halogen(s), Y=S/O, Z=NH/O).

A novel QSAR approach based on a whole set of topological indices as theoretical molecular descriptors was applied to these molecules by Besalú *et al.* The topological descriptors calculated for the set of molecules comprised 2D descriptors like the Wiener index, Wiener Path Number, Randic number, Schultz, Balaban and Hosoya indices, Harary Number, Kier and Hall generalized connectivity indices, and Gálvez charge indices. Furthermore, a 3D-topological descriptor named as Topological Quantum Similarity Indices (TQSI) was also computed. A total of 162 descriptors for each individual compound were generated. These descriptors were filtered to reduce their number and extract the best amongst them through the Unsupervised Forward Selection (UFS) algorithm [151]. Multiple linear regression analysis was used to build the models, and the models were validated with LOO cross-validation and the Internal Test Sets (ITS) Method. Based on extensive statistical treatment, the best equation (equation 49) with 5-terms was identified.

$$\log MIC_{Mt} = -(0.757 \pm 0.243)[2GS] + (2.919 \pm 0.680)[7GS] - (2.068 \pm 0.252)[3\chi cC] + (4.640 \pm 0.773)[7\chi chC] + (4.886 \pm 0.501)[9\chi chC] - 16.299\ (\pm 2.133) \qquad \ldots \text{(equation 49)}$$

During this study special attention was paid towards avoiding over-fitting of models and the development of models with some degree of flexibility in terms of prediction and extrapolation. The predictive powers of the models were gauged through the *Q-probability* test. The resulting model was subsequently applied to a wider virtual molecular library comprising around 576 molecules and the activities of these molecules were predicted. Depending on the predicted activity some new promising candidates for active antitubercular agents were taken up for further investigation [152].

It should be mentioned that the QSAR models developed by Besalú *et al.* were of the predictive type, to get a better perspective of QSAR, Gallegos *et al.* applied fragment-similarity based approach (**Scheme 37b**) to the same set of compounds (**Scheme 37a**).

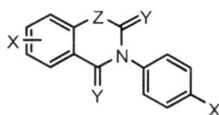

Scheme 37a: (X=halogen(s), Y=S/O, Z=NH/O).

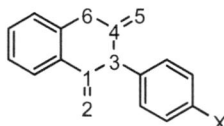

Scheme 37b: Fragmentation pattern, Numbering of atoms for the definition of molecular fragments considered as potential pharmacophores in compounds drawn in (**a**).

In this approach self-similarity measures of fragments were exploited as new universal molecular descriptors such that the models generated were themselves universal QSAR models which could describe the activity of all 39 molecules contained in the six series. In this project, the research group calculated the descriptors through application of the Molecule Quantum Similarity Measure (MQSM) method to the fragments of the molecules which is actually an extension of an older methodology where molecular similarities were judged based on the corresponding density function. These density functions were weighted by the Dirac delta function. For each fragment, the electron density was calculated from the electron density function of the whole molecule. Since the objective of this project was to dig out the essential features of the molecules responsible for activity, seven fragments were systematically defined with increasing number of atoms in one direction so that all permutations and combinations could be considered. The models were generated through statistical treatment of the MQSM fragments in the molecules and the respective $\log(1/MIC)$ values. Equation 50 was one of the best models obtained through Hansch analysis which was derived for one fragment $C_1=O_2$.

$$\log(1/MIC_{Mt}) = 0.826\sigma + 0.273\pi - 1.873 \qquad \ldots \text{(equation 50)}$$

A similar equation derived from all fragments is equation 51.

$$\log(1/MIC_{Mt}) = 0.0292 Z_{AA}{}^{C1=O2} - 0.0303 Z_{AA}{}^{O6-C4=O5} + 4.030 \qquad \ldots \text{(equation 51)}$$

Important contributions to activity were from fragments that represented the thioxo functionality and thus indicated that replacement of the -oxo group by the thioxo group could increase the potency of the molecules. The models built by this approach were compared with traditional Hansch and Futija QSAR scheme and it was found that the MQSM models were superior and universal in nature [153].

Very recently, a QSAR study has been reported for benzylsulfanyl benzoxazoles whose antitubercular activities were described at the beginning of this decade [154]. Klimešová *et al.* have put in some innovative efforts to extract structure-activity relationships about this series of molecules (**Scheme 38**) for combating MDR-TB.

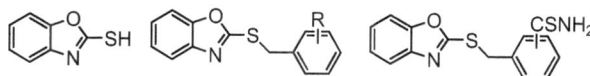

Scheme 38: Benzylsulfanyl benzoxazoles with antitubercular activity.

This study related the MIC of these series of molecules with descriptors like physico-chemical and quantum-chemical parameters, such as $\log P$ (hydrophobic effects), σ (Hammet constant) and energy of the HOMO and LUMO orbitals (ε_{HOMO}, ε_{LUMO}) as the electronic influences and *MR* (molar refractivity) for the steric effects, equations 52 and 53.

$$\log(1/C_{Mt}) = 0.06\ (\pm0.01)MR - 0.32\ (\pm0.11)\log P - 1.51\ (\pm0.45)HOMO - 12.00\ (\pm3.35)$$
$$\ldots \text{(equation 52)}$$

$$\log(1/C_{Mt}) = 0.05\ (\pm0.01)MR - 0.26\ (\pm0.11)\log P - 0.53\ (\pm0.14)LUMO + 0.68\ (\pm1.00)$$
$$\ldots \text{(equation 53)}$$

It was revealed through models (equation 52 and 53) that the molar refractivity and hence steric effects were quite significant for the potency of the compounds studied. It was interesting to note that in contrast to the earlier studied molecules, in this case, molecules with lesser lipophilicity were predicted superior in activity, and in line with the earlier report, the redox system based action mechanism of these antitubercular agents was affirmed through the presence of the E_{HOMO} and E_{LUMO} terms in the QSAR equations [155].

A related example which will be discussed in brief is the benzamidazole derivatives where activity contribution for the structural and substituent effects were determined from the correlation equations which were derived using stepwise regression technique. Geban *et al.* through these QSAR models, showed that the activity of benzimidazoles depends on the size and the field effect of the substituents at the various positions. In addition, a structural parameter IY which indicated the presence of oxygen between the benzimidazole and benzyl or phenyl groups was also anticipated to be contributing positively to the activity [156].

One of the recent examples, are morpholine derivatives which have been explored by Raparti *et al.* This group of scientists developed 4-(morpholin-4-yl)-*N*-(arylidene)benzohydrazides as antitubercular compounds (**Scheme 39**).

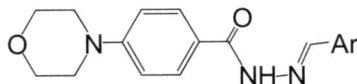

Scheme 39: 4-(Morpholin-4-yl)-*N*-(arylidene)benzohydrazides with antitubercular activity.

Both 2D- and 3D-QSAR studies (equation 54) were carried out for this set of molecules. For 2D-QSAR studies various experimental or calculated physico-chemical parameters were considered for generating QSAR models through Multiple Linear Regression (MLR).

$$pA_{logit(Mt)} = -0.473(T\_2\_N\_5) + 3.119 \qquad \text{... (equation 54)}$$

This model (equation 54) indicates that an increase in the number of double bonded atoms would decrease the potency; more precisely any double bonded atom separated from nitrogen atom by 5 bonds would be more detrimental.

The 3-D QSAR studies were carried out using the stepwise K nearest neighbour molecular field analysis [(SW) kNN-MFA]. The KNN-MFA models were generated using 10 molecules aligned by the field fit method. The models generated had good statistics and predictivity and could be useful for future development. The 3D-QSAR models pointed out that electrostatic contribution from the aromatic side chain would be the determining factor for activity of this series of molecules [157].

Furan Derivatives

It is noteworthy that until this point only nitrogen containing heterocycles had been studied but notably other heterocycles like furan and pyranes/pyrones have also been explored for antitubercular activity. There are many reports of nitrofuran derivatives as antitubercular agents in recent times (**Scheme 40**) [158-165].

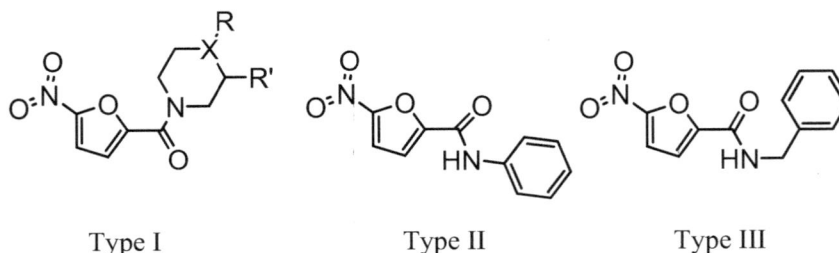

Type I   Type II   Type III

Scheme 40: Furan derivatives with antitubercular activity.

Nearly 110 molecules were used by Hevener *et al.* to investigate the QSAR of this class of compounds. The series contained three types of 5-nitro-2-carboxamides. Type I compounds were carboxamides of nitrogen containing saturated heterocycles, type II were N-aryl carboxamides and type III were N-methyl-aryl carboxamides. This is a good example of a systematic and rationale QSAR study. A number of 3D-QSAR models were built by CoMFA and CoMSIA methods. The ionized or non-ionized forms of the molecules were taken as two different sets and studied separately. Two different alignment methodologies were followed for these compounds. Likewise, the use of some selected descriptors like clog*P*, log*D*, molar refractivity (CMR), Polar Surface Area (PSA) along with the steric and electrostatic fields added to the number of models that were developed. Thus, 23 models were developed and compared and the best model was selected based on the statistical results as well as on the number of outliers. The models were further improved by the region focusing method and validated through bootstrapping and the y-scrambling techniques. The results revealed that molecules need sufficient polarity and hydrophobicity to exhibit antitubercular activity. There is yet scope for addition of bulky substituents on side chain at position-5. The electrostatic properties of the molecules are mainly due to the nitrofuryl group and it is sufficient enough for activity without further need for improvement [166].

Fused Heterocyclic Compounds

With the existence of extensive work on simple heterocyclic compounds as antitubercular agents, it was quite obvious that medicinal chemists would search for better, safer and specific agents among the fused heterocycles. One good example of such an effort, is the work by Bukowski and Kaliszan *et al.* who in the year 1989 synthesized and tested imidazo[4,5-b]pyridines (**Scheme 41**) as potent antitubercular agents.

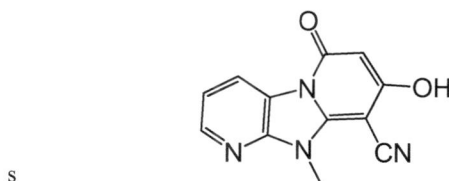

Scheme 41: Imidazo[4,5-b]pyridines with antitubercular activity.

In this study, many empirical descriptors were calculated, amongst them is the hydrophobicity parameter for fragments $\sum_f$, (calculated by the fragmental method of Hansch and Leo) [167]; also the indicator variable $\underline{D}$ (0 and 1 for methyl) was used in the study (equation 55).

$$\log(1/MIC_{Mt}) = -0.767\ (\pm0.222) + 0.298\ (\pm0.046)\sum f - 0.260\ (\pm0.192)D \qquad \ldots \text{(equation 55)}$$

Based on the QSAR equations (like equation 55) it was concluded that increasing hydrophobicity of the compounds would increase their *in vitro* antibacterial activity. A negative coefficient for $\underline{D}$ reflected that methyl substitution in the heterocyclic system decreases antibacterial activity [168,169].

Another good example of fused heterocycles, studied as antitubercular agents is thieno[2,3-d]pyrimidines which were synthesized and screened for activity by Chambhare *et al.*

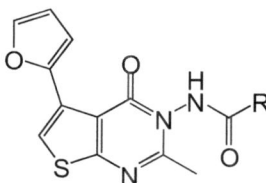

Scheme 42: Thieno[2,3-d]pyrimidines with antitubercular activity.

Two other series N-[5-(2-furanyl)-2-methyl-4-oxo-4H-thieno[2,3-d]pyrimidin-3-yl]-carboxamide and 3-substituted-5-(2-furanyl)-2-methyl-3H-thieno[2,3-d]pyrimidin-4-ones (**Scheme 42**) [170] were studied by Narasimhan *et al.* who calculated topological indices, molecular connectivity indices, valence order molecular connectivity indices, Kiers shape indices, Randic [R], Balban [J], and Wiener [W] topological indices as descriptors and tried to establish a correlation with logMIC for *M. tuberculosis* and *M. avium* strains (equations 56 and 57 respectively).

$$\log(1/MIC_{Mt}) = 1.622[^{3}\chi] - 0.990 \qquad \ldots \text{(equation 56)}$$

$$\log(1/MIC_{Ma}) = 1.645[^{3}\chi] - 1.019 \qquad \ldots \text{(equation 57)}$$

It was established from these studies that the third order molecular connectivity indices and Kiers shape indices correlated very well with antibacterial activity determined for several bacterial strains. The appearance of the $^3\chi$ term vividly indicated that the connectivity among the atoms and branching is sufficient to describe activity and there is some room for modification in the molecules to improve activity [171].

Like the aforementioned nitrofuranyl class of compounds, the nitroimidazoles also have been inspected by medicinal chemists for their antitubercular properties. Very recently, chiral 2-nitro-6-substituted 6,7-dihydro-5*H*-imidazo[2,1-*b*][1,3]oxazines/thiazines/diazines derivatives have been looked at for their antitubercular activity. It is worthwhile mentioning here that Kim *et al.* in this endeavor developed a purely pharmacophore model based on predictive models.

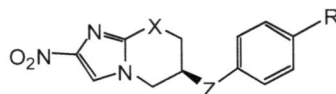

Scheme 43: (X = S, SO, SO$_2$, O, NR, Z = -OCH$_2$-, -NH$_2$-, -NHCOCH$_2$O-, -NH(CH$_2$)$_n$-, R = -OCF$_3$, -F,-OCH$_2$Ph(*p*-OCF$_3$), -^{n}Bu)

Though these molecules (**Scheme 43**) were designed to be inhibitors of *dependent nitroreductase (Ddn)* the enzyme inhibition assay was not performed for all molecules, hence the MIC values were explored for a possible correlation with the pharmacophoric features. Essential pharmacophoric features of the set of the molecules were first identified with a few highly active molecules and with it several *hypotheses* were generated. Each *hypothesis,* contained the distances, angles, planes *etc.,* between pharmacophoric features like H-donor, H-acceptor, hydrophobic features, aromatic nuclei and many others. These were used as descriptors and correlated with the activity. With the Fischer validation test, seven *hypotheses* were identified to be significant. These were further tested against an external data set of 42 compounds. From all significant models, it was explicated that oxygen atom at the 2-position of the imidazole ring was required for aerobic activity. For the other two series acetylating the amino group or oxidizing the thioether, or replacing the ether oxygen with carbon significantly reduced the potency of the compounds; this was backed by the *hypotheses* where H-acceptor features prevailed in the same position (oxazine oxygen). Similarly a hydrogen bond donor was positioned at the position-6 on the oxazine nucleus, indicating that the amino analog would be equipotent. Incidentally, replacement of the oxygen at 6-position by nitrogen slightly improved activity; this is probably due to the elevated solubility of the 6-amino compounds. Additionally the position and distances among the hydrophobes provided a good lead for understanding the receptor pocket and designing new molecules as well. Being a pharmacophore based approach, the models directly uncovered the binding pattern of these sets of ligands which was predicted to be a U-shaped conformation [172].

One more interesting example is the 6-acylamido-2-alkylthiobenzothiazoles which were tested *in vitro* for antitubercular activity towards *M. avium.* by Machcek *et al.* in the 1980s and relationships between the chemical structure and activity of these compounds were attempted by the Free-Wilson method. From this study, it was discovered that the substituents in positions 2 and 6 and the common benzothiazole part contributed significantly to the activity [173].

Xanthone Derivatives

In 1978 it was discovered that xanthone derivatives *i.e.* polyhydroxyxanthones possessed antitubercular activity [174]. After two years, Hambloch and Frahm through multiple linear analysis revealed the importance on activity, of the positions of -OH groups on polyhydroxyxanthones (**Scheme 44**) based on the ^{13}C chemical shifts [175].

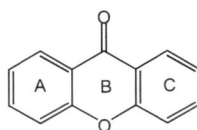

Scheme 44: Polyhydroxyxanthones with antitubercular activity.

Ungwitayatron *et al.* in the year 1996 took this work forward and developed the additivity rules for chemical shift calculations and used them as dependent variables for QSAR. The ^{13}C NMR data was taken from the literature. The chemical shift of atoms was assumed to be affected by the three neighboring atoms in the chain and this was the basis for formulation of the additivity rule. In this simple yet sophisticated research a correlation was sought between the charges on the atoms of the polyhydroxyxanthones and the ^{13}C NMR shifts. In combination with the hydroxyl chemical shift increment, equations like the one given below were deduced for a few representative compounds (equation 58).

$$\delta_{C1} = 153.46 + 270.47qC_1 - 45.09qC_2 + 185.12qC_{8b} + 24.59_{qOH1,H1} \qquad \ldots \text{(equation 58)}$$

These equations were able to predict the values of the chemical shift for different mono-, di- tri-, tetra- and pentahydroxyxanthone derivatives. As an extension of this project, Ungwitayatron *et al.* carried out a 3D-QSAR study for the same set of compounds using their antitubercular activity (1/*MICs*). The contour maps helped to identify that in order to improve the activity, the C_7 and C_8 positions should be occupied by bulky substituents while electropositive groups should be placed at the C_5 position of the polyhydroxyxanthone nucleus. These models were used to design a few molecules with improved activity [176].

Following a -hydroxy scan, Pickert *et al.* decided to study the effect of common substituents particularly – NO_2, –CH_3, –COOH, –COOR and –$CONH_2$ on the polyhydroxyxanthone nucleus [177]. The same series of 61 compounds were taken for QSAR studies in 1999. However, with multiple linear regression-based adaptive least squares analysis no statistically sound models could be generated, but with artificial neural network, nonlinear models could be developed between the activity and some of the physico-chemical parameters [178].

Some Assorted Organic Molecules

We will start with the most complex molecules - the phenanthrene derivatives. Diaryloxy-methanophenanthrenes were found to have antitubercular activity [179-181].

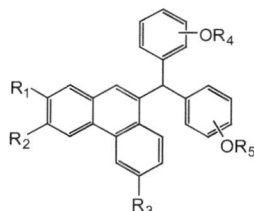

Scheme 45: Phenanthrene derivatives with antitubercular activity.

The same series (**Scheme 45**) were examined by the 3D-QSAR techniques of CoMFA and CoMSIA. The data set was aligned by the database alignment method (*atom fit*). One interesting feature in this study was that the research group tried to find the effect of the type of charges assigned to the atoms on the molecules on the quality of the models. Since both CoMFA and CoMSIA calculations involve calculation of interaction energies between the molecules and a probe atom, the type of charges will affect the model to a certain extent. The Gasteiger–Hückel charges were found to give the best models. The contour maps built using these charges indicated that there was still scope to add steric bulk in these molecules in certain positions. The active pocket for this class of molecules seems to be predominantly hydrophobic in nature, since the electrostatic contours were found make a lesser contribution to the overall activity [182].

For a medicinal chemist, the first priority is to make compounds with the simplest chemicals, methods, instruments and techniques. Since serendipity is rare, most of the medicinal chemistry research starts from the simplest organic compounds which then can be modified to improve selectivity, specificity and safety. QSAR and QSPR furnish this process of development quite efficiently and rationalistically. The esters of substituted phenylcarbamic acids represent a fine example of design and synthesis by Čižmárik, and Waisser *et al.* over the duration of almost five years, for antitubercular activity.

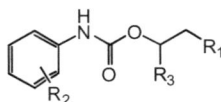

Scheme 46: (R_1 = piperidin-1-yl, pyrrolidin-1-yl or 4-methylpiperidin-1-yl; R_2 = *o-, m-*or *p-*$O(CH_2)_nCH_3$ (n = 0, 1, 2, 3, 4, 5, 6, 7, 8, 9), H, 4-isopropyl, 2-isopropoxy, 2-F, 2-I, 3-I, 2-CH_3 or 2-CH_2CH_3 group; R_3 = H, CH_3 or $CH_2OCH_2CH_3$).

These molecules (**Scheme 46**) were targeted and tested against *M. tuberculosis, M. avium and M. kansasii* and MDR-TB. The QSAR study with these chemical species involved a simple goal to establish a relation between activity and the lipophilic characteristic of the molecules (equations 59 – 67).

$$\log(MIC_{Mt}) = -0.571(\pm0.034)\log P + 3.774(\pm0.127) \qquad \text{... (equation 59)}$$

$$\log(MIC_{Ma}) = -0.557(\pm0.033)\log P + 3.871(\pm0.123) \qquad \text{... (equation 60)}$$

$$\log(MIC_{Mk}) = -0.698(\pm0.033)\log P + 4.356(\pm0.124) \qquad \text{... (equation 61)}$$

$$\log(MIC_{Mt}) = 0.169\,(\pm0.028)(\log P)^2 - 1.810\,(\pm0.210)(\log P) + 5.893\,(\pm0.372) \qquad \text{... (equation 62)}$$

$$\log(MIC_{Ma}) = 0.079\,(\pm0.029)(\log P)^2 - 1.121\,(\pm0.210)(\log P) + 4.801\,(\pm0.362) \qquad \text{... (equation 63)}$$

$$\log(MIC_{Mk}) = 0.071\,(\pm0.032)(\log P)^2 - 1.215\,(\pm0.237)(\log P) + 5.235\,(\pm0.418) \qquad \text{... (equation 64)}$$

$$\log(MIC_{Mt}) = \frac{-1.773}{1-e^{-(2.485\log P-7.842)}} + 2.847 \qquad \text{... (equation 65)}$$

$$\log(MIC_{Ma}) = \frac{-2.108}{1-e^{-(-2.108\log P-4702)}} + 3.166 \qquad \text{... (equation 66)}$$

$$\log(MIC_{Mk}) = \frac{-2.445}{1-e^{-(1.572\log P-5.367)}} + 3.184 \qquad \text{... (equation 67)}$$

The results from linear regression (best equations 59-61), parabolic regression (best equations 62-64) and sigmoidal regression (best equations 65-67) revealed that the sigmoidal relationships were most significant *i.e.* the activity would rise with the lipophilicity up to a certain extent and then fall [183].

The non-linear relationship between activity and lipophilicity (equations 68 and 69) was first reported by Waisser *et al.* who undertook an investigation on quaternary ammonium salts of piperidinylpropyl esters of alkoxy-substituted phenylcarbamic acids (**Scheme 47**).

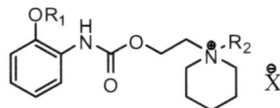

Scheme 47: Quaternary ammonium salts of piperidinylpropyl esters of alkoxy-substituted phenylcarbamic acids with antitubercular activity.

$$\log(MIC) = aN - bI + c \qquad \text{... (equation 68)}$$

$$\log(MIC) = d \log P + e \qquad \text{... (equation 69)}$$

It was postulated based on equations 68-69 that activity should increase with increasing lipophilicity of the compounds [184,185].

Another example which can be quoted is that of salicylanilide congeners (**Scheme 48**). The QSAR study was a combination of the Free–Wilson and the Hansch approaches for activities against *M. tuberculosis, M. avium* and two strains of *M. kansasii*. The Free–Wilson method categorized the molecules based on two moieties.

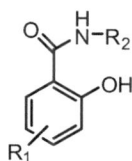

Scheme 48: Salicylanilide congeners with antitubercular activity.

Matyk *et al.* smartly used experimental values from physical measurements as electronic parameters. They assumed that the ^{1}H-NMR chemical shift of the proton on the nitrogen atom would describe the influence of electronic properties of the heterocyclic moiety. In addition, the partition coefficients were also included in the study to describe hydrophobicity. It was also suggested that probably electron donating substituents in positions 4- and 5- on the salicyl moiety increases antimycobacterial activity against all strains of *Mycobacterium*. Among the compounds that were studied were N-heterocyclic salicylanilide derivatives; in particular, N-benzimidazolyl derivatives were realized to have no effect on the antimycobacterial activity. The triazole ring was predicted to diminish the activity against all the three strains of mycobacterium while the isoxazolyl group would increase activity against *M. tuberculosis* [186].

A recent example of salicylanilide derivatives is the work by Imramovský *et al.*, which in one sense can be considered as an extension of the phenylcarbamate research. Precisely speaking, they synthesized N- and O=C-protected amino acids where N- was protected with -Cbz and the O=C- was protected with the salicylanilide functionality (**Scheme 49**).

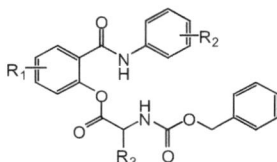

Scheme 49: Amino acids with antitubercular activity.

Activities against all three microorganisms, as discussed under salicylinides, were analyzed against 17 descriptors belonging to different categories like the Free-Wilson binary (presence/absence) descriptors, lipophilic, electronic, and steric properties. The simplified mathematical formulation for a three-layer ANN was as follows (equation 70).

$$y_i = f^3(W^{3,2}f^2(W^{2,1}f^1(W^{1,1}x_i + b^1) + b^2) + b^3) \qquad \qquad \text{... (equation 70)}$$

where (f) is the activation function of neurons $f^1 = f^2 = \tanh$ (input), $f^3 =$ identity (input), y_i is the antimycobacterial activity of compound *i*, $\mathbf{W}^{kj}$ is the weight matrix and $\mathbf{b}^k$ is the bias vector.

With the statistically sound models (superior overall r^2, overall s^p, r^2_{test} and s^p_{test}) a sensitivity analysis was done for log*k*, the Hammett substituent σ constant and the Taft inductive σ^* values of the R_1, R_2 and R_3 substituents on the *N*-Cbz-amino acid esters. This analysis verified the effects of log*P*, σ and σ^* constants and some Free-Wilson binary descriptors on the activity and the attributes they would confer on the molecules. Lipophilicity was found to have a direct correlation with the activity against *M. kansasii 235/80* strain. Specifically the Free-Wilson descriptor of R_2 (3-Cl) was found significant for activity against *M. tuberculosis* 331/88 strain; Free-Wilson descriptor of R_2 (4-Cl) for activity against *M. avium* 330/88 strain; Free-Wilson descriptor of R_2 (4-Br) for activity against *M. kansasii 235/80* strain and Free-Wilson descriptor of R_2 (3-Cl), for activity against *M. kansasii 6509/96* strain. The R_2 substituents were found to have a greater influence on the activity than the R_1 or R_3 substituents [187].

Meanwhile, Dolezal *et al.,* also scanned simpler salicylanilides (**Scheme 50**) for antitubercular activity along with their QSAR.

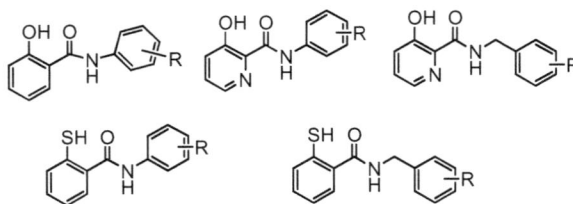

Scheme 50: Some simple salicylanilides with antitubercular activity.

The 14 and 21 day activities against *M. tuberculosis, M. kansasii*, and *M. avium* were analyzed by QSAR. The descriptors for the QSAR were priory examined and selected based on the statistical analyses (equations 71 – 76) .

$$\log(MIC_{Mt}^{14D}) = -58.655 - 11.1436\varepsilon_{HOMO} - 7.083C_{-Se} - 13.485D_{-Se} + 0.119H_{-am-Sn} + 0.0003H_{-am-Sn-w}$$
$$\text{... (equation 71)}$$

$$\log(MIC_{Mt}^{21D}) = -61.968 - 14.806\varepsilon_{HOMO} - 7.427C_{-Se} - 14.090D_{-Se} + 0.121H_{-am-Sn} + 0.0003H_{-am-Sn-w}$$
$$\text{... (equation 72)}$$

$$\log(MIC_{Ma}^{14D}) = -8.081 - 0.389Softness - 3.481A_{-Fe} + 0.448B_{-Fn} + 6.606G_{-char} - 4.142G_{-Se}$$
$$\text{... (equation 73)}$$

$$\log(MIC_{Ma}^{21D}) = -8.027 - 0.414Softness - 3.071A_{-Fe} + 0.477B_{-Fn} + 6.415G_{-char} - 4.328G_{-Se}$$
$$\text{... (equation 74)}$$

$$\log(MIC_{Mk}^{14D}) = -7.399 - 0.0001El.ext. - 0.228O_{-am-Fn} - 0.334Rnl_{-Se-sum} + 0.095H_{-am-Sn} + 0.0001Sum_{-Sn-w}$$
$$\text{... (equation 75)}$$

$$\log(MIC_{Mk}^{21D}) = -7.1413 - 0.0002El.ext. - 0.265O_{-am-Fn} - 0.361Rnl_{-Se-sum} + 0.0824H_{-am-Sn} + 0.000Sum_{-Sn-w}$$
$$\text{... (equation 76)}$$

It was established with such efforts (equations 71-76) that in general, activity increased with increasing lipophilicity and electron donating effect of the substituents in the acyl moiety, while a diminishing activity was observed with the electrophilic super-delocalizability of the molecules. More precisely, for *M. tuberculosis*, the QSAR model was observed to contain only descriptors relating to electronic properties of the molecules. The biological activity would increase with electron-donating substituents. In the QSAR model for *M. avium* and *M. kansasii* the electronic descriptors were important. Particularly in the latter case, the positive regression coefficient of the descriptor *Sum-Sn-w* meant that a higher uniformity of distribution of virtual molecular orbitals would lower the biological activity [188].

On the similar lines Dolezal *et al.,* further studied molecules of same origin but also some N-benzylsalicylthioamides (**Scheme 51**) [189].

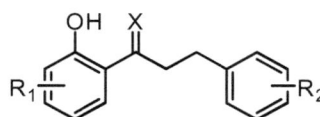

Scheme 51: (X=O/S).

A set of 177 descriptors was calculated and used for QSAR analysis. The best three-parameter QSAR models were sought using MLR technique (equations 77-83).

$$\log(MIC_{Mt}^{14d}) = -0.478\ (\pm0.107)\log P + 1.551\ (\pm0.206)\sigma_{R1} + 0.048\ (\pm0.009)SumS_e^w - 2.46\ (\pm0.646)$$
$$\text{... (equation 77)}$$

$$\log(MIC_{Mt}^{21d}) = -0.491 \ (\pm 0.097)\log P + 1.581 \ (\pm 0.187)\sigma_{R1} + 0.043 \ (\pm 0.008)SumS_e^w - 1.725 \ (\pm 0.586)$$
$$\text{... (equation 78)}$$

$$\log(MIC_{Ma}^{14d}) = 1.819 \ (\pm 0.163)\sigma_{R1} - 0.013 \ (\pm 0.002)S_n3' + 0.043 \ (\pm 0.007)SumS_e^w + 2.118 \ (\pm 0.765)$$
$$\text{... (equation 79)}$$

$$\log(MIC_{Ma}^{21d}) = 1.611 \ (\pm 0.19)\sigma_{R1} - 0.012 \ (\pm 0.002)S_n3' + 0.041 \ (\pm 0.008)SumS_e^w + 2.101 \ (\pm 0.887)$$
$$\text{... (equation 80)}$$

$$\log(MIC_{Mk}^{14d}) = -0.472 \ (\pm 0.106)\log P - 1.948 \ (\pm 0.204)\sigma_{R1} + 0.04 \ (\pm 0.009)SumS_e^w - 1.27 \ (\pm 0.64)$$
$$\text{... (equation 81)}$$

$$\log(MIC_{Mk}^{21d}) = 39.247 \ (\pm 10.225)\text{charge1} - 98.36 \ (\pm 9.011)S_{eH} - 3.271 \ (\pm 0.43)Sumf_e - 13.978 \ (\pm 2.456)$$
$$\text{... (equation 82)}$$

The descriptors for the QSAR were priory examined and selected based on the statistical analyses. It was established that in general, activity increased with increasing lipophilicity and electron donating effect of the substituents in the acyl moiety; while a diminishing activity was observed with the electrophilic super-delocalizability of the molecules. More precisely, for *M. tuberculosis*, the QSAR model was observed to contain only descriptors relating to the electronic properties of the molecules. The biological activity would increase with electron-donating substituents. For the QSAR model for *M. avium* and *M. kansasii* the electronic descriptors were important. Particularly in the latter case the positive regression coefficient of the Sum-Sn-w descriptor meant that a higher uniformity of distribution of virtual molecular orbitals would lower the biological activity [188,189].

At this point we would like to quote one very old but related example of thiobenzamides which was studied by Waisser *et al.* This was a unique example where it was claimed from the QSAR models that a strongly polarized C=S bond of the thiocarbamido group is necessary, as well as the value of the Hammett constant that must be positive (to prevent hepatotoxicity). Ideally, to meet such a requirement was difficult. Though the conjugated system can be extended to reduce the excitation energy of the eta-eta electronic transition, the lipophilicity should not increase (to avoid the risk of increased antimitotic activity and acute toxicity) [190].

Some scientists have shown that even simple substituted derivatives of alkanes may have potent activities. For instance, Pathak *et al.*, have explored some functionalized chiral heptenol and octenol derivatives (**Scheme 52**) and then later on their close associates Gupta *et al.*, probed the QSAR of these molecules.

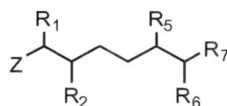

Scheme 52: Nitro/acetamido alkenols: R_1 = H, CH_3; R_2=-H, -OH, -OAc. Chloro/amino alkenols: Z = -CH_2Cl; R_1 =-Ac, -CN; R_2 = -H,-OH; R_6= -OH, -OAc, -OBn; R_7=H, - CH_2OBn.

They identified the suitable descriptors based on the CP-MLR method and then built models for antitubercular activities with the chosen descriptors (equations 83 and 84).

$$\log(1/MIC_{Mt}^9) = 11.281 - 21.071 \ (\pm 6.443)ATS8P - 15.806 \ (\pm 2.566)MATS7m \qquad \text{... (equation 83)}$$

$$\log(1/MIC_{Mt}^{10}) = 7.482 - 11.098 \ (\pm 3.112)ATS7v - 3.716 \ (\pm 0.785)GATS4m \qquad \text{... (equation 84)}$$

Primarily, one of the topological descriptors TOPO and a more complex 2D autocorrelation descriptor 2DAUTO were identified as more significant than empirical, constitutional, molecular walk counts, modified Burden eigen values, and Galvez topological charge indices for activity. The models made a clear distinction between the two types of compounds subjected for analysis *i.e.* nitro/acetamido alkenol derivatives and chloro/amino alkenol derivatives; the two classes of derivatives had different factors which could modulate the activity (equations 83 and 84, respectively) [191].

Another recent example of simple organic molecules studied for antitubercular activity is the α,ω-diaminoalkanes (**Scheme 53**) evaluated against *M. tuberculosis H37Rv* by Vergara *et al.* These were demonstrated to be comparable to the first line antitubercular agents *in vitro*.

$$H_2N(CH_2)_nNH_2$$

Scheme 53: (n = 9–12).

A simple and preliminary QSAR study suggested that both lipophilicity and the free *bis* amine functions were crucial for activity. This fact was also proved through observations as well as experimentations [192].

Molecules of Natural/Semisynthetic Origin

The compounds extracted from naturally occurring flora and fauna have good potential as drugs. These agents from natural origin may not have the desired pharmacokinetics or pharmacodynamic features; hence they may need synthetic modifications to optimize the properties. Moreover, for a detailed study, the medicinal chemist needs a substantial amount of the active compounds which may not be available because it may be cumbersome to extract from its natural source. Such amounts are often obtained through intervention of synthetic chemistry approaches and while doing so this can also be used to generated congeners of these molecules. There are innumerable examples of such synthetic and semisynthetic efforts in the history of pharmaceutical industry. *Mycobacteria* have also been found to be susceptible to many naturally occurring agents but issues of high virulence, intracellular growth and resistance opens the door for semisynthetic modifications in these compounds.

Sugars and their derivatives have been studied for a long time for their antitubercular activities as a dense network of cross-linked sugar residues esterified with mycolic acid at their ends have been identified as an integral part of mycobacterial cell walls. Interference by modified sugars and their derivatives offers an interesting strategy to fight TB [193-201].

A recent study of C_3 alkyl/arylalkyl-2,3-dideoxyhex-2-enopyranosides (**Scheme 54**) provides a good example of the development of sugar derivatives as antitubercular agents.

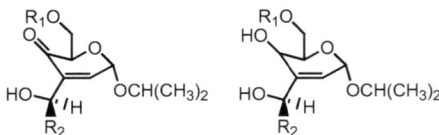

Scheme 54: Alkyl/arylalkyl-2,3-dideoxyhex-2-enopyranosides as promising sugar derivatives with antitubercular activity.

Saquib *et al.* also carried out QSAR studies on these newly discovered molecules. The QSARs were performed on two different conformations of the molecules. One of them was identical to the conformation of a representative molecule identified by X-ray diffraction while the second conformation was obtained through an energy minimization protocol. The molecules undertaken for this study were made to adopt these two conformations and then alignment free 3D descriptors were calculated (681 and 679 for the two groups respectively). A Combinatorial Protocol in Multiple Linear Regression (CP-MLR) in conjunction with a three-stage descriptor classification protocol was used to build the models (equations 85 and 86). The former model, built from the conformation in the X-ray analysis as the template, identified the most significant descriptors and classified them accordingly in three different classes. Unfortunately, the XRD conformation based approach could not build good enough models.

$$\log(1/MIC_{Mt}) = -3.667 \, (\pm 0.558)PJI3 - 5.204 \, (\pm 2.110)E3v + 11.832 \, (\pm 3.735)Rlu^+ + 7.654$$

... (equation 85)

$$\log(1/MIC_{Mt}) = -3.571 \ (\pm 0.518)PJI3 - 14.986 \ (\pm 4.940)ISH + 40.044 \ (\pm 10.726)HATS2p - 59.740$$
$$(\pm 17.397)R6e^+ + 19.290 \qquad \text{... (equation 86)}$$

With the second set (minimum energy conformation) under identical conditions, three 3D-descriptors classes *viz.*GEO (geometrical descriptors), WHIM (weighted holistic invariant molecular descriptors), and GETAWAY (geometry, topology, and atom-weights assembly descriptors) were found to be good contributors to the activity (-log *MIC*). The GEO descriptors depicted that that molecular structures with compact conformational features are favorable for activity. The WHIM descriptors suggested reduced axial density for enhancement in activity. GETAWAY descriptors pointed that symmetric molecules would have better activity [202].

Similarly, Sivakumar *et al.* explored chalcones, chalcone-like compounds, flavones and flavanones (**Scheme 55**) through QSAR methodology. The antitubercular activities of these set of the compounds were collected from previous published reports [203].

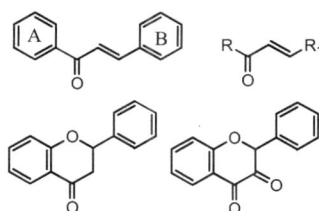

Scheme 55: Flavonoid derivatives with antitubercular activity.

Separate models were built for each category of compounds and the descriptors found in the best models were the Jurs descriptors, surface-weighted charged partial surface (WNSA-3), atomic charge weighted negative surface area: sum of the product of solvent-accessible surface X partial charge for all negatively charged atoms (PNSA-3) and relative positive charge surface area: solvent-accessible surface area of the most positive atom divided by descriptor (RPCS), Principle moments of inertia (PMI-mag), dipole (magnetic), Kier and Hall-molecular connectivity index valence modified CHI-2 (CHI-V-2), order 1 chi index, related to the number of edges and rings order: number of skeleton atoms in the subgraphs considered (CHI-1), valence modified CHI-3_C, C: cluster (CHI-V-3_C), valence-modified CHI-3_P, V: valence-modified connectivity index (CHI-V-3_P), conformational energy (Energy), H-bond donor and HOMO energy. The robust statistical technique such as Genetic Function Algorithm (GFA) was used to generate the QSAR models (equations 87-90)

For chalcones

$$Activity_{Mt} = 3.45 + 38.14 \ (CHI_{-V-2}) - 31.93 \ (H_{-bond\ donor}) - 0.11 \ (PMI_{-mag}) - 1.11 \ (Jurs_{WNSA-3})$$
$$\text{... (equation 87)}$$

For chalcone like compounds

$$Activity_{Mt} = 186.53 - 2.33 \ (Jurs_{PNSA-3}) - 2.62 \ (CHI_{-V-3-C}) + 1.34 \ (energy) + 17.85 \ (CHI_{-1})$$
$$\text{... (equation 88)}$$

For flavones

$$Activity_{Mt} = 56.90 - 19.87 \ (ADME\_BB\_level\_2D) + 7.81 \ (Jurs_{RPCS}) + 0.39 \ (energy) - 11.13 \ (CHI_{-V-3-p})$$
$$\text{... (equation 89)}$$

For flavanones

$$Activity_{Mt} = 190.16 - 0.91 \ (energy) - 19.37 \ (dipole - mag) - 2.70 \ (HOMO) \qquad \text{... (equation 90)}$$

The results of the QSAR were in conjugation with Lipinski's rule and previous studies which pointed to the need of hydrophobicity in antitubercular agents. In addition, the contributions of ADME properties and connectivity indices for chalcones and flavones were also portrayed [204].

Inspired by the activity of licochalcone, a natural product obtained from *Glycyrrhiza inflate* [205] and the earlier reports of chalcones as antitubercular agents [206], Sivakumar *et al.* developed some chalcone derivatives (**Scheme 56**).

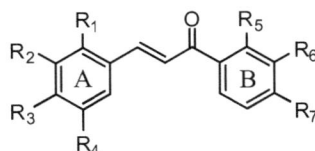

Scheme 56: Chalcones as antitubercular agents.

With 21 derivatives, QSAR models were developed between the observed activity [calculated as log(p/100-p) with p as percentage inhibition from Relative Light Units (RLU) data on a luminometer at 50 μg/ml concentration] and the spatial, topological, and ADME descriptors using multiple linear regression. Descriptors with small cross-correlated values were considered for the QSAR study; thus ADME_solubility_level, CHI-V-1 and Shadow-Zlength descriptors were found in the best equation 91.

$$Activity_{Mt}^{50\mu g/ml} = -1.398 + 1.609\ (ADME\_solubility\_level) - 0.624\ (CHI_{-V-1}) + 0.424\ (Shadow\text{-}Zlength)$$
$$\text{... (equation 91)}$$

Thus contributions of the ADME properties and the connectivity indices toward TB activity have been observed in this QSAR. The positive contribution of shadow indices in the QSAR equations revealed that increasing the length of the molecule (also related to hydrophobicity) would increase the antimycobacterial activity of the compounds [207].

Coumarin derivatives have been reported with a broad spectrum of activities. The most famous of these are their blood thinning capabilities and antiplatelet activity. Currently, there is a growing interest in these molecules for their anticancer and antitubercular activity. Very recently Manver *et al.* have studied several coumarin analogs as antitubercular agents. Their research on coumarin derivatives began with the discovery of calanolides, found in the extracts of the calophyllum seed, which are coumarin derivatives and posses weak antitubercular activity.

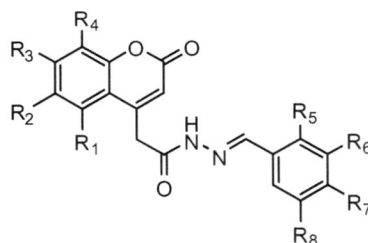

Scheme 57: Coumarin-4-acetic acid benzylidene hydrazides as antitubercular agents.

Around 25 coumarin-4-acetic acid benzylidene hydrazides (**Scheme 57**) were synthesized and monitored for activity against M. tuberculosis *H37Rv* strain. The coumarins yield variable and interesting activity and with this data, QSAR studies were evoked. 3D-QSAR studies were executed using the CoMFA methodology. Several statistically significant CoMFA models were generated with variable grid sizes, the orientation of the molecules and the alignments (atom/database fit and field fit). It must be considered that any alteration in the three aspects just now mentioned would lead to differences in the probe-molecule interaction energies which are used as descriptors in the 3D-QSAR. One of the models generated with *database/atom fit alignment* was found to be the best in terms of overall statistics and predictive powers. The contour maps indicated that enhancement in the activity could be obtained if steric bulk as added on the benzylidene side chain at the 4-position and in addition, these groups need to possess an electropositive character [208].

One more series of compounds which are structurally close to flavones, coumarins and the xanthones are the juglone derivatives which were first identified in the South African medicinal plant, *Euclea natalensis*. Diospyrin is the constituent active against drug-resistant strains of *M. tuberculosis* [209]. Mahapatra and co-workers prepared several juglone derivatives (**Scheme 58**) and evaluated them against *M. tuberculosis* [210] which was further exploited by Sharma *et al.* for QSAR investigation.

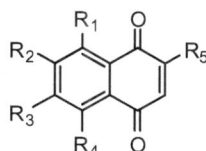

Scheme 58: Juglone derivatives as antitubercular agents.

Descriptors belonging to the constitutional, charge, physico-chemical, topological, geometrical and quantum chemical categories were computed and QSAR models were constructed based on those descriptors identified through a variable selection approach CP-MLR (Combinatorial Protocol in Multiple Linear Regression) approach. Partial least square analysis between the descriptors and the biological activity represented as $\log(1/MIC)$ and $\log(1/IC_{50})$ for the cytotoxic activity. The models were validated through cross-validation and y-scrambling protocols. The equation 92 corresponds to the best model generated by CP-MLR analysis while equation 93 is one of the best equations obtained with the PLS technique.

$$\log(1/MIC_{Mt}) = 2.919 - 0.071\ (\pm0.019)MSWHIM:\hat{E}3,\text{-}(SAS) + 42.565\ (\pm9.525)FPSA_3 - 0.794$$
$$(\pm0.226)Mor18u \qquad\qquad\qquad \text{... (equation 92)}$$

$$\log(1/MIC_{Mt}) = 0.705 - 2.178\ (\pm0.157)MATS2r - 0.205\ (\pm0.043)GATS4m - 0.033\ (\pm0.155)MSWHIM:\hat{E}3,\text{-}$$
$$(SAS) - 57.028\ (\pm0.391)FPSA_3 - 0.068\ (\pm0.097)RDFgu(2.0) - 0.216\ (\pm0.018)RDFgu(3.0) - 0.353$$
$$(\pm0.138)Mor18u \qquad\qquad\qquad \text{... (equation 93)}$$

Since the objective of this research was to delineate the factors responsible for the activity and the toxicity, Sharma *et al.* suggested that GATS4m, MS-WHIM|Ê3,-(SAS) and FPSA_3 are good for modulating the activity of the compounds (equation 94).

$$\log(1/MIC_{Mt}) = 4.249 - 0.032\ (\pm0.005)MSWHIM:\tilde{A}1,\text{-}(VdW) - 3.191\ (\pm1.295)Mor21v + 5.242$$
$$(\pm1.826)HATS1m \qquad\qquad\qquad \text{... (equation 94)}$$

On the other hand, the parameters MS- WHIM|Ã₁,-(VDW), E1s and HATS0v were found to be good for modulating the toxicity of the compounds (equation 95). These features collectively suggest that the substitution with bulky groups which are electropositive while truncating the electron rich groups may enhance the activity of the compounds as well as reduce their toxicity.

$$\log(1/MIC_{Mt}) = 3.938 - 0.016\ (\pm0.333)MSWHIM:\tilde{A}1,\text{-}(VdW) + 0.568\ (\pm0.085)E1s + 0.884\ (\pm0.246)Mor22m$$
$$- 0.561\ (\pm0.048)Mor21v + 1.345\ (\pm0.066)\ HATS1m + 3.699\ (\pm0.222)HATS0v \qquad \text{... (equation 95)}$$

Like phenanthrene derivatives discussed above, the steroids which are perhydrophenathrene derivatives have also been tested as antitubercular agents. It had been established through extensive research that progesterone, cholesterol, stigmasterol, ergosterol, sitosterol, chondrilasterol and ergosterol were active against *M. tuberculosis H37Rv*. It had also been put forward that the antimycobacterial activity is dependent on hydrophobicity and the type of substituents on the phytyl moiety on the steroidal backbone. In order to rationalize the activity, Rugutt *et al.* calculated several molecular descriptors including van der Waals surface area (*Vdw*, A), van der Waals volume (*Vdw*, v), polarizability, dipole moment, log *P*, and the differences between the Highest Occupied Molecular Orbital and the Lowest Unoccupied Molecular Orbital (HOMO-LUMO gap) for generating QSAR models [211].

QSAR Studies Involving Diverse Classes of Compounds

Until now we have been discussing QSAR studies where compounds were from the same series. Rationally speaking, the QSAR studies helped to understand the structure activity relationships for a set of molecules evaluated under the same conditions. For a long time in the history of QSAR and molecular modeling, QSARs were performed for molecules belonging to the same chemical class and thus it was just a tool to understand the effect of substitutions. Currently, the scenario has changed and researchers have started believing that for QSAR studies the core structures need not be similar. This revolution is probably due to exposure to the fact that molecules belonging to different categories have a common pharmacophore and it is the pharmacophores which are recognized by the receptor. As a consequence, there are now a number of examples where QSAR studies have been carried out on diverse classes of compounds. This plays a vital role in Virtual Ligand Screening (VLS), where hits are identified through pharmacophore mapping and/or docking protocols and the activity of the hits can be predicted using QSAR.

One of the good examples of such a study is that carried out by García-García *et al.* in 2004. In this study they considered 29 structurally heterogeneous drugs and applied Linear Discriminant Analysis (LDA) and multilinear regression analysis (MLR) to construct models with Topological Indices (TI) as structural descriptors with activity against *M. avium complex*. Thirty two of these descriptors (Topological Indices) were the Randić-Kier-Hall connectivity indices, and differences and quotients between them; 20 were topological charge indices and other discrete invariants comprised 10 in all. For the LDA, the descriptors were selected based on the Fisher-Snedecor *F* parameter. The independent variables were TIs while the discrimination property was the presence or absence of anti-MAC activity (equation 96).

$$DF = 0.47 - 1.26\,^{3}\chi_{c}^{v} + 0.22G_{1}^{v} - 14.53J_{3}^{v} - 0.22\,^{4}C_{c} \qquad \text{... (equation 96)}$$

The models (discrimination function; DF) obtained by LDA (equation 96 being the best) were able to predict the existence of antimycobacterial activity for a given structure and, consequently, were useful in selecting new candidate drugs. These DFs were validated using the LOO algorithm. The topological models obtained were able to classify over 80% of compounds as active or inactive. The MLR analyses also generated QSAR models (equation 97 being the best) which were validated with the LOO algorithm and y-scrambling, see Equaiton 97.

$$P = -0.59 + 43.94\,^{4}\chi_{c}^{v} - 2.91G_{4}^{v} + 7.50G_{S} + 9.29\,^{2}C - 2.43\,^{4}C_{c} - 0.93PR3 \qquad \text{... (equation 97)}$$

One direct use of this model was in carrying out a VLS on an inbuilt database of 20,000 compounds [212].

Another report of a QSAR study with diverse set of molecules was in 2006 where Prakash *et al.* used Hologram QSAR (H-QSAR) and clustering. The H-QSAR relates biological activity to structural molecular composition where molecular composition is described in terms of patterns of sub-structural fragments. The goal of their work was to identify the minimum common sub-structure (MCS) for antitubercular agents. Furthermore, such models could be used in a data mining operation to search for the motif responsible for the activity of a diverse class. In this scheme around 847 diverse molecules were considered and their antitubercular activities were collected from literature. Based on the activities, the molecules were divided into three classes as active, moderately active and inactive. Structurally, the data set contained about 28 diverse classes of compounds. Since only 2D structures were considered in this research no consideration was made of the stereoisomerism of the compounds. Three hierarchical clustering experiments (with Tanimoto similarity of 50%) were used to arrive at the final clustering and the clustering at a 40-s 'time out' and a minimum of 5 atoms in MCS with 4 target groups was taken for further study, on the grounds that they were closest to the natural cluster specifications. This resulted in a dendrogram where the molecules were reflected to belong to different classes through colour codes. With H-QSAR the substructural fragments for each of the molecules in a data set were generated and then these fragments were encoded in holograms followed by correlation with the available biological data. For one H-QSAR model the clustering information was used and only actives identified previously were considered; also another clustering-independent model was built. The models were validated by the cross-validation method. The worth of this study was clearly reflected in results where MCS for each class of compounds were identified and moreover, the fragments that are important for delimiting for activity were revealed. The effectiveness of such models were judged by searching an external dataset for potential antitubercular agents and it was heartening to note that rifampicin was found among the hits which corroborated the utility and scope of this study [213].

One more interesting link in the chain is the work of Prathipati *et al.* who chose a data set of around four thousand antitubercular compounds with *M. tuberculosis H37Rv* activities from the NIAID website. Due to the large size, structural diversity, and "noise" in the activity data, a method like Bayesian modeling was thought to be the most appropriate for this dataset. Bayesian statistics considers both the good and bad compounds while generating models. As a screen, hierarchical cluster was applied to select the best amongst the molecules. Different types of finger prints were used for cluster analysis and the finger print which best described the pharmacophoric similarity of the compounds was identified. The fingerprints belonging to the circular substructure class of fingerprints *i.e.* Extended-Connectivity Fingerprints (ECFP), Functional Class Fingerprints (FCFP) and *Sybyl* atom types Extended-Connectivity Fingerprints (SCFP) were used for development of the Bayesian model. With this, the best Bayesian model was developed on a training set obtained from diversity analysis of FCFP_4 fingerprints and derived using the ECFP_12 fingerprints plus Lipinski-type descriptors (Alog*P*, molecular weight, number of hydrogen bond donors and acceptors, number of rotatable bonds) and molecular fractional polar surface area, and the best model was chosen based on the optimum prediction accuracy for actives (total positive recall) and total accuracy values. The aim of Prathipati *et al.* in this project was to build a global predictive QSAR model based on 2D-fingerprints which would be competent enough against its 3D-QSAR and pharmacophore based approaches. In addition, they also tried to identify essential features in terms of abstract substructural features that could discriminate active from inactive molecules. They identified some scaffolds which could be further worked out to reach better antitubercular compounds which included substituted imidazooxazoles, pyrroles, furans, and oxazoles; substructures such as hydroxyl ketones, chalcones, phenyl ethylenes, tertiary amides, dienes, were predicted to have substantial influence on the antitubercular activities [214].

Hybridization of existing molecules was then a booming trend in medicinal chemistry. In this approach through connection of two active molecules by a bridge that is easily released *in vivo* new types of potentially active molecules can be created. Imramovský *et al.* designed, synthesized and analyzed by QSAR some new antitubercular compounds based on the original compounds isonicotinyl hydrazide, pyrazinamide, p-aminosalicylic acid (PAS), ethambutol, and ciprofloxacin [215].

6. QSAR STUDIES ON TARGETED ANTITUBERCULAR DRUGS

The advancements in the fields of biotechnology and biochemistry have propelled medicinal chemists to use rational methods to design and synthesize molecules. Since the biological response in the true sense is the effect of drug-receptor interactions, an activity value obtained at the receptor level would be more useful for QSAR analysis. In this scenario, the embedded ADME effect could be separated from the QSAR model and this could delineate between QSAR and QSPR models. A number of validated targets are known for development of antitubercular drugs. Many of them have been validated and for each target multiple classes of organic, inorganic and organometallic compounds have been designed. QSAR methodologies ever so often have been applied to study molecules against these targets. A few targets which have been investigated by QSAR are *thymidine monophosphate kinase, enoyl-ACP-reductase, dihydrofolate reductase, DNA gyrase, ribonucleotide reductase, salicyl AMP ligase, etc.*

Dihydrofolate Reductase

Dihydrofolate reductase (DHFR) plays a crucial role in one carbon metabolism by mediating the biosynthesis of DNA, RNA, and the essential amino acid methionine.

To analyze the structural requirements responsible for the enhanced activity of the highly active benzylpyrimidines (**K130**), Czaplinski *et al.* synthesized a series of trimethoprim analogs (**Scheme 59**) with various 4-anilinoalkoxy moieties and screened them against dihydrofolate reductase (DHFR) derived from various species *(M. lutu, E. coli, C. albicans* and rats).

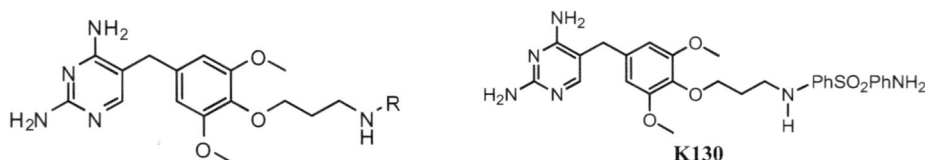

Scheme 59: Trimethoprim analogs as antitubercular agents.

The importance of the secondary amino group for binding affinity was investigated by varying the substituent on the *sec-amino* moiety of K-130. Multiple linear regression and CoMFA studies were performed to further support these findings. The study showed that the polarized SO_2 group in K-130, which was assumed to be responsible for the improved antitubercular activity, was not the only group important for receptor binding. Additionally, Principle Component Analysis (PCA) was carried out to investigate the high selectivity of some new candidates against the DHFR of various bacterial species and *C. albicans* and DHFR derived from rat liver.

X-ray crystallographic studies of DHFR from *E. coli* in complex with 3'-substituted analogs of trimethoprim (TMP) bearing a carboxylic acid moiety attached *via* an alkoxy side chain have shown that an ionic interaction of this carboxyl group with the guanidinium moiety of Arg57. The computational studies supported the assumption of this additional binding with Arg57 for the benzylpyrimidines. One of the benzylpyrimidines derivatives (K-130) showed an improved inhibitory activity compared to TMP, for DHFR derived from *M. lufu*.

The most appropriate indicator of the electronic influence of the *para-substituent* on the -NH group seems to be the NMR chemical shift of the -NH proton. This descriptor together with other physico-chemical parameters [the van der Waals volume (*VdW*) and *MR*] were used in the final multiple linear regression analysis. The following two best models obtained were used for understanding the QSAR of this series of compounds (equations 98 and 99).

$$\log(1/IC_{50M}{}^{DHFR}) = 0.58\ (\pm0.33)\delta H^{3,5} - 0.021\ (\pm0.332)MR_{R4} \qquad \ldots \text{(equation 98)}$$

$$\log(1/IC_{50M}{}^{DHFR}) = 0.75\ (\pm0.21)\delta H^{3,5} - 0.29\ (\pm0.34)MR_{R4} - 0.077\ (\pm0.34)\ . \qquad \text{.. (equation 99)}$$

However, the electronic substituent effects explained a major fraction of the variance. Descriptors beyond $\delta H^{3,5}$ and MR, such as total lipophilicity (log *P*), did not improve the correlations. Only poor correlations were found between the IC_{50} values of rat liver with all chemical descriptors used. The results obtained from the CoMFA study showed that the activities of the derivatives against *E. coli*-derived DHFR could be described by the electrostatic field at the *para-substituent,* probably indicating a dipole interaction with the receptor site.

Furthermore, CoMFA also indicated the importance of the *sec*-amino group for enhanced activity. Even Multiple linear regression analysis showed that the *sec*-amino moiety is important for the binding affinity of benzylpyrimidines to DHFR derived from *E coli* and to some extent to DHFR derived from *M. lufu* and *C. albicans*. The increased activity of benzylpyrimidines against DHFR derived from *E. coli* despite the negative steric effect contributed by the *para-* substituents may be explained by the compensation of this negative steric effect by an additional binding of the extended structures of this series. In addition, a principle component analysis (PCA) was performed to support the results of the MLR and CoMFA-analysis, *i.e.* different structural dependences of binding to DHFR derived from various species especially *M. lufu* [216].

Jain *et al.* established a quantitative structure activity relationship for inhibition of DHFR in various opportunistic pathogens like *P. carinii, T. gonidii, M. avium* and rat liver by substituted 2,4-diaminopyrido[2,3-d]pyrimidines and 2,4-diaminopyrrolo[2,3-d]pyrimidines (**Scheme 60**). This study was aimed at identifying those physico-chemical parameters of DHFR inhibitors that could differentiate activity against the DHFR of *P. carinii, T. gonidii, M. avium vs.* mammalian DHFR.

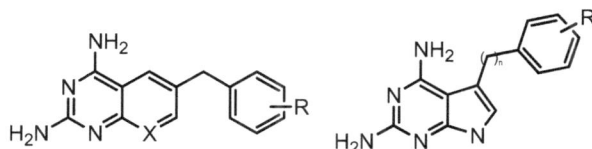

Scheme 60: Substituted 2,4-diaminopyrido[2,3-d]pyrimidines and 2,4-diaminopyrrolo[2,3-d]pyrimidines as antitubercular agents.

Correlations were sought between DHFR inhibitory activity and various substituent constants of the molecules like hydrophobic (π), steric (molar refractivity, MR), hydrogen acceptor (HA), hydrogen donor (HD) and electronic parameters (field effect, resonance effect and Hammett's constant) in the QSAR models. Thermodynamic descriptors *viz.* critical temperature (Tc), ideal gas thermal capacity (Cp), critical pressure (Pc), boiling point (BP), Henry's law constant (H), stretch bend energy (SBE), bend energy (Eb) and log *P* were calculated for all the molecules. Steric descriptors namely Connolly accessible area (CAS), Connolly molecular area (CMA), Connolly solvent excluded volume (CSEV), exact mass (EM), molecular weight (MW), Principle moment of inertia-X component (PMIX), principle moment of inertia-Y component (PMIY), principle moment of inertia-Z component (PMIZ), molar refractivity (MR) and ovality (OVAL) were also derived. Electronic descriptors which were derived included electronic energy (ElcE), highest occupied molecular orbital energy (HOMO), lowest occupied molecular orbital energy (LUMO), X- component of dipole moment (DPL_1), Z- component of dipole moment (DPL_3), resultant dipole moment (DPL), repulsion energy (NRE), Vdw-1,4 energy (E14), non-1,4-Vdw energy (E_v) and total energy (E).

Sequential Multiple Linear Regression (SMLR) was used to perform 2D-QSAR analysis with the VALSTAT program wherein the inhibition data of DHFR in various opportunistic pathogens was used as the dependent variables and the substituent constants as independent variables. Statistically significant correlations (for *e.g.,* equation 100) were obtained for models using the IC_{50} data of various pathogens. A correlation was also established between the substituent constants and the DHFR inhibitory activity in rat liver to identify those descriptors, which were responsible for toxicity in the hematopoietic system in mammalians.

$$pIC_{50Ma}^{DHFR} = 2.244\ (\pm 0.786)LUMO + 3.049\ (\pm 0.386)SBE - 0.339\ (\pm 0.092)DPL_3 + 3.031$$

... (equation 100)

2D-QSAR analyses suggested that absence of the nitrogen atom in the ring system is favorable for antifolate activity. MR was found to contribute positively to the DHFR inhibitory activity against all three species. Also the electronic parameters *i.e.* field effect and Hammet's constant showed positive correlation with antifolate activity of rat liver DHFR. However, field effect was found to contribute negatively to the antifolate activity of *T. gondii and M. avium* DHFR. These results suggested that electronic parameters may contribute significantly in the development of potent and selective DHFR inhibitors [217].

Thymidine Monophosphate Kinase Inhibitors

M. tuberculosis thymidine monophosphate kinase (TMPK) catalyses the phosphorylation of deoxythymidine monophosphate (*d*TMP) to deoxythymidine diphosphate (*d*TDP) utilizing ATP as a phosphoryl donor. This reaction is at the junction of the *de novo* and salvage pathways of thymidine triphosphate (TTP) metabolism and is the last specific enzyme for TPP synthesis. The sequence of TMPK when compared with the human isozyme shows only 22% sequence homology making it one of the potential targets for the design of new antitubercular drugs.

Aparna *et al.* have carried out a 3D-QSAR study employing Molecular Field Analysis (MFA) on *d*TMP analogs (**Scheme 61**) which inhibit *thymidine monophosphate kinase* (TMPK) in *M. tuberculosis*, to gain further insight into the key structural features required to design potential drug candidates of this class. Since the reliability and the efficiency of 3D-QSAR models depend on the orientation of the molecule, three different alignment techniques, *viz.*, least squares, pharmacophore based and receptor based methods were adopted to develop the MFA models. It was found that the receptor based MFA model could better explain the structure activity relationship when compared with the least squares and pharmacophore based models, equations 101 and 102.

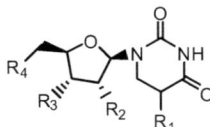

Scheme 61: *d*TMP analogs as antitubercular agents.

$$pK_{i\,Mt}^{TMPK} = 4.196 + 0.038\,(H_2O/218) - 0.026\,(CH_3/685) - 0.031\,(NH_2/398) + 0.023\,(CH_3/675) - 0.016$$
$$(H_2O/476) - 0.027\,(NH_2/298) - 0.023\,(H_2O/605) \qquad \text{... (equation 101)}$$

$$pK_{i\,Mt}^{TMPK} = 2.371 - 0.070\,(H_2O/463) + 0.054\,(CH_3/534) + 0.036\,(H_2O/348) + 0.043\,(CH_3/311) + 0.029$$
$$(CH_3/339) + 0.045\,(CH_3/125) - 0.011\,(NH_2/331) \qquad \text{... (equation 102)}$$

Though the disposition of the 3D-descriptors obtained from the three QSAR models was similar, the receptor based MFA model showed higher predictivity compared to the other two models. Also the 3D-descriptors obtained from the receptor based models (equations 101 and 102) complement well with the active site residues involved in ligand binding. It was found that the -Br group at the C_5 position satisfies the electronic requirements at this position while substituents at $C_{3'}$ interact with the acidic residues (Asp9 and Asp163) to increase the inhibitory activity. Also it was concluded that electron-donating substituents at $C_{5'}$ that can displace the water molecules from the co-ordination sites of Mg^{2+} ion may be entropically favorable for the drug-receptor interaction. The results provided useful information about the chemical and structural features of TMPK inhibitors and guidelines to design novel and potent antitubercular agents [218].

Andrade *et al.*, reported a receptor-independent (RI) 4D-QSAR formalism to develop QSAR models and corresponding 3D-pharmacophores for a set of 5'-thiourea-substituted α-thymidine inhibitors (**Scheme 62**) with significant inhibitory activity against *M. tuberculosis* TMPK and low human cytotoxicity.

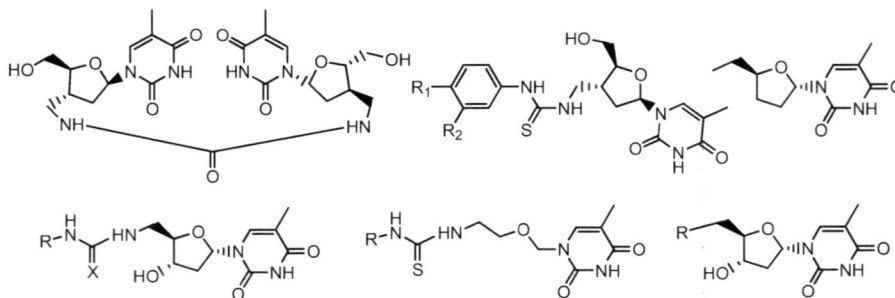

Scheme 62: 5'-Thiourea-substituted α-thymidine as antitubercular agents.

QSAR models were derived for the entire training set and also for a subset consisting of the most potent inhibitors. Statistically significant correlation models were obtained which were also found to have good predictivity based on test set predictions. The most and least potent inhibitors were docked in the active site of the TMPK crystallographic structure, in their respective postulated active conformations derived from the models. The objective of these docking-relaxation studies was to probe the possible types of interactions between the inhibitors with the surrounding amino acid residues lining the active site. A strong harmony was observed between the 3D-pharmacophore sites defined by the QSAR models and interactions with binding site residues. Equation 103 was derived based on the full set of molecules, while equation 104 was built based upon the most potent molecules in the set.

$$pK_{i\,Mt}^{TMPK} = 5.40GC1(any) - 4.77GC2(P+) + 2.18GC3(any) + 14.11GC4(any) + 1.14GC5(any) +$$
$$0.17(ClogP)^2 + 2.71 \qquad \text{... (equation 103)}$$

$$pK_{i\,Mt}^{TMPK} = -2.33GC1(P+) + 3.67GC2(any) - 8.25GC3(any) + 2.79GC4(HA) + 2.28GC5(any) +$$
$$0.40ClogP + 4.28 \qquad \text{... (equation 104)}$$

It is worth mentioning that in both models, increase in lipophilicity has a favorable contribution to pK_i showing that the α-thymidine derivatives bind in the active site in an 'upside down' fashion compared to the natural substrate. The tails (5′-arylthiorea moieties) of these molecules are oriented to the exterior of the enzyme through a channel, which is surrounded by nonpolar and aromatic residues that includes Ala35, Phe36, Pro37, and Arg160. The lipophilic substituents present on the 5′-aryl moiety are an important feature of the α-thymidine derivatives possibly representing an additional pharmacophore site contributing to the higher inhibition potency of these derivatives. The model could also identify new regions of the inhibitors that contain pharmacophore sites, such as the sugar-pyrimidine ring structure and the region of the 5′-arylthiourea moiety. These new ligand regions could be utilized to identify novel, and, perhaps, better TMPK inhibitors. Furthermore, the 3D-pharmacophores defined by these models can be used as structural design templates for future receptor dependent antitubercular drug design as well as to elucidate candidate sites for substituent addition to optimize ADMET properties of these analogs [219].

InhA, Enoyl Acyl Carrier Protein Reductase (EACP Reductase)

InhA, the *enoyl acyl carrier protein reductase* (EACP reductase) from *M. tuberculosis*, is one of the key enzymes involved in the mycobacterial fatty acid elongation cycle and has been validated as an attractive target for the design of new antibacterial agents.

Fatty acid biosynthesis in bacteria is catalyzed by a set of distinct, monofunctional enzymes collectively known as the type II FAS (FASII). These enzymes differ significantly from the type I FAS (FAS-I) in mammals, where all of the enzymatic activities are encoded in one or two multifunctional polypeptides. This distinctive difference in the FAS molecular organization between most bacteria and mammals makes possible the design of specific inhibitors of increased selectivity and lower toxicity. *M. tuberculosis* contains some signature fatty acids, the mycolic acids that are unusually long chain alkyl, β-hydroxy fatty acids of 60–90 carbons. The TB-specific drugs isoniazid (INH) and ethionamide have been shown to target the synthesis of these mycolic acids, which are central constituents of the mycobacterial cell wall. Among the enzymes involved in FASII, the NADH-dependent *enoyl-ACP reductase* encoded by the mycobacterium gene *InhA* is a key catalyst in mycolic acid biosynthesis. Studies over the years have established that *InhA* is the primary molecular target of INH, the drug that for the past 40 years has been, and continues to be, the frontline agent for the treatment of TB.

Ashutosh *et al.* has reported a CoMFA study on pyrrolidine carboxamides (**Scheme 63**), which have been reported as selective inhibitors of *EACP reductase* from *M. tuberculosis*.

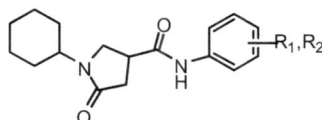

Scheme 63: Pyrrolidine carboxamides as antitubercular agents.

The CoMFA model produced statistically significant results with good predictability. Subsequently based on the important features observed in the CoMFA model, new pyrrolidine carboxamide analogs were proposed with the LeapFrog program. The designed molecules were predicted with higher activity and binding energy, suggesting that the newly proposed molecules in this series of compounds may be more potent and selective toward *EACP reductase* inhibition.

It was found that the preferred area for addition of steric bulk is located near the hydrophobic amino acids Phe149, Leu218, Trp222, Ile202, Trp230 and Met232 suggesting that more bulky aromatic substitutions on the inhibitors can interact better with the side chains of these residues *via* hydrophobic interactions. On the other hand, the area disfavored sterically is around Ile215, Trp230, Phe97 and Ile95; thus, bulkier substituents in these regions could cause steric hindrance.

The electrostatic contour maps suggests the presence of a negatively charged environment around the side chain oxygen of Tyr 158 and the nicotinamide ribose group of the cofactor NAD; as these can form favorable hydrogen bonds. A negatively charged favored region was also observed near Arg195 implying that electron rich groups on the inhibitors may interact with the side chain of Arg195 and therefore increase the inhibitory potencies of pyrrolidine carboxamides. However, it was observed from the statistical analysis that contribution of the electrostatic component to the variance was relatively smaller compared to the steric component. This study provided a good insight into the structure requirement for pyrrolidine carboxamides to design better EACP inhibitor as antitubercular agents [220].

Continuing the search for *InhA* inhibitors Xiao-Yun *et al.* developed an efficient approach for discovering new direct inhibitors.

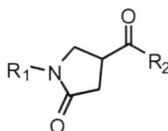

Scheme 64: Other pyrrolidine carboxamides as antitubercular agents.

In this study an efficient virtual screening model based on the *InhA* bound conformation of a pyrrolidine carboxamide inhibitor (**Scheme 64**) was built using LigandScout 2.0. The pharmacophore model generated by the LigandScout 2.0 program included six features: two Hydrogen Bond Acceptors (HBA) and four hydrophobic groups. Besides, a series of excluded volumes in the pharmacophore model were also defined. Both HBA features characterize the carbonyl group of the ligand which forms two hydrogen bonds with Tyr158 and the 20-hydroxyl group of the nicotinamide ribose of the nucleotide. One of the hydrophobic features is located on the cyclohexyl and the other three hydrophobic features are located on the 3- and 5-chloro-substituted phenyl group, respectively. This pharmacophore model was successfully used as a query to screen compound libraries (SPECS database) to retrieve potential hits against *InhA*.

It was observed that strong hydrogen bonding interactions occur between the carbonyl group of the pyrrolidine ring present in the ligands and the protein catalytic residue (Tyr158) as well as its co-substrate (the 20-hydroxyl group of the nicotinamide ribose of the nucleotide). Several other reported potent *InhA* inhibitors have a carbonyl group or a hydroxyl group which might also form hydrogen bonds with the Tyr158 residue and NAD co-substrate. Therefore, more attention was paid to the two HBA features of the pharmacophore model. In addition, the lactam ring of the ligand shows π-π interactions with the NAD$^+$ nicotinamide ring which did not show up in the pharmacophore model.

In order to incorporate the structural information of the receptor and identify the bioactive conformations of the pyrrolidine carboxamide inhibitors of *InhA*, CoMFA and CoMSIA analysis was performed based on the pharmacophore alignment. A statistically valid 3D-QSAR model was obtained which described the steric, electrostatic, and hydrogen bond donor requirements for the *InhA* inhibitory activity.

Screening of the ZINC database produced 30 hits based on the conserved interactions between *InhA* and the inhibitors and which also exhibited good activities as predicted by the 3D-QSAR models. The hits belonged to two classes, namely, thiazolones and nitrobenzene carboxamides. Also some other new structures were identified. The 3D-QSAR model also showed that the introduction of larger hydrophobic substituents to the hits might provide more hydrophobic contacts with the active site of *InhA* and improve their inhibitory activities. In general, the nitrobenzene carboxamide series showed better activity than the thiazolones.

To sum up, the whole procedure of pharmacophore modeling, 3D-QSAR study, *in silico* screening, and molecular docking resulted in the identification of some potential candidates as direct inhibitors of *InhA*.

MbtA (A Salicyl AMP Ligase)

The iron concentration in serum and human body fluids is approximately 10^{-24} M, which is too low to support bacterial colonization and growth. In order to overcome this iron limitation, *M. tuberculosis*, synthesizes iron chelators called siderophores for iron acquisition. As a result, siderophore biosynthesis has emerged as a potential biochemical pathway for the development of antitubercular agents. *Aryl acid adenylation enzyme* (AAAE) MbtA, which catalyzes the first step in aryl-capped siderophore biosynthesis, has proved as an attractive target for intervening the growth of *M. tuberculosis*. 5'-*O*-[*N*-(salicyl)sulfamoyl]adenosine (Sal-AMS), a rationally designed bisubstrate analog, is the prototypical AAAE inhibitor. It comprises four modules: the aryl group, the linker, the glycosyl moiety, and the nucleobase.

There are a number of reports in the literature where a predictive binding models has been generated to weight theoretical interaction energies using experimental binding affinities. Within such Linear Interaction Energy (LIE) approximations, binding affinities are estimated after only one ligand-receptor and one ligand-solvent simulation for each additional compound. The resulting models often display high correlation and, in contrast to some activity relationship models, have the advantage of being structure-based and therefore serve as guidelines for rational drug design. Labello *et al.* have reported one such structure-based model for the Sal-AMS scaffold (**Scheme 65**) to predict the binding affinities of aryl acid-AMP bisubstrate inhibitors of MbtA employing Linear Interaction Energy (LIE) technique to derive linear equations relating ligand structure to function (Equation 105).

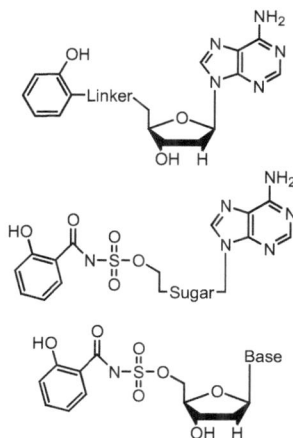

Scheme 65: 5'-*O*-[*N*-(salicyl)sulfamoyl]adenosine as antitubercular agents.

$$A = \alpha\Delta G_{vdW} + \beta\Delta G_{El} + \gamma \qquad \qquad ...\text{(equation 105)}$$

The LIE model (equation 105) was developed with emphasis on providing a quantitative model for predicting binding affinities and a grounded physical interpretation of the SAR to guide future synthesis. It was found that modifications of the nucleobase are of particular interest, as this moiety represents the best opportunity for improving potency and increasing specificity and lipophilicty. The linker and glycosyl regions were also examined, with a view that variation of these functionalities may be required to modify the number of hydrogen bond donors and acceptors or otherwise optimize the pharmacokinetic properties. With only two parameters derived from molecular dynamics simulations, good correlation ($R^2 = 0.70$) was achieved for a set of 31 inhibitors with binding affinities spanning 6 orders of magnitude. The results were applied to understand the effect of steric and heteroatom substitutions on bisubstrate ligand binding and to predict second generation inhibitors of MbtA. As a final check of the applicability of the model, the resulting model was further validated by chemical synthesis of a novel inhibitor, *N*-6-cyclopropyl-2-phenyl-Sal-AMS which was predicted with a LIE binding affinity of 1.6 nM but subsequently experimentally determined (K_i^{app}) as 0.7 nM [221].

Ribonucleotide Reductase (RNR)

M. tuberculosis ribonucleotide reductase (RNR) has emerged as yet another potential target for new antitubercular drugs which catalyzes the reduction of ribonucleotides to the corresponding deoxyribonucleotides and is therefore an essential enzyme for DNA synthesis. The active enzyme is a tetramer composed of two large subunits (R_1) and two small subunits (R_2). The R_1 subunit possesses the substrate and effector binding sites while R_2 harbors a tyrosine radical essential for catalytic activity. The catalytic mechanism involves electron transfer between the radical in R_2 and the active site in R_1. The association of the subunits is therefore crucial for enzymatic activity. Several different approaches for inhibiting RNR have been explored, and one possible approach is to inhibit the association of the R_1 and R_2 subunits. Studies have shown that peptides corresponding to the C-terminal end of the R_2 subunit can compete for the R_2 binding site of R_1 and thus inhibit RNR activity.

Johanna *et al.* have described the synthesis and evaluation of peptide inhibitors of RNR derived from the C-terminus of the small subunit of *M. tuberculosis* RNR. An N-terminal truncation, an alanine scan and a novel Statistical Molecular Design (SMD) approach based on the heptapeptide Ac-Glu-Asp-Asp-Asp-Trp-Asp-Phe-OH were applied in this study. The alanine scan showed that Trp5 and Phe7 were important for inhibitory potency. A series of peptides was synthesized based on the classical approach of identifying the minimal active sequence of the peptide. Based on the inhibitory potency of these peptides, a decision was made to explore the SAR of the heptapeptide in greater detail. A systematic variation of each position quickly led to a large number of peptides for synthesis and evaluation. This together with the enormous peptide space available (207 using only coded amino acids) encouraged them to explore this space using the FHDoE approach [222]. One of the benefits of using Statistical Molecular Design (SMD) [223] is that information-rich datasets can be generated from few experiments.

A Quantitative Structure Activity Relationship (QSAR) model was developed based on the synthesized heptapeptide analogs which highlighted the importance of the individual amino acid positions in the peptide. It was found that anionic amino acids in positions 2, 3, and 6 would be preferred for inhibitory potency. While, in position 5 the model coefficients indicated that there is room for a larger side chain, which was also supported from inspection of the *S. typhimurium* complex structure. The RNR of *M. tuberculosis* has high sequence identity to its *S. typhimurium* counterpart [224].

7. CONCLUSION

The studies discussed above indicate that there exists a great deal of research employing QSAR techniques for the development and improvement of known antitubercular agents along with the search for novel antitubercular agents; work is going on at a brisk pace. At this rate, it could be surely expected that sooner or later some new antitubercular agents could be introduced in the market which would not only have better activity but also superior pharmacokinetic and pharmacodynamic profiles. QSAR models can be put forth with more confidence if the receptors are known and well characterized. As mentioned earlier, many classes of antitubercular agents are known, however the action mechanisms have not been revealed. The advancements in biochemistry, biophysics and pharmacology will make it possible to determine the mechanism or target of action, and this information, if incorporated in the QSAR studies, would definitely make a difference in the final outcome. All said and done, there is still a tremendous scope for development of novel antitubercular agents and QSAR can serve as a valuable tool to achieve the goal of eradication of tuberculosis.

REFERENCES

[1] Crum-Brown A, Fraser TR. On the connection between chemical constitution and physiological action I. On the physioligical action on the salts of the ammonium basis, derived from strychnia, brucia, thebaia, codeia, morphia, and nicotia. Trans Roy Soc Edinburgh 1868; 25: 151.

[2] Hansch C. On The Structure of medicinal chemistry. J Med Chem 1976; 19: 1.

[3] Meyer H. Zur Theorie der Alkoholnarkose - Erste Mittheilung. Welche Eigenschaft der Anästhetica bedingt ihre narkotische Wirkung? Arch Exp Pathol Pharmak 1899; 42: 109.

[4] Overton CE., Studien uber die Narkose Zugleich ein Beitrag zur Allgemeinen Pharmakologie Jena, Switzerland: Verlag von Gustav Fischer 1901.

[5] Hansch C, Fujita T. ρ-σ-π Analysis. A method for the correlation of biological activity and chemical structure. J Am Chem Soc 1964; 86: 1616.

[6] Free SM Jr, Wilson JW. A mathematical contribution to structure-activity studies. J Med Chem 1964; 7: 395.

[7] Hansch C. A quantitative approach to biochemical structure-activity relation-ships. Acc Chem Res 1969; 2: 232.

[8] Hansch C, Gao H. Comparative QSAR: radical reactions of benzene derivatives in chemistry and biology. Chem Rev 1997; 97: 2995.

[9] Hansch C, Hoekman D, Gao H. Comparative QSAR: toward a deeper understanding of chemicobiological interactions. Chem Rev 1996; 96: 1045.

[10] Hansch C, Kurup A, Garg R, Gao H. Chem-Bioinformatics and QSAR. A review of QSAR lacking positive hydrophobic terms. Chem Rev 2001; 101: 619.

[11] Hansch C, Kutter E, Leo A. Homolytic constants in the correlation of chloramphenicol structure with activity. J Med Chem 1969; 12: 746.

[12] Hansch C, Maloney PP, Fujita T, Muir RM. Correlation of biological activity of phenoxyacetic acids with hammett substituent constants and partition coefficients. Nature 1962; 194: 178.

[13] Hansch C, Muir RM, Fujita T, Maloney PP, Geiger F, Streich M. The correlation of biological activity of plant growth regulators and chloromycetin derivatives with hammett constants and partition coefficients. J Am Chem Soc 1963; 85: 2817.

[14] Hansch C. A quantitative approach to biochemical structure-activity relation-ships. Acc Chem Res 1969; 2: 232.

[15] Hansche PE. A theoretical basis for the entrainment of chemostat populations. J Theor Biol 1969; 24: 335.

[16] Fujita T, Ban T. Structure-activity relation. 3. Structure-activity study of phenethylamines as substrates of biosynthetic enzymes of sympathetic transmitters. J Med Chem 1971; 14: 148.

[17] Fujita T, Hansch C. Analysis of the structure-activity relationship of the sulfonamide drugs using substituent constants. J Med Chem 1967; 10: 991.

[18] Fujita T, Iwasa J, Hansch C. A new substituent constant, π, derived from partition coefficients. J Am Chem Soc 1964; 86: 5175.

[19] Cramer III RD, Patterson DE, Bunce JD. Comparative molecular field analysis (CoMFA). 1. Effect of shape on binding of steroids to carrier proteins. J Am Chem Soc 1988; 110: 5959.

[20] Cramer III RD, Patterson DE, Bunce JD. Recent advances in comparative molecular field analysis (CoMFA). Prog Clin Biol Res 1989; 291: 161.

[21] Hillebrecht A, Supuran CT, Klebe G. Integrated approach using protein and ligand information to analyze selectivity- and affinity-determining features of carbonic anhydrase isozymes.Chem Med Chem 2006; 1: 839.

[22] Radestock S, Bohm M, Gohlke H. Improving binding mode predictions by docking into protein-specifically adapted potential fields. J Med Chem 2005; 48: 5466.

[23] Kovatcheva A, Golbraikh A, Oloff S, Xiao YD, Zheng W, Wolschann P, Buchbauer G, Tropsha A. Combinatorial QSAR of ambergris fragrance compounds. J Chem Inf Comput Sci 2004; 44: 582.

[24] Silverman BD. Three-dimensional moments of molecular property fields. J Chem Inf Comput Sci 2000; 40: 1470.

[25] Silverman BD, Platt DE. Comparative molecular moment analysis (CoMMA): 3D-QSAR without molecular superposition. J Med Chem 1996; 39: 2129.

[26] Klebe G. Comparative molecular similarity indices analysis: CoMSIA. Persp Drug Discov Des 1998; 12: 87.

[27] Klebe G, Abraham U, Mietzner T. Molecular similarity indices in a comparative analysis (CoMSIA) of drug molecules to correlate and predict their biological activity. J Med Chem 1994; 37: 4130.

[28] Niedbala H, Polanski J, Gieleciak R, Musiol R, Tabak D, Podeszwa B, Bak A, Palka A, Mouscadet JF, Gasteiger J, Le Bret M. Comparative molecular surface analysis (CoMSA) for virtual combinatorial library screening of styrylquinoline HIV-1 blocking agents. Comb Chem High Throughput Screen 2006; 9: 753.

[29]   Polanski J, Gieleciak R. Comparative molecular surface analysis: a novel tool for drug design and molecular diversity studies. Mol Divers 2003; 7: 45.

[30]   Polanski J, Gieleciak R. The comparative molecular surface analysis (CoMSA) with modified uniformative variable elimination-PLS (UVE-PLS) method: application to the steroids binding the aromatase enzyme. J Chem Inf Comput Sci 2003; 43: 656.

[31]   Polanski J, Gieleciak R, Bak A. The comparative molecular surface analysis (COMSA)--a nongrid 3D QSAR method by a coupled neural network and PLS system: predicting pK(a) values of benzoic and alkanoic acids. J Chem Inf Comput Sci 2002; 42: 184.

[32]   Polanski J, Gieleciak R, Bak A. Probability issues in molecular design: predictive and modeling ability in 3D-QSAR schemes. Comb Chem High Throughput Screen 2004; 7: 793.

[33]   Polanski J, Gieleciak R, Magdziarz T, Bak A. GRID formalism for the comparative molecular surface analysis: application to the CoMFA benchmark steroids, azo dyes, and HEPT derivatives. J Chem Inf Comput Sci 2004; 44: 1423.

[34]   Polanski J, Gieleciak R, Wyszomirski M. Comparative molecular surface analysis (CoMSA) for modeling dye-fiber affinities of the azo and anthraquinone dyes. J Chem Inf Comput Sci 2003; 43: 1754.

[35]   Polanski J, Walczak B. The comparative molecular surface analysis (COMSA): a novel tool for molecular design. Comput Chem 2000; 24: 615.

[36]   Jain AN, Harris NL, Park JY. Quantitative Binding Site Model Generation: Compass Applied to Multiple Chemotypes Targeting the 5-HT1A Receptor. J Med Chem 1995; 38: 1298.

[37]   Jain AN, Koile K, Chapman D. Compass: predicting biological activities from molecular surface properties. Performance comparisons on a steroid benchmark. J Med Chem 1994; 37: 2315.

[38]   Datar PA, Khedkar SA, Malde AK, Coutinho EC. Comparative residue interaction analysis (CoRIA): a 3D-QSAR approach to explore the binding contributions of active site residues with ligands. J Comput Aided Mol Des 2006; 20: 343.

[39]   Khedkar SA, Malde AK, Coutinho EC. Design of Inhibitors of the MurF Enzyme of Streptococcus pneumoniae Using Docking, 3D-QSAR, and *de Novo* Design. J Chem Inf Model 2007; 47: 1839.

[40]   Verma J, Khedkar VM, Prabhu AS, Khedkar SA, Malde AK, Coutinho EC. A comprehensive analysis of the thermodynamic events involved in ligand-receptor binding using CoRIA and its variants. J Comput Aided Mol Des 2008; 22: 91.

[41]   Asikainen A, Ruuskanen J, Tuppurainen K. Spectroscopic QSAR methods and self-organizing molecular field analysis for relating molecular structure and estrogenic activity. J Chem Inf Comput Sci 2003; 43: 1974.

[42]   Vedani A, Dobler M. Quasar, version 5.0. Basel, Switzerland: Biographics Laboratory 3R 2002.

[43]   Vedani A, Dobler M. Multi-dimensional QSAR in drug research. Predicting binding affinities, toxicity and pharmacokinetic parameters. Prog Drug Res 2000; 55: 105.

[44]   Vedani A, Dobler M. 5D-QSAR: the key for simulating induced fit? J Med Chem 2002; 45: 2139.

[45]   Vedani A, Dobler M, Lill MA. A safe identification of environmental toxins *in silico* - Approaching receptor-mediated endocrine disruption by 6D-QSAR. Med Chem 2005; 48: 3700.

[46]   Dunteman GH. An introduction to generalized linear models. In: Dunteman GH, London: SAGE Publications Ltd 1989; pp. 15.

[47]   Dunteman GH. An introduction to generalized linear models. In: Dunteman GH, London: SAGE Publications Ltd 1989; pp. 65.

[48]   Wold S, Johansson E, Cocchi M. 3D-QSAR in Drug Design. Theory, Methods and Applications In: Kubinyi MH, The Netherlands: ESCOM Lieden 1993; pp. 523.

[49]   Rogers D, Hopfinger AJ. Application of genetic function approximation to quantitative structure-activity relationships and quantitative structure-property relationships. J Chem Inf Comp Sci 1994; 34: 854.

[50]   Baskin II, Palyulin VA, Zefirov NS. Neural networks in building QSAR models. Meth Mol Biol 2008; 458: 137.

[51]   Dunn III WJ, Rogers D. Neural networks in QSAR and drug design (principles of QSAR and drug design) In: Devillers J, London: Academic Press 1996; pp. 109.

[52]   Hemmateenejad B, Akhond M, Miri R, Shamsipur M. Genetic algorithm applied to the selection of factors in principal component-artificial neural networks: application to QSAR study of calcium channel antagonist activity of 1,4-dihydropyridines (nifedipine analogous). J Chem Inf Comput Sci 2003; 43: 1328.

[53]   Gramatica P. Principles of QSAR models validation: internal and external. QSAR Comb Sci 2007; 26: 694.

[54]   Richard D, Cramer III RD, Bunce JD, Patterson DE, Frank IE. Crossvalidation, Bootstrapping, and Partial Least Squares Compared with Multiple Regression in Conventional QSAR Studies. Quant Struct-Act Relat 1988; 7: 18.

[55]   Stone M. Cross-validatory choice and assessment of statistical predictions (with discussion). J Roy Stat Soc B 1974; 36, 111.

[56]   Archdeacon TJ. In: Archdeacon, T. J. Regression and explained variance. In: Correlation and Regression Analysis: a Historian's Guide; Archdeacon, T.J., Ed, USA: Univ of Wisconsin Press 1994; pp. 178.

[57]   Campbell MK, Torgerson DJ. Bootstrapping: estimating confidence intervals for cost-effectiveness ratios. QJM 1999; 92: 177.

[58]   Conrad M. Bootstrapping on the adaptive landscape. Biosystems 1979; 11: 167.

[59]   Conrad M. Bootstrapping model of the origin of life. Biosystems 1982; 15: 209.

[60]   Franke J, Neumann MH. Bootstrapping neural networks. Neural Comput 2000; 12: 1929.

[61]   Gunter B. Bootstrapping. Infect Control Hosp Epidemiol 1994; 15: 543.

[62]   Hunt CA, Givens GH, Guzy S. Bootstrapping for Pharmacokinetic Models: Visualization of Predictive and Parameter Uncertainty. Pharm Res 1998; 15: 690-697.

[63]   Kerr MK, Churchill GA. Bootstrapping cluster analysis: Assessing the reliability of conclusions from microarray experiments Proc Natl Acad Sci USA 2001; 98: 8961.

[64]   Loughin TM, Koehler KJ. Bootstrapping regression parameters in multivariate survival analysis. Lifetime Data Anal 1997; 3: 157.

[65]   Shao J. Bootstrap model selection. J Am Stat Assoc 1996; 91: 655.

[66]   Mullner M. Doctors and managers. BMJ 2003; 326: 911.

[67]   Archdeacon TJ. Proceedings of the Symposium on Archdeacon In: Archdeacon TJ, USA: Ed. Univ of Wisconsin Press 1994; pp. 160.

[68]   Rucker C, Rucker G, Meringer M. y-Randomization and Its Variants in QSPR/QSAR. J Chem Inf Model 2007; 47: 2345.

[69]   Klopman G. The MultiCASE program II. Baseline activity identification algorithm (BAIA). J Chem Inf Comput Sci 1998; 38: 78.

[70]   Klopman G, Chakravarti SK. Screening of high production volume chemicals for estrogen receptor binding activity (II) by the MultiCASE expert systemChemosphere 2003; 51: 461.

[71]   Klopman G, Chakravarti SK. Structure–activity relationship study of a diverse set of estrogen receptor ligands (I) using MultiCASE expert system. Chemosphere 2003; 51: 445.

[72]   Rosenkranz HS, Cunningham AR, Zhang YP, Klopman G. Applications of the CASE/MULTICASE SAR method to environmental and public health situations. SAR QSAR Environ Res 1999; 10: 263.

[73]   Klopman G, Wang S, Jacobs MR, Bajaksouzian S, Edmonds K, Ellner JJ. Anti-Mycobacterium avium activity of quinolones: *in vitro* activities. Antimicrob Agents Chemother 1993; 37: 1799.

[74]   Klopman G, Wang S, Jacobs MR, Ellner JJ. Anti-Mycobacterium avium activity of quinolones: structure-activity relationship studies. Antimicrob Agents Chemother 1993; 37: 1807.

[75]   Klopman G, Fercu D, Li JY, Rosenkranz HS, Jacobs MR. Antimycobacterial quinolones: a comparative analysis of structure-activity and structure-cytotoxicity relationships. Res Microbiol 1996; 147: 86.

[76]   Klopman G, Fercu D, Renau TE, Jacobs MR. N-1-tert-butyl-substituted quinolones: *in vitro* anti-Mycobacterium avium activities and structure-activity relationship studies Antimicrob Agents Chemother 1996; 40: 2637.

[77]   Klopman G, Li JY, Wang S, Pearson AJ, Chang K, Jacobs MR, Bajaksouzian S, Ellner JJ. *In vitro* anti-Mycobacterium avium activities of quinolones: predicted active structures and mechanistic considerations. Antimicrob Agents Chemother 1994; 38: 1794.

[78]   Gozalbes R, Brun-Pascaud M, Garcia-Domenech R, Galvez J, Girard PM, Doucet JP, Derouin F. Prediction of quinolone activity against Mycobacterium Avium by molecular topology and virtual computational screening. Antimicrob Agents Chemother 2000; 44: 2764.

[79]   Agrawal VK, Bano S., Mathur KC, Khadikar PV. The study of inhibition of DNA synthesis by hydeoxyureaq(s). Proc Indian Acad Sci (Chem Sci) 2000; 112: 137.

[80]   Ghosh P, Thanadath M, Bagchi MC. On an aspect of calculated molecular descriptors in QSAR studies of quinolone antibacterials. Mol Divers 2006; 10: 415.

[81]   Bagchi MC, Mills D, Basak SC. Quantitative structure-activity relationship (QSAR) studies of quinolone antibacterials against *M. fortuitum* and *M. smegmatis* using theoretical molecular descriptors. J Mol Model 2007; 13: 111.

[82]   Gozalbes R, Brun-Pascaud M, Garcia-Domenech R, Galvez J, Girard PM, Doucet JP, Derouin F. Anti-Toxoplasma Activities of 24 Quinolones and Fluoroquinolones *in Vitro*: prediction of activity by molecular topology and virtual computational techniques. Antimicrob Agents Chemother 2000; 44: 2771.

[83]   Anquetin G, Greiner J, Mahmoudi N, Santillana-Hayat M, Gozalbes R, Farhati K, Derouin F, Aubry A, Cambau E, Vierling P. Design, synthesis and activity against *Toxoplasma gondii*, Plasmodium spp., and *Mycobacterium tuberculosis* of new 6-fluoroquinolones. Eur J Med Chem 2006; 41: 1478.

[84]   Jain R, Vaitilingam B, Nayyar A, Palde PB. Substituted 4-methylquinolines as a new class of anti-tuberculosis agents. Bioorg Med Chem Lett 2003; 13: 1051.

[85]   Vangapandu S, Jain M, Jain R, Kaur S, Singh PP. Ring-substituted quinolines as potential anti-tuberculosis agents. Bioorg Med Chem 2004; 12: 2501.

[86]   Vaitilingam B, Nayyar A, Palde PB, Monga V, Jain R, Kaur S, Singh PP. Synthesis and antimycobacterial activities of ring-substituted quinolinecarboxylic acid/ester analogues. Part 1. Bioorg Med Chem 2004; 12: 4179.

[87]   Monga V, Nayyar A, Vaitilingam B, Palde PB, Jhamb SS, Kaur S, Singh PP, Jain R. Ring-substituted quinolines. Part 2: Synthesis and antimycobacterial activities of ring-substituted quinolinecarbohydrazide and ring-substituted quinolinecarboxamide analogues. Bioorg Med Chem 2004; 12: 6465.

[88]   Kellogg GE, Semus SF, Abraham DJ. HINT - a new method of empirical hydrophobic field calculation for CoMFA. J Comput-Aided Mol Des 1991; 5: 545.

[89]   Nayyar A, Malde A, Jain R, Coutinho E. 3D-QSAR study of ring-substituted quinoline class of anti-tuberculosis agents. Bioorg Med Chem 2006; 14: 847.

[90]   Nayyar A, Malde A, Coutinho E, Jain R. Synthesis, anti-tuberculosis activity, and 3D-QSAR study of ring-substituted-2/4-quinolinecarbaldehyde derivatives. Bioorg Med Chem 2006; 14: 7302.

[91]   Nayyar A, Monga V, Malde A, Coutinho E, Jain R. Synthesis, anti-tuberculosis activity, and 3D-QSAR study of 4-(adamantan-1-yl)-2-substituted quinolines. Bioorg Med Chem 2007; 15: 626.

[92]   Nayyar A, Patel SR, Shaikh M, Coutinho E, Jain R. Synthesis, anti-tuberculosis activity and 3D-QSAR study of amino acid conjugates of 4-(adamantan-1-yl) group containing quinolines. Eur J Med Chem 2009; 44: 2017.

[93]   Jaso A, Zarranz B, Aldana I, Monge A. Synthesis of New Quinoxaline-2-carboxylate 1,4-Dioxide Derivatives as Anti-Mycobacterium tuberculosis Agents. J Med Chem 2005; 48: 2019.

[94]   Vicente E, Perez-Silanes S, Lima LM, Ancizu S, Burguete A, Solano B, Villar R, Aldana I, Monge A. Selective activity against *Mycobacterium tuberculosis* of new quinoxaline 1,4-di-N-oxides. Bioorg Med Chem 2009; 17: 385.

[95]   Vicente E, Duchowicz PR, Castro EA, Monge A. QSAR analysis for quinoxaline-2-carboxylate 1,4-di-N-oxides as anti-mycobacterial agents. J Mol Graph Model 2009; 28: 28.

[96]   Ghosh P, Vracko M, Chattopadhyay AK, Bagchi MC. On Application of constitutional descriptors for merging of quinoxaline data sets using linear statistical methods. Chem Biol Drug Des 2008; 72: 155.

[97]   Seydel JK, Schaper KJ, Wempe E, Cordes HP. Mode of action and quantitative structure-activity correlations of tuberculostatic drugs of the isonicotinic acid hydrazide type. J Med Chem 1976; 19: 483.

[98]   Bagchia MC, Maitib BC, Bose S. QSAR of anti tuberculosis drugs of INH type using graphical invariants. J Mol Struct (Theochem) 2004; 679: 179.

[99]   Bagchi MC, Maiti BC, Mills D, Basak SC. Usefulness of graphical invariants in quantitative structure-activity correlations of tuberculostatic drugs of the isonicotinic acid hydrazide type. J Mol Model 2004; 10: 102.

[100]   Klopman G, Fercu D, Jacob J. Computer-aided study of he relationship between structure and antituberculosis activity of a series of isoniazid derivatives. Chem Phys 1996; 204: 181.

[101]   Pasqualoto KF, Ferreira EI, Santos-Filho OA, Hopfinger AJ. Rational design of new antituberculosis agents: receptor-independent four-dimensional quantitative structure–activity relationship analysis of a set of isoniazid derivatives. J Med Chem 2004; 47: 3755.

[102]   Ventura C, Martins F. Application of quantitative structure–activity relationships to the modeling of antitubercular compounds. 1. The hydrazide family. J Med Chem 2008; 51: 612.

[103]   Andrade CH, Salum LB, Castilho MS, Pasqualoto KF, Ferreira EI, Andricopulo AD. Fragment-based and classical quantitative structure–activity relationships for a series of hydrazides as antituberculosis agents. Mol Divers 2008; 12: 47.

[104]   Quinn FR, Driscoll JS, Hansch C. Structure-activity correlations among rifamycin B amides and hydrazides. J Med Chem 1975; 18: 332.

[105]   Kubinyi H. Quantitative structure-activity relationships. 2. A mixed approach, based on Hansch and Free-Wilson analysis. J Med Chem 1976; 19: 587.

[106]   Kiritsy JA, Yung DK, Mahony DE. Synthesis and quantitative structure-activity relationships of some antibacterial 3-formylrifamycin SV N-(4-substituted phenyl)piperazinoacethydrazones. J Med Chem 1978; 21: 1301.

[107]   Maggi N, Pallanza R, Sensi P. New derivatives of rifamycin SV. Antimicrob Agents Chemother 1965; 5: 765.

[108] Dimcho D, Zoya N, Svetla H, Gerrit S, Ovanes M. QSAR modeling of antimycobacterial activity and activity against other bacteria of 3-formyl rifamycin SV derivatives. Quant Struct-Act Relat 2001; 20: 298.

[109] Deeb O, Singh J, Varma RG, Khadikard PV. Topological modeling of antimycobacterial activity of 3-formyl rifamycin SV derivatives. Arkivoc 2007; XIV: 141.

[110] Wachter GA, Davis MC, Martin AR, Franzblau SG. Antimycobacterial activity of substituted isosteres of Pyridine- and Pyrazinecarboxylic acids. J Med Chem 1998; 41: 2436.

[111] Desai B, Sureja D, Naliapara Y, Shah A, Saxena AK. Synthesis and QSAR Studies of 4-Substituted phenyl-2,6-dimethyl-3, 5-bis-N-(substituted phenyl)carbamoyl-1,4-dihydropyridines as potential antitubercular agents. Bioorg Med Chem 2001; 9: 1993.

[112] Kharkar PS, Desai B, Gaveria H, Varu B, Loriya R, Naliapara Y, Shah A, Kulkarni VM. Three-dimensional quantitative structure–activity relationship of 1,4-dihydropyridines as antitubercular agents. J Med Chem 2002; 45: 4858.

[113] Manvar AT, Pissurlenkar RR, Virsodia VR, Upadhyay KD, Manvar DR, Mishra AK, Acharya HD, Parecha AR, Dholakia CD, Shah AK, Coutinho EC. Synthesis, *in vitro* antitubercular activity and 3D-QSAR study of 1,4-dihydropyridines. Mol Divers 2010; 14: 285-305.

[114] Klimesova V, Palat K, Waisser K, Klimes J. Combination of molecular modeling and quantitative structure-activity relationship analysis in the study of antimycobacterial activity of pyridine derivatives. Int J Pharm 2000; 207: 1.

[115] Coleman MD, Tims KJ, Rathbone DL. The use of computational QSAR analysis in the toxicological evaluation of a series of 2-pyridylcarboxamidrazone candidate anti-tuberculosis compounds. Environ Toxicol Pharmacol 2003; 14: 33.

[116] Virsodia V, Pissurlenkar RR, Manvar D, Dholakia C, Adlakha P, Shah A, Coutinho EC. Synthesis, screening for antitubercular activity and 3D-QSAR studies of substituted N-phenyl-6-methyl-2-oxo-4-phenyl-1,2,3,4-tetrahydro-pyrimidine-5-carboxamides. Eur J Med Chem 2008; 43: 2103.

[117] Kaliszan R, Foks H, Janowiec M. Studies on the quantitative structure-activity relationships in pyrazine carbothioamide derivatives. Pol J Pharmacol Pharm 1978; 30: 579-583.

[118] Bergmann KE, Cynamon MH, Welch JT. Quantitative structure–activity relationships for the *in vitro* antimycobacterial activity of pyrazinoic acid esters. J Med Chem 1996; 39: 3394.

[119] Ragno R, Marshall GR, Di Santo R, Costi R, Massa S, Rompei R, Artico M. Antimycobacterial pyrroles: synthesis, anti-Mycobacterium tuberculosis activity and QSAR studies. Bioorg Med Chem 2000; 8: 1423.

[120] Biava M, Porretta GC, Poce G, Supino S, Deidda D, Pompei R, Molicotti P, Manetti F, Botta M. Antimycobacterial agents. Antimycobacterial agents. Novel diarylpyrrole derivatives of BM212 endowed with high activity toward mycobacterium tuberculosis and low cytotoxicity. J Med Chem 2006; 49: 4946.

[121] Biava M, Porretta GC, Poce G, Supino S, Manetti F, Forli S, Botta M, Sautebin L, Rossi A, Pergola C, Ghelardini C, Norcini M, Makovec F, Giordani A, Anzellotti P, Cirilli R, Ferretti R, Gallinella B, La Torre F, Anzini M, Patrignani P. Synthesis, *in vitro*, and *in vivo* biological evaluation and molecular docking simulations of chiral alcohol and ether derivatives of the 1,5-diarylpyrrole scaffold as novel anti-inflammatory and analgesic agents. Bioorg Med Chem 2008; 16: 8072.

[122] Biava M, Porretta GC, Poce G, De Logu A, Saddi M, Meleddu R, Manetti F, De Rossi E, Botta M. 1,5-diphenylpyrrole derivatives as antimycobacterial agents. Probing the influence on antimycobacterial activity of lipophilic substituents at the phenyl rings. J Med Chem 2008; 51: 3644.

[123] Biava M, Porretta GC, Manetti F. New derivatives of BM212: A class of antimycobacterial compounds based on the pyrrole ring as a scaffold. Mini Rev Med Chem 2007; 7: 65.

[124] Biava M, Porretta GC, Deidda D, Pompei R. New trends in development of antimycobacterial compounds. Infect Disord Drug Targets 2006; 6: 159.

[125] Biava M, Porretta GC, Poce G, Deidda D, Pompei R, Tafi A, Manetti F. Antimycobacterial compounds. Optimization of the BM 212 structure, the lead compound for a new pyrrole derivative class. Bioorg Med Chem 2005; 13: 1221.

[126] Biava M, Porretta GC, Deidda D, Pompei R, Tafi A, Manetti F. Antimycobacterial compounds. New pyrrole derivatives of BM212. Bioorg Med Chem 2004; 12: 1453.

[127] Biava M, Porretta GC, Deidda D, Pompei R, Tafi A, Manetti F. Importance of the thiomorpholine introduction in new pyrrole derivatives as antimycobacterial agents analogues of BM 212. Bioorg Med Chem 2003; 11: 515.

[128] Biava M. BM 212 and its derivatives as a new class of antimycobacterial active agents. Curr Med Chem 2002; 9: 1859.

[129] Manetti F, Corelli F, Biava M, Fioravanti R, Porretta GC, Botta M. Building a pharmacophore model for a novel class of antitubercular compounds. Farmaco 2000; 55: 484.

[130] Biava M, Fioravanti R, Porretta GC, Deidda D, Maullu C, Pompei R. New pyrrole derivatives as antimycobacterial agents analogs of BM212. Bioorg Med Chem Lett 1999; 9: 2983.

[131] Biava M, Porretta GC, Poce G, De Logu A, Meleddu R, De Rossi E, Manetti F, Botta M. 1,5-Diaryl-2-ethyl pyrrole derivatives as antimycobacterial agents: Design, synthesis, and microbiological evaluation. Eur J Med Chem 2009; 44: 4734.

[132] Karali N, Gursoy A, Kandemirli F, Shvets N, Kaynak FB, Ozbey S, Kovalishyn V, Dimoglo A. Synthesis and structure–antituberculosis activity relationship of 1H-indole-2,3-dione derivatives. Bioorg Med Chem 2007; 15: 5888.

[133] Vicini P, Geronikaki A, Incerti M, Busonera B, Poni G, Cabras CA, La Colla P. Synthesis and biological evaluation of benzo[d]isothiazole, benzothiazole and thiazole Schiff bases. Bioorg Med Chem 2003; 11: 4785.

[134] Sivakumar PM, Geetha Babu SK, Doble M. Impact of topological and electronic descriptors in the QSAR of pyrazine containing thiazolines and thiazolidinones as antitubercular and antibacterial agents. Chem Biol Drug Des 2008; 71: 447.

[135] Desai NC, Bhavsar AM, Shah MD, Saxena AK. Synthesis and QSAR studies of thiosemicarbazides, 1,2,4-triazoles, 1,3,4-thiadiazoles and 1,3,4-oxadiazoles derivatives as potential antibacterial agents. Ind J Chem 2008; 47B: 579.

[136] Desai NC, Shah MD, Bhavsar AM, Saxena AK. Synthesis and QSAR studies of 4-oxo-thiazolidines and 2-oxo-azetidines as potential antibacterial agents. Ind J Chem 2008; 47B: 1135.

[137] Thaker KM, Kachhadia VV, Joshi HS. Synthesis of 4-thiazolidinones and 2-azetidinones bearing benzo (b) thiophene nucleus as potential antitubercular and antimicrobial agents. Ind J Chem 2003; 42B: 1544.

[138] Narute AS, Khedekar PB, Bhusari KP. QSAR studies on 4-thiazolidinones and 2-azetidinones bearing benzothiophene nucleus as potential anti-tubercular agents. Ind J Chem 2008; 47B: 586.

[139] Foroumadi A, Mirzaei M, Shafiee A. Antituberculosis agents, I: Synthesis and antituberculosis activity of 2-aryl-1,3,4-thiadiazole derivatives. Pharmazie 2001; 56: 610.

[140] Hemmateenejad B, Miri R, Jafarpour M, Tabarzad M, Foroumadi A. Multiple liner regression and principal component analysis-based prediction of the anti-tuberculosisactivity of some 2-aryl-1,3,4-thiadizole derivatives. QSAR Comb Sci 2006; 25: 56.

[141] Foroumadi A, Sakhteman A, Sharifzadeh Z, Mohammadhosseini N, Hemmateenejad B, Moshafi MH, Vosooghi M, Amini M, Shafiee A. Synthesis, antituberculosis activity and QSAR study of some novel 2-(nitroaryl)-5-(nitrobenzylsulfinyl and sulfonyl)-1,3,4-thiadiazole derivatives. DARU 2007; 15: 218.

[142] Macaev F, Rusu G, Pogrebnoi S, Gudima A, Stingaci E, Vlad L, Shvets N, Kandemirli F, Dimoglo A, Reynolds R. Synthesis of novel 5-aryl-2-thio-1,3,4-oxadiazoles and the study of their structure–anti-mycobacterial activities. Bioorg Med Chem 2005; 13: 4842.

[143] Sriram D, Yogeeswari P, Thirumurugan R, Pavana RK. Discovery of new antitubercular oxazolyl thiosemicarbazones. J Med Chem 2006; 49: 3448.

[144] Ban T, Fujita T. Mathematical approach to structure-activity study of sympathomimetic amines. Norepinephrine uptake inhibition. J Med Chem 1969; 12: 353.

[145] Fujita T, Ban T. Structure-activity relation. 3. Structure-activity study of phenethylamines as substrates of biosynthetic enzymes of sympathetic transmitters. J Med Chem 1971; 14: 148.

[146] Gupta RA, Gupta AK, Soni LK, Kaskhedikar SG. Rationalization of physicochemical characters of oxazolyl thiosemicarbazones analogs towards multi-drug resistant tuberculosis: A QSAR approach. Eur J Med Chem 2007; 42: 1109.

[147] Waisser K, Macháček M, Dostál H, Gregor J, Kubicová L, Klimešová V, Kuneš Jr J, Palát K, Hladůvková J, Kaustová J, Möllmann U. Relationships between the chemical structure of substances and their antimycobacterial activity against atypical strains. Part 18. 3-phenyl-2H-1,3-benzoxazine-2,4(3H)-diones and isosteric 3-phenylquinazoline-2,4 (1H, 3H)-diones. Collect Czech Chem Commun 1999; 64: 1902.

[148] Waisser K, Hladuvkova J, Gregor J, Rada T, Kubicova L, Klimesova V, Kaustova J. Relationships between the chemical structure of antimycobacterial substances and their activity against atypical strains. Part 14: 3-Aryl-6,8-dihalogeno-2H-1,3-benzoxazine-2,4(3H)-diones. Arch Pharm 1998; 331: 3.

[149] Waisser K, Gregor J, Dostal H, Kunes J, Kubicova L, Klimesova V, Kaustova J. Influence of the replacement of the oxo function with the thioxo group on the antimycobacterial activity of 3-aryl-6,8-dichloro-2H-1,3-benzoxazine-2,4(3H)-diones and 3-arylquinazoline-2,4(1H,3H)-diones. Farmaco 2001; 56: 803.

[150] Waisser K, Gregor J, Kubicova L, Klimesova V, Kunes J, Machacek M, Kaustova J. New groups antimycobacterial agents: 6-chloro-3-phenyl-4-thioxo-2H-1,3-benzoxazine -2(3H)-ones and 6-chloro-3-phenyl-2H-1,3-benzoxazine-2,4(3H)-dithiones. Eur J Med Chem 2000; 35: 733.

[151] Whitley DC, Ford MG, Livingstone DJ. Unsupervised forward selection: a method for eliminating redundant variables. J Chem Inf Comput Sci 2000; 40: 1160.

[152] Besalu E, Ponec R, de Julian-Ortiz JV. Virtual generation of agents against *Mycobacterium tuberculosis*. A QSAR study. Mol Divers 2003; 6: 107.

[153] Gallegos A, Carbo-Dorca R, Ponec R, Waisser K. Similarity approach to QSAR. Application to antimycobacterial benzoxazines. Int J Pharm 2004; 269: 51.

[154] Koci J, Klimesova V, Waisser K, Kaustova J, Dahse HM, Mollmann U. Heterocyclic benzazole derivatives with antimycobacterial *in vitro* activity. Bioorg Med Chem Lett 2002; 12: 3275.

[155] Klimesova V, Koci J, Waisser K, Kaustova J, Mollmann U. Preparation and *in vitro* evaluation of benzylsulfanyl benzoxazole derivatives as potential antituberculosis agents. Eur J Med Chem 2009; 44: 2286.

[156] Geban O, Ertepinar H, Ozden S. QSAR analysis of a set of benzimidazole derivatives based on their tuberculostatic activities. Pharmazie 1996; 51: 34.

[157] Raparti V, Chitre T, Bothara K, Kumar V, Dangre S, Khachane C, Gore S, Deshmane B. Novel 4-(morpholin-4-yl)-N'-(arylidene)benzohydrazides: Synthesis, antimycobacterial activity and QSAR investigations. Eur J Med Chem 2009; 44: 3954.

[158] Tangallapally RP, Sun D, Budha NR, Lee RE, Lenaerts AJ, Meibohm B. Discovery of novel isoxazolines as anti-tuberculosis agents. Bioorg Med Chem Lett 2007; 17: 6638.

[159] Tangallapally RP, Yendapally R, Daniels AJ, Lee RE. Nitrofurans as novel anti-tuberculosis agents: identification, development and evaluation. Curr Top Med Chem. 2007; 7: 509.

[160] Tangallapally RP, Yendapally R, Lee RE, Lenaerts AJ. Synthesis and evaluation of cyclic secondary amine substituted phenyl and benzyl nitrofuranyl amides as novel antituberculosis agents. J Med Chem 2005; 48: 8261.

[161] Tangallapally RP, Yendapally R, Lee RE, Hevener K, Jones VC, Lenaerts AJ, McNeil MR, Wang Y, Franzblau S. Synthesis and evaluation of nitrofuranylamides as novel antituberculosis agents. J Med Chem 2004; 47: 5276.

[162] Tangallapally RP, Lee RE, Lenaerts AJ. Synthesis of new and potent analogues of anti-tuberculosis agent 5-nitro-furan-2-carboxylic acid 4-(4-benzyl-piperazin-1-yl)-benzylamide with improved bioavailability. Bioorg Med Chem Lett 2006; 16: 2584.

[163] Budha NR, Mehrotra N, Tangallapally R, Qi J, Daniels AJ, Lee RE, Meibohm B. Pharmacokinetically-guided lead optimization of nitrofuranylamide anti-tuberculosis agents. AAPS J 2008; 10: 157.

[164] Hurdle JG, Lee RB, Budha NR, Carson EI, Qi J, Scherman MS, Cho SH, McNeil MR, Lenaerts AJ, Franzblau SG, Meibohm B, Lee RE. A microbiological assessment of novel nitrofuranylamides as anti-tuberculosis agents. J Antimicrob Chemother 2008; 62:1037.

[165] Budha NR, Lee RE, Meibohm B. Biopharmaceutics, pharmacokinetics and pharmacodynamics of antituberculosis drugs. Curr Med Chem 2008; 15: 809.

[166] Hevener KE, Ball DM, Buolamwini JK, Lee RE. Quantitative structure–activity relationship studies on nitrofuranyl anti-tubercular agents. Bioorg Med Chem 2008; 16: 8042.

[167] Hansch CH, Leo AJA. Substituent constants for correlation analysis in chemistry and biology. New York: Wiley 1979.

[168] Bukowski L, Kaliszan R. Imidazo[4,5-b]pyridine derivatives of potential tuberculostatic activity. Part 1: Synthesis and quantitative structure-activity relationships. Arch Pharm 1991; 324: 537.

[169] Bukowski L, Kaliszan R. Imidazo[4,5-b]pyridine derivatives of potential tuberculostatic activity. Part 1: Synthesis and quantitative structure-activity relationships. Arch Pharm 1991; 324: 121.

[170] Chambhare RV, Khadse BG, Bobde AS, Bahekar RH. Synthesis and preliminary evaluation of some N-[5-(2-furanyl)-2-methyl-4-oxo-4H-thieno[2,3-d]pyrimidin-3-yl]-carboxamide and 3-substituted-5-(2-furanyl)-2-methyl-3H-thieno[2,3-d]pyrimidin-4-ones as antimicrobial agents. Eur J Med Chem 2003; 38: 89.

[171] Narasimhan B, Kumari M, Jain N, Dhake A, Sundaravelan C. Correlation of antibacterial activity of some N-[5-(2-furanyl)-2-methyl-4-oxo-4H-thieno[2,3-d]pyrimidin-3-yl]-carboxamide and 3-substituted-5-(2-furanyl)-2-methyl-3H-thieno[2,3-d]pyrimidin-4-ones with topological indices using Hansch analysis. Bioorg Med Chem Lett 2006; 16: 4951.

[172] Kim P, Kang S, Boshoff HI, Jiricek J, Collins M, Singh R, Manjunatha UH, Niyomrattanakit P, Zhang L, Goodwin M, Dick T, Keller TH, Dowd CS, Barry CE. Structure–activity relationships of antitubercular nitroimidazoles. 2. Determinants of aerobic activity and quantitative structure–activity relationships. J Med Chem 2009; 52: 1329.

[173] Machacek M, Kunes J, Sidoova E, Odlerova Z, Waisser K. [Relation between the chemical structure of substances and their antimicrobial action against atypical strains. II. 6-acycloamido-2-alkylthiobenzothiazoles, quantitative relation to their effectiveness spectrum]. Cesk Farm 1989; 38: 9.

[174] Ghosal S, Biswas K, Chaudhuri RK. Chemical constituents of Gentianaceae XXIV: Anti-Mycobacterium tuberculosis activity of naturally occurring xanthones and synthetic analogs. J Pharm Sci 1978; 67: 721.

[175] Frahm AW, Hambloch H. Computer supported structure elucidation of polymethoxy- and polyacetoxyxanthones. V†—^{13}C NMR spectroscopy of substituted xanthones. Org Magn Reson 1982; 19: 43.

[176] Ungwitayatorn J, Pickert M, Frahm AW. Quantitative structure-activity relationship (QSAR) study of polyhydroxyxanthones. Pharm Acta Helv 1997; 72: 23.

[177] Pickert M, Schaper KJ, Frahm AW. Substituted xanthones as antimycobacterial agents, Part 2: Antimycobacterial activity. Arch Pharm 1998; 331: 193.

[178] Schaper KJ, Pickert M, Frahm AW. Substituted xanthones as antimycobacterial agents. Part 3: QSAR investigations. Arch Pharm 1999; 332: 91.

[179] Panda G, Shagufta, Mishra JK, Chaturvedi V, Srivastava AK, Srivastava R, Srivastava BS. Diaryloxy methano phenanthrenes: a new class of antituberculosis agents. Bioorg Med Chem 2004; 12: 5269.

[180] Panda G, Shagufta, Srivastava AK, Sinha S. Synthesis and antitubercular activity of 2-hydroxy-aminoalkyl derivatives of diaryloxy methano phenanthrenes. Bioorg Med Chem Lett 2005; 15: 5222.

[181] Siddiqi S. Clinical Microbiology Handbook, ASM Press: Washington DC, 1992.

[182] Shagufta, Kumar A, Panda G, Siddiqi MI. CoMFA and CoMSIA 3D-QSAR analysis of diaryloxy-methano-phenanthrene derivatives as anti-tubercular agents. J Mol Model 2007; 13: 99.

[183] Čižmárik J, Waisser K, Doležal R. QSAR study of antimycobacterial activity of esters of substituted phenylcarbamic acids. Acta Facult Pharm Univ Comenianae 2008; 55: 90.

[184] Waisser K, Dolezal R, Palat Jr K, Cizmarik J, Kaustova J. QSAR study of antimycobacterial activity of quaternary ammonium salts of piperidinylethyl esters of alkoxysubstituted phenylcarbamic acids. Folia Microbiol 2006; 51: 21.

[185] Waisser K, Drazkova K, Cizmarik J, Kaustova J. Antimycobacterial activity of piperidinylpropyl esters of alkoxy-substituted phenylcarbamic acids. Folia Microbiol 2003; 48: 585.

[186] Matyk J, Waisser K, Drazkova K, Kunes J, Klimesova V, Palat Jr K, Kaustova J. Heterocyclic isosters of antimycobacterial salicylanilides. Farmaco 2005; 60: 399.

[187] Imramovsky A, Vinsova J, Ferriz JM, Dolezal R, Jampilek J, Kaustova J, Kunc F. New antituberculotics originated from salicylanilides with promising *in vitro* activity against atypical mycobacterial strains. Bioorg Med Chem 2009; 17: 3572.

[188] Dolezal R, Van Damme S, Bultinck P, Waisser, K. QSAR analysis of salicylamide isosteres with the use of quantum molecular descriptors. Eur J Med Chem 2009; 44: 869.

[189] Dolezal R, Waisser K, Petrlikova E, Kunes J, Kubicova L, Machacek M, Kaustova J, Dahse HM. N-benzylsalicylthioamides: highly active potential antituberculotics. Arch Pharm 2009; 342: 113.

[190] Waisser K, Celadnik M, Palat K, Karlicek R, Odlerova Z, Bartos F, Drsata J. [Quantitative structure-activity analysis of thiobenzamides]. Pharmazie 1983; 38: 874.

[191] Gupta MK, Sagar R, Shaw AK, Prabhakar YS. CP-MLR directed QSAR studies on the antimycobacterial activity of functionalized alkenols—topological descriptors in modeling the activity. Bioorg Med Chem 2005; 13: 343.

[192] Vergara FM, Henriques MG, Candea AL, Wardell JL, De Souza MV. Antitubercular activity of alpha,omega-diaminoalkanes, H2N(CH2)nNH2. Bioorg Med Chem Lett 2009; 19: 4937.

[193] Lee RE, Smith MD, Nash RJ, Griffiths RC, McNeil M, Grewal RK, Yan W, Besra GS, Brennan PJ, Fleet GWJ. ChemInform Abstract: inhibition of udp-gal mutase and mycobacterial galactan biosynthesis by pyrrolidine analogues of galactofuranose. Tetrahedron Lett 1997; 28: 6733.

[194] Cren S, Gurcha SS, Blake AJ, Besra GS, Thomas NR. Synthesis and biological evaluation of new inhibitors of UDP-Galf transferase--a key enzyme in *M. tuberculosis* cell wall biosynthesis. Org Biomol Chem 2004; 2: 2418.

[195] Centrone CA, Lowary TL. Sulfone and phosphinic acid analogs of decaprenolphosphoarabinose as potential anti-tuberculosis agents. Bioorg Med Chem 2004; 12: 5495.

[196] Wen X, Crick DC, Brennan PJ, Hultin PG. Analogues of the mycobacterial arabinogalactan linkage disaccharide as cell wall biosynthesis inhibitors. Bioorg Med Chem 2003; 11: 3579.

[197] Pathak R, Shaw AK, Bhaduri AP, Chandrasekhar KV, Srivastava A, Srivastava KK, Chaturvedi V, Srivastava R, Srivastava BS, Arora S, Sinha S. Higher acyclic nitrogen containing deoxy sugar derivatives: A new lead in the generation of antimycobacterial chemotherapeutics. Bioorg Med Chem 2002; 10: 1695.

[198] Reynolds RC, Bansal N, Rose J, Friedrich J, Suling WJ, Maddry JA. Ethambutol-sugar hybrids as potential inhibitors of mycobacterial cell-wall biosynthesis. Carbohydrate Res 1999; 317: 164.

[199] Maddry JA, Suling WJ, Reynolds RC. Glycosyltransferases as targets for inhibition of cell wall synthesis in *M. tuberculosis* and *M. avium*. Res Microbiol 1996; 147: 106.

[200] Wojtowicz M, Wieniawski W. [N-glycosides of nitrogen heterocyclic compounds. VI. Synthesis of N-D-glucopyranosides of 2-amino-5-(2-pyridyl)-1,3,4-oxadiazole] Acta Pol Pharm 1977; 34: 575.

[201] Wojtowicz M, Wieniawski W. [N-glycosides of nitrogen heterocyclic compounds. VII. Synthesis of N-(D-galactopyranoside) of 2-amino-5-(pyridyl)-1,3,4-oxadiazole]. Acta Pol Pharm 1978; 35: 37.

[202] Saquib M, Gupta MK, Sagar R, Prabhakar YS, Shaw AK, Kumar R, Maulik PR, Gaikwad AN, Sinha S, Srivastava AK, Chaturvedi V, Srivastava R, Srivastava BS. C-3 Alkyl/Arylalkyl-2,3-dideoxy Hex-2-enopyranosides as Antitubercular Agents: Synthesis, Biological Evaluation, and QSAR Study. J Med Chem 2007; 50: 2942.

[203] Lin YM, Zhou Y, Flavin MT, Zhou LM, Nie W, Chen FC. Chalcones and flavonoids as anti-Tuberculosis agents. Bioorg Med Chem 2002; 10: 2795.

[204] Sivakumar PM, Geetha Babu SK, Mukesh D. QSAR studies on chalcones and flavonoids as anti-tuberculosis agents using genetic function approximation (GFA) method. Chem Pharm Bull 2007; 55: 44.

[205] Kanokmedhakul S, Kanokmedhakul K, Phonkerd N, Soytong K, Kongsaeree P, Suksamrarn A. Antimycobacterial anthraquinone-chromanone compound and diketopiperazine alkaloid from the fungus Chaetomium globosum KMITL-N0802. Planta Med 2002; 68: 834.

[206] Ballell L, Field RA, Duncan K, Young RJ. New small-molecule synthetic antimycobacterials. Antimicrob Agents Chemother 2005; 49: 2153.

[207] Sivakumar PM, Seenivasan SP, Kumar V, Doble M. Synthesis, antimycobacterial activity evaluation, and QSAR studies of chalcone derivatives. Bioorg Med Chem Lett 2007; 17: 1695.

[208] Manvar A, Malde A, Verma J, Virsodia V, Mishra A, Upadhyay K, Acharya H, Coutinho E, Shah A. Synthesis, anti-tubercular activity and 3D-QSAR study of coumarin-4-acetic acid benzylidene hydrazides. Eur J Med Chem 2008; 43: 2395.

[209] Lall N, Meyer JJ. Inhibition of drug-sensitive and drug-resistant strains of *Mycobacterium tuberculosis* by diospyrin, isolated from Euclea natalensis. J Ethnopharmacol 2001; 78: 213.

[210] Mahapatra A, Mativandlela SP, Binneman B, Fourie PB, Hamilton CJ, Meyer JJ, van der Kooy F, Houghton P, Lall N. Activity of 7-methyljuglone derivatives against *Mycobacterium tuberculosis* and as subversive substrates for mycothiol disulfide reductase. Bioorg Med Chem 2007; 15: 7638.

[211] Rugutt JK, Rugutt KJ. Relationships between molecular properties and antimycobacterial activities of steroids. Nat Prod Lett 2002; 16: 107.

[212] Garcia-Garcia A, Galvez J, de Julian-Ortiz JV, Garcia-Domenech R, Munoz C, Guna R, Borras R. Search of chemical scaffolds for novel antituberculosis agents. J Biomol Screen 2005; 10: 206.

[213] Prakash O, Ghosh I. Developing an antituberculosis compounds database and data mining in the search of a motif responsible for the activity of a diverse class of antituberculosis agents. J Chem Inf Model 2006; 46: 17.

[214] Prathipati P, Ma NL, Keller TH. Global Bayesian models for the prioritization of antitubercular agents. J Chem Inf Model 2008; 48: 2362.

[215] Imramovsky A, Polanc S, Vinsova J, Kocevar M, Jampilek J, Reckova Z, Kaustova J. A new modification of anti-tubercular active molecules. Bioorg Med Chem 2007; 15: 2551.

[216] Czaplinski K-H, Hänsel W, Wiese M, Seydel JK. New benzyl pyrimidines: inhibition of DHFR from various species. QSAR, CoMFA and PC analysis. Eur J Med Chem 1995; 30: 779.

[217] Jain P, Soni LK, Gupta AK, Kashkedikar SG. Jain P, Soni LK, Gupta AK, Kashkedikar SG. Ind J Biochem Biophys 2005; 42: 315. Ind J Biochem Biophys 2005; 42: 315.

[218] Aparna V, Jeevan J, Ravi M, Desiraju GR, Gopalakrishnan B. 3D-QSAR studies on antitubercular thymidine monophosphate kinase inhibitors based on different alignment methods. Bioorg Med Chem Lett 2006; 16: 1014.

[219] Andrade CH, Pasqualoto KF, Ferreira EI, Hopfinger AJ. Rational design and 3D-pharmacophore mapping of 5'-thiourea-substituted alpha-thymidine analogues as mycobacterial TMPK inhibitors. J Chem Inf Model 2009; 49: 1070.

[220] Kumar A, Siddiqi MI. CoMFA based *de novo* design of pyrrolidine carboxamides as inhibitors of enoyl acyl carrier protein reductase from *Mycobacterium tuberculosis*. J Mol Model 2008; 14: 923.

[221] Labello NP, Bennett EM, Ferguson DM, Aldrich CC. Quantitative three dimensional structure linear interaction energy model of 5'-O-[N-(Salicyl)sulfamoyl]adenosine and the aryl acid adenylating enzyme MbtA. J Med Chem 2008; 51: 7154.

[222] Daniel M, Per ML, Johanna N, Anders K, Torbjörn L. Focused hierarchical design of peptide libraries—follow the lead. J Chemom 2007; 21: 486.

[223] Linusson A, Gottfries J, Lindgren F, Wold S. Statistical molecular design of building blocks for combinatorial chemistry. J Med Chem 2000; 43: 1320.

[224] Johanna N, Annette KR, Daniel M, Erik W, Daniel JE, Torbjörn L, Torsten U, Anders K. Design, synthesis and evaluation of peptide inhibitors of *Mycobacterium tuberculosis* ribonucleotide reductase. J Pep Sci 2007; 13: 822.

CHAPTER 5

Antileishmaniasis Agents: Molecular Dynamics Simulations

Tanos Celmar Costa França[1]* and Alan Wilter Sousa da Silva[2]

[1]Chemical Engineering Department, Military Institute of Engineering, Pça General Tibúrcio, 80, Urca, 22290-270, Rio de Janeiro/RJ, Brazil and [2]Department of Biochemistry, University of Cambridge, 80 Tennis Court Road, CB2 1GA, United Kingdom

Abstract: Molecular dynamics simulations have showed to be powerful tools when applied to the preliminary investigations of the interactions of potential molecular targets with its natural subtracts or, eventually, with their potential inhibitors. When the 3D structure of a molecular target is yet unknown, sometimes it is possible to build a very consistent model using one of the several softwares available today for this purpose (see chapter on homology modeling) and, further, use it to analyze the overall structure of the target, the active site residues and their potential interactions with potential ligands, by performing MD simulations studies in order to afford additional information towards the rational design of inhibitors to the molecular target in focus. Literature has reported a few interesting studies using this approach on leishmaniasis. Those studies have afforded useful information for the experimentalists on new drug targets for the rational design of new, more selective and powerful antileishmiasis agents.

Keywords: Molecular dynamics simulation, proteins, antileishmaniasis agents.

1. INTRODUCTION

1.1. Molecular Dynamics Simulations

Molecular Dynamics (MD) is a technique developed to study the movements of a system of particles by simulation. This technique can be employed to electrons, atoms or molecules as well as to macromolecular systems. Its essential elements are the knowledge of the potential of interaction among the particles and the classic equations of movement ruling the dynamic of those particles. The potential can change from simple, like the gravitational star interactions, to the complex, with several terms, like the description of the interactions between atoms and molecules. For many systems, including the biomolecular, the equations of classic dynamics are suitable. However, for some problems, like the galaxy evolution, relativist terms are included, while for others, like chemical reactions involving tunneling effects, quantum mechanical corrections are needed [1].

The microscopic state of a system can be specified in terms of the positions and moments of its particles. Therefore, in classic mechanics, one can write the Hamiltonian **H** of a classic molecular system as the sum of the kinetic (C) and potential (U) energies in function of the series of generalized coordinates q_i and generalized moments p_i of all N atoms of the system, like in Eq. 1:

$$H(q_i, p_i) = C(p_i) + V(q_i) \qquad \text{(equation 1)}$$

Where:

$$q_i = q_1, q_2, ., q_{Nat} \text{ and } p_i = p_1, p_2, \ldots, p_{Nat}$$

The potential energy $V(q_i)$ contains the short and long distance inter and intramolecular interaction terms and can be replaced by the potential $V(r_i)$ in a way that the coordinates q_i become the Cartesian coordinates and r_i and p_i its conjugated moments. The kinetic energy, on the other hand, can assume the form of Eq. 2:

*Address correspondence to Tanos Celmar Costa França: Chemical Engineering Department, Military Institute of Engineering, Pça General Tibúrcio, 80, Urca, 22290-270, Rio de Janeiro/RJ, Brazil; Tel: +00552125467195; E-mail: tanos@ime.eb.br

Teodorico C. Ramalho, Matheus P. Freitas and Elaine F. F. da Cunha (Eds)

$$C(p_i) = \sum_{i=1}^{Nat} \frac{p_i^{\,2}}{2m_i}$$ (equation 2)

where m_i is the mass of the atom i.

From Eq. 1 it is possible to construct the equations of movement ruling the temporal evolution of the system and their dynamical properties. As the potential energy is independent from velocities and time, **H** is the total energy of the system and Hamilton's equations of motion (Eqs. 3 and 4):

$$\dot{q}_i = \frac{\partial H}{\partial p_i}$$ (equation 3)

$$\dot{p}_i = -\frac{\partial H}{\partial q_i}$$ (equation 4)

conduct to the Newton equations of motion (Eq. 5 and 6):

$$\dot{r}_i = \frac{p_i}{m_i} = v_i$$ (equation 5)

$$\dot{p}_i = m_i \ddot{r}_i = -\frac{\partial V(r_i)}{\partial r_i} = F_i$$ (equation 6)

In Eqs. 5 and 6, $\dot{r}_i$ (or v_i) and $\dot{p}_i$ are the velocity and the acceleration of the atom i, while F_i is the force on the atom i [2-4].

The MD consists, therefore, in the numeric resolution of Eqs. 5 and 6 and their integration, step by step, in an efficient and accurate way. As a result, we have the energies and trajectories of all the particles (or atoms) for the system as a whole and from which several properties can be calculated. In systems with hydrogen, a time step of 5.0 x 10^{-16} seconds is usually applied. In this procedure it is essential that the potential energy be a continuum function of the particle positions and that the positions change smoothly with time. The F_i forces on each atom, obtained from the spatial derivative of the potential energy, as shown in Eq. 6, can be considered constant in the gap between two steps. Once the dynamic stability is favored, the particles follow their classic trajectories more accurately and the total system energy tends to be conserved.

A limitation of the MD simulation resides, therefore, in the fact that for each nanosecond of simulation, 2 million of steps are needed with the time step above. A simulation of 1 ns for a macromolecule with 200 atoms can take hours of CPU time in a supercomputer, when using an efficient algorithm. A description and analysis of the efficiency of algorithms for simulation and molecular dynamics can be found in Berendsen & van Gunsteren [3] and Allen & Tildesley [4]. The latter includes routines in FORTRAN for some simulation methods.

1.1.2. Molecular Mechanics

Most of the systems in molecular modeling are too big to be treated by quantum mechanics or semi empirical methods. Because these methods consider the movements of the electrons and a large number of other particles in the system, they are so time consuming and unreliable for big systems. The force field methods or the molecular Mechanics Methods (MM) ignore the electron movements and calculate the energy of the system as a function of only the nuclear positions. This makes MM a suitable method to deal with systems containing a large number of atoms. In some cases, force fields can lead to answers as accurate as the top level of quantum mechanics, in much less computational time [5]. MM, however, is not able to predict properties depending on the electronic distribution in a molecule, like transition states or charge distributions but, still, it is a very useful and powerful technique for drug design.

MM is based on a simpler model of the interactions inside a system with contributions of processes like bond stretching, the opening-closing of angles and the rotations around single bonds. However, there are limitations inherent in this methodology. For instance, when single functions, like Hooke's law, are employed to describe these contributions, the force field may not respond very well, producing values inconsistent with experimental results. Transferability is also a key property of a force field because it permits that a set of parameters developed and tested for a relatively small number of cases could be applied to a wider range of problems. Furthermore, parameters developed from data of small molecules could be used to study much large molecules like polymeric structures for example.

Most of the force fields in molecular modeling used today for molecular systems, GROMOS [6], AMBER [7,8], MM2/MM3/MM4 [9-17], CHARMM [18] could be interpreted in terms of a single set of four components corresponding to the intra and intermolecular forces inside the system. Energetic penalties are associated to the lift of angles and bonds from their reference or equilibrium values. There is a function describing how energy changes when bonds are turned and, finally, the force field has terms describing the interaction between non linked parts of the system. More sophisticated force fields can include additional terms but usually contain these four components. An interesting feature in this representation is that several terms can be related to changes in specific internal coordinates as to bond lengths, bond rotations or atom movements related to other atoms (see Table **1**). This facilitates the understanding of how changes in the force field parameters affect its performance and also helps in the parameterization process.

Table 1: Illustrations of the contributing terms for the potentials in a force field.

Terms for the interactions between pairs of bonded atoms (Bond potential)		
$$\upsilon_l = \frac{1}{2}\sum_{n=1}^{N_b} k_{bn}(r_n - r_{i,0})^2$$		Describes the interaction between N_b pairs of bonded atoms through a harmonic potential (Hooke's law).
Term for the angular potential energy		
$$\upsilon_\theta = \frac{1}{2}\sum_{n=1}^{N_\theta} k_\theta(\theta_n - \theta_{0n})^2$$		The deviation of the angles from their reference values is, also, usually described using the Hooke's law.
Term for the torsional potential		
$$\upsilon(\varphi) = \sum_{n=0}^{N} \frac{V_{\varphi n}}{2}\left[1 + \cos(m\varphi_n - \gamma_n)\right]$$		The torsional potentials are usually expressed as an expansion in cosines series where ω is the torsion angle, n the multiplicity and γ the phase factor determining where the torsion angle is minimum and $V_{\phi n}$ the constant defining the torsion barrier.
Term for the improper torsional potential and angular movements out of the plane		
$$\upsilon(\omega) = k(1 - \cos 2\omega)$$		Vibrations out of the plane can be dealt with by a torsional potential maintaining the improper torsional angle between 0° and 180° and also to conserve the structure of chiral centres

Term for the potential energy of Lennard-Jones		
$$v_{vdW} = \sum_{i<j}^{N^*} \left(\frac{A_{ij}}{r_{ij}^{12}} - \frac{B_{ij}}{r_{ij}^{6}} \right)$$	R_{vdw} (×2)	Non-bonded terms are usually modeled using a term of Lennard-Jones for electrostatic interactions.
Term for the electrostatic potential energy		
$$v_{el} = \sum_{i<j}^{N^*} \frac{q_i q_j}{4\pi\varepsilon_0 r_{ij}}$$		Non-bonded terms are usually modeled using a term of the Coulomb potential for electrostatic interactions.

A functional form of a force field is shown in Eq. 7:

$$v(r^N) = \sum_{bonds} \frac{k_i}{2}(r_i - r_{i,0})^2 + \sum_{angles} \frac{k_i}{2}(\theta_i - \theta_{i,0})^2 + \sum_{torsions} \frac{V_n}{2}[1 + \cos(n\omega - \gamma)]$$

$$+ \sum_{i=1}^{N} \sum_{j=i+1}^{N} \left(4\varepsilon_{ij} \left[\left(\frac{\sigma_{ij}}{r_{ij}}\right)^{12} \left(\frac{\sigma_{ij}}{r_{ij}}\right)^{6} \right] + \frac{q_i q_j}{4\pi\varepsilon_0 r_{ij}} \right)$$

$$v(r^N) = \sum_{bonds} \frac{k_i}{2}(r_i - r_{i,0})^2 + \sum_{angles} \frac{k_i}{2}(\theta_i - \theta_{i,0})^2 + \sum_{torsions} \frac{V_n}{2}[1 + \cos(n\omega - \gamma)]$$

$$+ \sum_{i=1}^{N} \sum_{j=i+1}^{N} \left(4\varepsilon_{ij} \left[\left(\frac{\sigma_{ij}}{r_{ij}}\right)^{12} \left(\frac{\sigma_{ij}}{r_{ij}}\right)^{6} \right] + \frac{q_i q_j}{4\pi\varepsilon_0 r_{ij}} \right)$$

(equation 7)

Where $v(r^N)$ is the total potential energy as a function of the positions (r) of N particles (usually atoms). The first term models the interactions between pairs of bonded atoms, modeled here by the harmonic potential affording the energy increment when the length r_i deviates from the reference value $r_{i,0}$. The second term is the summation on all the valence angles (angles A-B-C) in the molecule, again modeled by the harmonic potential. The third term is the torsional potential that modulates how the energy changes when bonds change. The dihedral or torsional terms typically have multiple minima and thus cannot be modeled as harmonic oscillators, though their specific functional form varies with the implementation. This class of terms may include "improper" dihedral terms, which function as correction factors for out-of-plane deviations (for example, they can be used to keep benzene rings planar). The fourth contribution is the non-bonded term. It is calculated among all pairs of atoms (i e j) in different molecules or in the same molecule, but at least at three bonds from each other (*i.e.*, they have a relation 1, n where n ≥ 4).

In a single force field the non bonded term is usually modeled using a term of the Coulomb potential for electrostatic interactions and a term of Lennard-Jones potential for the van der Waals, interactions (See Eq. 7). Table 1 illustrates each one of the terms discussed above.

1.1.3. Structure Optimization

The forces in an initial configuration for a 3D structure derived from Eq. 7 can be very large, leading to high local accelerations, high velocities and, consequently, large local displacements that do not satisfy the conditions for MD simulations. So, before performing a full MD simulation, an energy minimization is usually necessary that is also important to remove any local tension. Eq. 7 generates a hypersurface with local minima in which one can be the global minimum. The more degrees of freedom in the system the more local minima it has. This makes the work of scanning a multidimensional surface searching for the global minimum in a biological system such as a protein practically impossible. The solution is to scan a piece of the hypersurface of potential energy to find a point with the smallest local potential energy. The goal of getting into a minimum is to obtain a spatial conformation having relaxed the distortions in the chemical bonds, the angles between bonds and in the *van der Waals* contacts.

A minimum point (local or global) is determined when the forces acting on the systems are null. The global minimum would be a spatial configuration for the biological system in which the total potential energy would be the lowest among others for any point on the hypersurface.

Several methods for geometry optimization or energy minimization are available in the literature [19] but the most widely used are the steepest descent [19, 20] and conjugate gradients [19, 21].

1.1.3.1. Steepest Descent

The steepest descent [19, 20] is a first derivative method that converges, slowing when close to the minimum, but is powerful for configurations far from an energy minimum.

This method is useful to improve poorly refined crystallographic structures and to optimize graphically built, NMR or homology modeling structures.

In the steepest descent method, the resulting force on an atom i is obtained by the potential energy gradient in i given by Eq. 8.

$$F_i = -\nabla E(r_i) \qquad\qquad \text{(equation 8)}$$

The step or increment in the coordinates Δr_i, n of an atom i is given in the direction of the resulting force on this atom, as defined by Eq. 9.

$$\Delta \mathbf{r}_{i,n} = k_n \left(\frac{F_{i,n}}{|F_{i,n}|} \right) \qquad\qquad \text{(equation 9)}$$

Where k_n is an adjustment parameter of the step size and $\dfrac{F_{i,n}}{|F_{i,n}|}$ is the unitary vector in the direction of the resulting force on i in the step n. The minimization algorithm for the steepest descent method is, therefore, defined in Eq. 10.

$$\mathbf{r}_{i,n+1} = \mathbf{r}_{i,n} + k_n \left(\frac{F_{i,n}}{|F_{i,n}|} \right) \qquad\qquad \text{(equation 10)}$$

Where $\mathbf{r}_{i,n+1}$ affords the new position of atom i in the step n + 1.

The result after minimization is a set of coordinates (r_i) that, when applied to Eq. 8, makes the force F_i approximate to zero, thus reaching a local minimum. Fig. **1** illustrates the method for two dimensions.

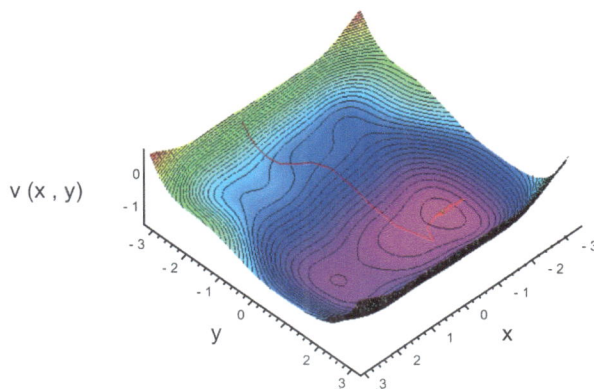

Figure 1: Representation of the steepest descent method for two dimensions.

1.1.3.2. Conjugate Gradient

The conjugate gradient [19, 21] is a more sophisticated method for searching minima in an energy function than steepest descent because, besides utilizing information on the first derivative (gradient), this method also takes into consideration the path already performed in the minimum search. This permits a faster convergence than the steepest descent method.

Conjugate gradient [19, 21] uses for determination of the following step, besides the gradient value in the current point, the gradient value obtained from the last step. In this method the coordinate shift ($\Delta r_{i,n}$) is given by Eq. 11.

$$\Delta r_{i,n} = \alpha_n \delta_{i,n}$$
(equation 11)

where:

$$\delta_{i,n} = \left(\frac{F_{i,n}}{|F_{i,n}|} \right) + \delta_{i,n-1} \left(\frac{|F_{i,n}|^2}{|F_{i,n-1}|^2} \right)$$
(equation 12)

and α_n is an adjustable parameter at each step. The ideal value for α_n is obtained by a minimization process of the potential in the direction defined by the vector $\delta_{i,n}$ [($F_i = -\Delta V(r_i)$)].

The advantage of using conjugate gradients come from the fact that the mathematical expressions in Eq. 11 and 12 guarantee that the gradient direction in the new point ($r_{i,n} + 1 = r_{i,n} + \Delta r_{i,n}$) will be orthogonal to the gradient direction in the last point (r_1, n), and consequently, to all the former points. So the direction δ_i,n will be "conjugated" to all the former directions. This property determines a more directly path into the end of the potential energy well, avoiding the return over paths already performed. Steepest descent is not able to exclude this possibility. A graphic representation of the conjugate gradient algorithm in a two dimensional space is presented in Fig. **2**.

In practice, however, for macromolecular systems, the steepest descent algorithm is more effective and faster in the starting points when the system is far from the minimum. However, when the system approaches the minimum, the conjugate gradient is much faster and precise. So it is possible to combine both methods using, at first, steepest descent with a weak convergence criteria and, the conjugate gradients method for the final refinement.

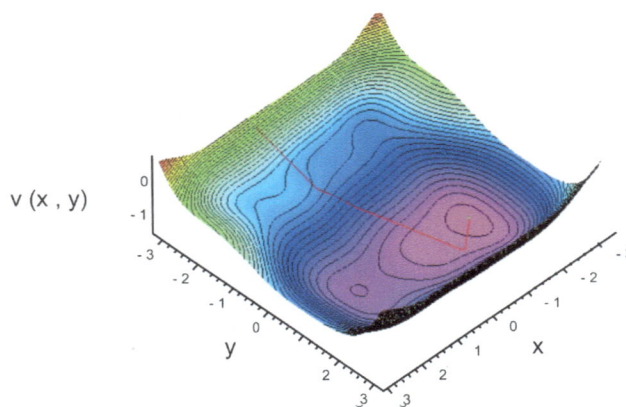

Figure 2: Representation of the conjugate gradients algorithm in a two dimensional space.

1.1.3.3. Statistical Analysis

MD simulations generate a huge amount of data that should be carefully analyzed in order to guarantee the success of the method. The most important data are the files with the positions and velocities of each atom in the system along time obtained after the simulations. From the data in these files all the further analysis can be performed. However, it is also practical, and very relevant, to record all the energies present in the system including the general energies and the interaction energies between atoms and/or groups in the system as well as temperature, pressure, forces and other macroscopic physical properties.

The Root Mean Square Deviation (RMSD) or Root Mean Square Fluctuation (RMSF) [22] is the statistic method usually applied for comparison between structures and, in the case of MD simulation results, can be extended to the comparison among the several spatial conformations assumed by the system along time related to the average structure of the cluster and excluding the solvent. This parameter can afford important information on how much one structure fluctuated along the MD simulation time as well as to permit the observation of local fluctuations as, for example, which amino acid residues presented higher mobility along the MD simulation in case of proteins. The RMSD can be calculated by the Eq. 13.

$$RMSD = \sqrt{\frac{1}{n}\sum_{i=1}^{n}(X_i - \overline{X})^2}$$

(equation 13)

Where n is the total number of samples, X_i is the tridimensional coordinate of each atom in sample i and X is the average position of the atom among all the samples. A good reference for this kind of statistical analysis can be found in Box *et al.*[23] and Zar [24]. As an example of the RMSD analysis, Fig. **3** shows a qualitative illustration, in tube representation, of the conformations adopted by a model of the enzyme serine hydroximethyltransferase from *P. falciparum* [25] along 1 ns of MD simulation. In order to build the illustration in Fig. **3**, one conformation (frame) was extracted from the MD output file at each 40th frame of simulation, totaling 25 conformations. Analyzing Fig. **3**, the most mobile regions along the MD simulations can be determined as the regions with high values of the spatial RMSD art. These regions correspond to the residues near the two terminus of each monomer and to the loop regions. On the other hand, the residues at the active site regions, as well as those at the α-helix and β-sheet regions, present lower spatial RMSD values, revealing to be the most stable regions in the system.

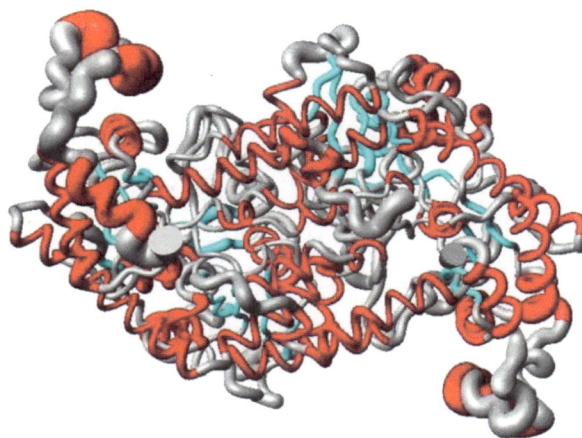

Figure 3: Qualitative illustration of the spatial RMSD for Serine Hydroxymethyltransferase from *P. falciparum.* Red = α-helices, blue = β-sheets and light gray = loops.

2. MOLECULAR DYNAMIC SIMULATIONS APPLIED TO THE STUDY OF MOLECULAR TARGETS AND DRUG DESIGN AGAINST LEISHMANIASIS

Today the molecular dynamic simulation techniques are well established and widely employed by the theoretical medicinal chemists in drug design. However, there are yet few studies applying these techniques on the investigation of molecular targets in the parasites responsible for neglected diseases. Specifically talking about leishmaniasis we can cite, as examples, the MD simulation of *Leishmania major* surface metalloprotease GP63 (leishmanolysin) by Bianchini *et al.* [26], the MD studies on actin from *Leishmania genus* by Kapoor *et al.* [27] and the design of inhibitors for Nucleoside Hydrolase (NH) from *L. donovani* using MD simulations by França *et al.* [28]. Some relevant aspects of those works illustrating interesting applications of the MD studies will be discussed below.

2.1. MD Simulation of *Leishmania major* GP63 (Leishmanolysin)

Bianchini *et al.* [26] have performed 9.2 ns of MD simulations in water, at pH 7, of the metalloprotease GP63 from *Leishmania major* in order to analyze its basic mechanical-structural features with the ultimate goal of quantifying the balance between rigidity and flexibility of the structural changes conceivably relevant in substrate recognition and (pro)enzyme activation. This was probably the first MD-based study on a whole metalloprotease.

GP63 (EC 3.4.24.36) is the major surface protease of *Leishmania* genus. This enzyme is a metalloprotease belonging to the themetzincin class and is evolved in the resistance of promastigotes to complement-mediated lysis and in receptor-mediated uptake of *leishmania* sp. [29]. Literature reports GP63 as a promising drug target and vaccine candidate [29, 30-32] because metalloproteases are believed to be quite important to the *Leishmania* sp. life cycle and to its relationship with the host in several ways. These enzymes are crucial to the penetration of the parasite into host cells, participate in the nutrition of the parasite by the host and are involved in the parasite escape mechanisms from the host immune system. Inhibition of GP36 block in *Leishmania* sp. can provide an alternative to traditional therapy against drug resistant parasites [29, 32, 33-39].

When placed in water, GP63 undergoes a sharp structural relaxation in solvent-exposed nonstructured regions while its active site turns out to be rigid [30]. MD simulation results of Bianchini *et. al.* [26] showed that these GP63 fluctuations are practically characterized by the motion of a large part of the N-terminal domain that, also, is evolved in substrate recognition and (pro) enzyme activation. Furthermore, an analysis of 10 GP63 homologs showed that the residues involved in the interdomain bending of GP63 are highly conserved, suggesting a possible relationship between the maintenance of proteolytic activity and the similarity of the dynamical properties of the related enzymes [26].

The software used for all the simulations performed by Bianchini *et al.* [26] was the GROMACS software modified for using the isokinetic temperature coupling [40] and the rototranslational constraint to stop the molecule at the center of the box during the simulation [41]. The protein structures were recovered from the Protein Data Bank (PDB) [42]. They heated up the system slowly until 300 K and, after a solvent relaxation at constant pressure and temperature, performed a MD simulation at constant volume and temperature (canonical ensemble) for 9.2 ns. To analyze the conformational changes occurring in GP63 in terms of interdomain movements they used the program DynDom [43].

RMSD and RMSF fluctuations on the Cα atoms, with respect to those of the crystal structure [29], obtained by Bianchini *et al.* [26], after MD simulation, are reported in Fig. **4**. As can be seen, large values affect the loop regions and the largest ones fall in the C-terminal domain. The RMSF results show that the α-helice and β-strand results as fairly rigid, in good agreement with the low thermal factors found for these regions in the crystal structure [29]. In addition, Essential Dynamics (ED) analysis [44] of the equilibrated portion of the trajectory shows that more than 70% of the overall motion is confined in the first three eigenvectors. According to Bianchini *et al.* [26], this result indicates that the conformational fluctuations of GP63 are mainly confined in the N- and C-terminal domains, whereas the central domain appears to be more rigid.

Results reported by Bianchini *et al.* [26] showed that, as a whole, GP63 is characterized by an overall rigid body movement of domains that can be described as a combination of two orthogonal motions of a large part of the N-terminal domain relatively to the central and C-terminal domains. It was also observed that the residues constituting the active site remain essentially rigid being positioned in the interfacial edge of the fixed domain. Also, the MD simulations performed by Bianchini *et al.* [26] indicated that GP63 shows, in water solution at 300 K, a rather fixed catalytic center with significant fluctuations concentrated in a portion of the N-terminal domain, which is involved in substrate recognition [29] and in (pro)enzyme activation. Further comparison with GP63 homologs, performed by the authors, showed that more than 80% of the amino acids involved in the interdomain motion are highly conserved. According to Bianchini *et al.* this result further shows the peculiar feature of MD simulations and their ability in correlating the similarity of the mechanical-dynamical properties of an enzyme with the maintenance of its activity.

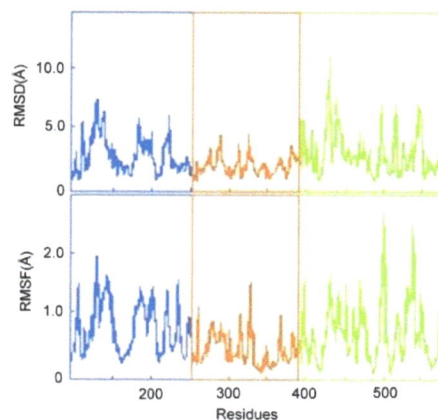

Figure 4: Reproduction of the RMSD and RMSF plots per residue of GP63, related to the crystal structure [29], obtained by Bianchini *et al.* [26]. The blue box indicates the N-terminal domain, the orange box the central domain and the green box the C-terminal domain.

2.2. MD Studies on Actin from *Leishmania donovani*

Kapoor *et al.* [27], after cloning and over expressing *L. donovani* actin (*Ld*ACT) in baculovirus insect cell system, observed that the purified protein polymerized optimally in the presence of Mg^{+2} and ATP, but differed from conventional actins in several ways. It did not polymerize in the presence of Mg^{+2}, but polymerized in a restricted range of pH 7.0-8.5. The critical concentration for polymerization was found to be 3-4 fold lower than of muscle actin and predominantly formed bundles rather than single filaments at pH 8.0. In addition, it displayed considerably higher ATPase activity during polymerization and did not inhibit DNase-I activity, nor bound the F-actin-binding toxin phalloidin or the actin polymerization disrupting agent Latrunculin B. According to Kapoor *et al.* [27], these results suggest that *Ld*ACT may be significantly different from other eukaryotic actins in its three-dimensional structure. In order to check these unconventional behaviors they built a homolog model for *Ld*ACT and, further, performed MD simulations studies on it. These studies revealed that the unusual behavior is related to the diverged amino acid stretches in *Ld*ACT sequence, which may lead to changes in the overall charge distribution on the solvent-exposed surface, ATP binding cleft, Mg^{+2} binding sites, and the hydrophobic loop that is involved in monomer-monomer interactions. Kapoor *et al.* [27], also suggest based on these results, that actin in *Leishmania* may serve as a novel target for the design of new antileishmanial drugs.

In order to perform the MD simulation studies, first Kapoor *et al.* [27], built a three-dimensional model of the *Ld*ACT monomer based on the crystal structure templates of actin from *O. cuniculus* (PDB [42] entry 2A5X), *C. elegans* (PDB [42] entry 1D4X), *S. cerevisiae* (PDB [42] entry 1YAG) and *D. discoideum* (PDB [42] entry 1NM1). This model was subsequently submitted to energy minimization and MD simulations in the presence and absence of ATP. The model was energetically minimized using steepest descent algorithm [19, 20], followed by conjugate gradients [19, 21] to remove the geometrical strains, and manually docked with ATP, based on previous structural information [45-47], in order to investigate the ATP binding site. To examine the ATP-induced conformational changes in *Ld*ACT, Kapoor *et al.* [27], carried out two independent 500-ps MD simulations in the presence and absence of ATP with the CHARMM program using CHARMM force field [18] and solvated the systems using Explicit Spherical Boundary with harmonic restraints. They heated the systems from 50 to 300 K over a period of 50 ps with a time step of 1 fs and the velocities being reassigned in the systems every 0.05 ps, and further equilibrated with a 1 fs time step, for 100 ps so that its energy achieved complete stability. Production runs were performed at 300 K and carried out under a constant number of particles, volume, and temperature conditions for 500 ps with a 1 fs time step. All the bonds involving hydrogen atom were constrained using the SHAKE algorithm in all simulations [48].

Results obtained by Kapoor *et al.* [27], showed that all the key diverged amino acid residues were found to be concentrated on the surface of the *Ld*ACT monomer (Fig. **5**), especially including those regions that participate in filament formation, like DNase-I loop formed by residues 40-53. Moreover, apart from the DNase-I loop, a region in subdomain 4 also showed marked fluctuations as revealed by average RMSF values, indicating the retention of flexible regions involved in filament formation (Fig. **6**). Kapoor *et al.* [27], noticed that the ATP binding cleft in *Ld*ACT is formed by residues Gly14, Ser15, Gly16, Met17, Lys19, Gln138, Asp155, Asp158, Lys214, Glu215, Gly303, Ser304, Met306, Phe307, and Lys337 and lies between the lobes formed by subdomains 1 and 2 and subdomains 3 and 4. This information was consistent with that reported for the ATP binding with rabbit actin [47]. After MD simulations, it was observed that among the residues that interacted with ATP, Lys19 showed the most dramatic conformational change. In the initial "apo" structure, Lys19 pointed away from the ATP binding site, but in the MD simulations of the apoprotein, Lys19 showed considerable flexibility and changed orientation often. In the MD simulation of the binary complex (*Ld*ACT-ATP), they observed a very stable hydrogen bond between Lys19 and ATP throughout the trajectory contrasting with rabbit actin where the corresponding Lys18 residue did not show any direct interaction with ATP [47]. Another conserved residue observed by Kapoor *et. al.* [27] that underwent a major conformational change upon ATP binding, was Ser15. In MD simulations of the apoprotein, this residue never changed orientation to point to the ATP binding site. However, in the binary complex, Ser15 was found to move toward the ATP binding site before simulation started and stayed in this orientation throughout the simulations. Additionally, the backbone nitrogen atom of Gly303 formed a hydrogen bond with the O1 atom of the phosphate. Similarly, the β- and γ-phosphates of ATP formed hydrogen bonds with the main chain nitrogen atoms of Ser15, Met17, and Asp158. The α-and β-phosphates of ATP also interacted with *Ld*ACT by forming hydrogen bonds with the side chain nitrogen atoms of Lys19 and Lys214 of *Ld*ACT interacted, by forming hydrogen bonds, with the oxygen atoms of the sugar moiety of ATP.

The MD results obtained by Kapoor *et al.* [27], together with the analysis of the structural features of *Ld*ACT helped to explain why *Ld*ACT is a novel form of actin with unique biochemical properties that had not been reported for any other actin to date.

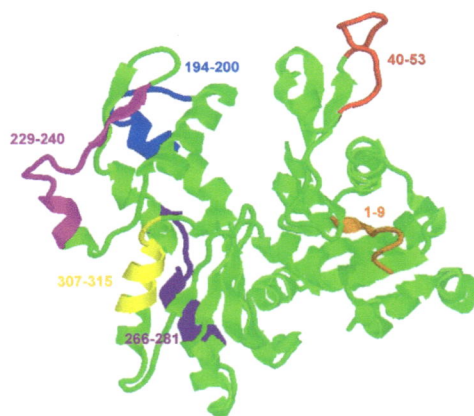

Figure 5: Reproduction of the model of *Ld*ACT proposed by Kapoor *et al.* [27] after MD simulations showing colored stretches of diverged amino acid residues (1-9 in subdomain 1, 40-53 in subdomain 2, 266-281 and 307-315 in subdomain 3, 194-200 and 229-240 in subdomain 4).

Furthermore, the results suggest that *Ld*ACT could serve as an attractive target for designing novel antileishmanial agents.

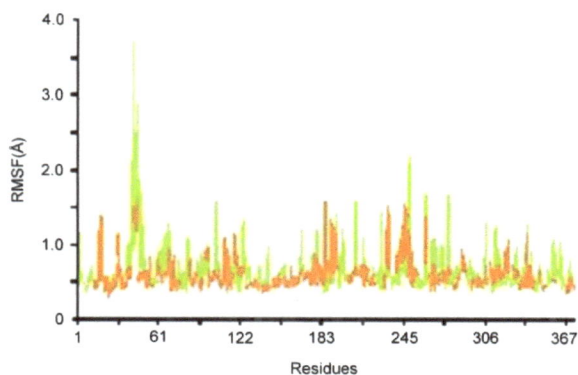

Figure 6: Adaptation of the plot reported by Kapoor *et. al.* [27], showing RMSF (measured in Å) as a function of amino acid residues observed along the MD simulation. Orange lines = without ATP; and green lines = with ATP.

2.3. MD Simulations of Nucleoside Hydrolase from *L. donovani* and *L. mayor*

França *et al.* [28] proposed the first homology model for NH from *Leishmania donovani* (*Ld*NH), built based on the crystallographic structures of *Crithidia fasciculata* and *Leishmania major* NHs (*Cf*NH and *Lm*NH). They also performed MD simulation studies, using the GROMACS 3.2 package [48, 49], of this model and its templates anchored with two potential inhibitors, analogues of the *Cf*NH inhibitor *p*-aminophenyliminoribitol (pAPIR), proposed based on the interaction information on the active site from the crystallographic structure of *Cf*NH in complex with pAPIR and Ca^{+2} (the natural cofactor of NHs) [50].

NH is a potential target for drug design because no NH activity or genes encoding this enzyme have been identified yet in mammals [51] and, also, because it is evolved in complex host purine salvage machinery essential to the protozoan, because they lack *de novo* biosynthetic pathways for purines [52]. NH catalyses the hydrolysis of nucleosides playing a crucial role in the purine salvage from the host DNA and RNA. The understanding of the binding mode of inhibitors and the participation of important active-site residues during the inhibition of NH would be useful for the drug design of antileishmaniasis agents [53].

França *et al.* [28] analyzed the pAPIR structure docked inside the active site of crystallographic *Cf*NH, and observed six residues close enough to pAPIR to interact with eventual pAPIR derivatives with substitutions in their aromatic base portion (see Fig. **7**). The analysis of the potential interactions with such residues lead to the proposition of the two new potential NH inhibitors showed in Fig. **8**. Further 2 ns of MD simulations of pAPIR and the proposed inhibitors inside the active sites of *Ld*NH, *Lm*NH and *Cf*NH, showed that those compounds, differently from pAPIR, remained tightly bound inside the active sites of the three enzymes during the whole simulation, interacting strongly with those residues of the hydrophobic pocket too far away to interact with pAPIR.

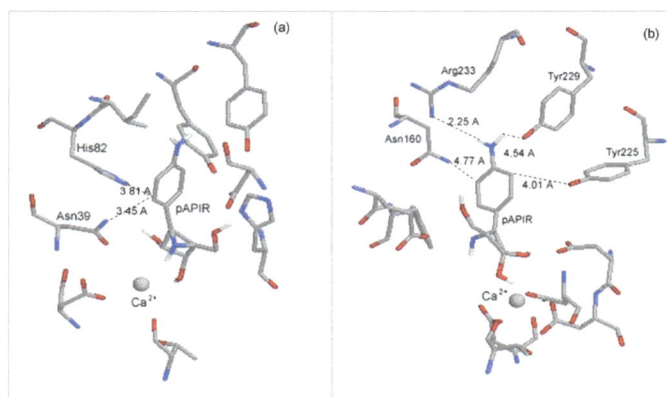

Figure 7: Potential residues in *Cf*NH active site able to interact with pAPIR derivatives. Available in: http://jbcs.sbq.org.br/jbcs/2008/vol19_n1/10-07126AR.pdf. Copyright of the Brazilian Chemical Society.

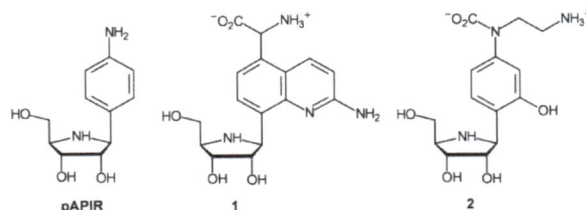

Figure 8: Structures of pAPIR and the proposed inhibitors. Available in: http://jbcs.sbq.org.br/jbcs/2008/vol19_n1/10-07126AR.pdf. Copyright of the Brazilian Chemical Society.

In order to perform the MD simulations it was necessary to first build the structures of *Ld*NH, *Lm*NH and *Cf*NH complexed with Ca^{+2} and pAPIR or the ligands. That was accomplished by the superposition of the backbone of *Ld*NH or *Lm*NH with the crystallographic structure of *Cf*NH complexed with Ca^{+2} and pAPIR (PDB [42] entry 2MAS).

The coordinates of Ca^{2+} and pAPIR were then copied into *Ld*NH or *Lm*NH active sites. The atomic partial charges of Ca^{2+}, pAPIR and the ligands were previously calculated at the Hartree Fock level with the 6-31G* basis set using the CHELPG approach of the Gaussian98 package [54], and their topological files, compatible with the format of the GROMACS 3.2 package [48, 49] databank, were generated at the Dundee PRODRG server [55]. The holoenzymes obtained by this procedure, were minimized with the GROMOS96 force field [6], implemented in the GROMACS 3.2 package. [48, 49] until an energy gradient of 20 kcal mol^{-1} nm^{-1}. The MD steps carried out were first, a 50 ps of MD at 300 K in the water molecules inside the box in order to allow for the equilibration of the solvent around the protein residues. In this simulation, all protein atoms had their positions restrained. Then, full MD simulations of 2,000 ps at 300 K were carried out with no restrictions for each system. As a whole, 1,000 conformations were stored during each simulation. In this step the pair lists were updated every 500 time steps and all the Lys and Arg residues were positively charged while the Glu and Asp residues were negatively charged. Also, in order to obtain a neutral net charge for all the systems, the charges were neutralized by the addition of Na^+ ions. Periodic Boundary Conditions (PBC) were employed and the water model used was the Simple Point Charge water (SPC), the default solvent of the GROMACS 3.2 package [48, 49].

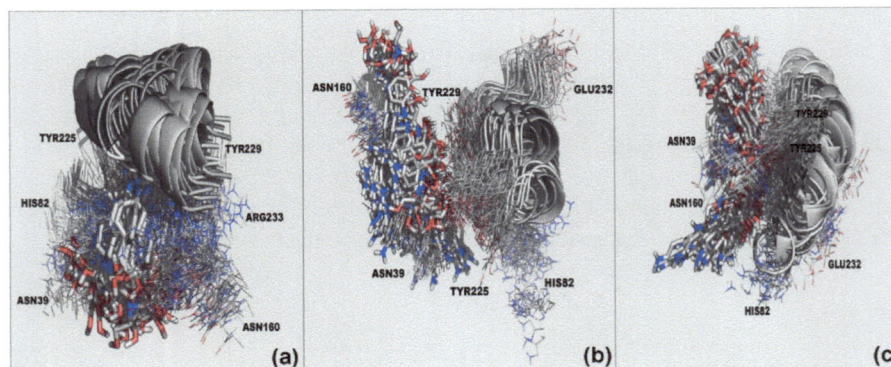

Figure 9: Dynamic Behavior of pAPIR inside the active sites of *Cf*NH (a), *Lm*NH (b) and *Ld*NH (c) along the 2,000 ps of MD simulations. Available in: http://jbcs.sbq.org.br/jbcs/2008/vol19_n1/10-07126AR.pdf. Copyright of the Brazilian Chemical Society.

Results of the MD simulations performed by França *et al.* [28] showed that the two proposed compounds stayed tightly bound to the active sites of all the enzymes along the 2 ns. The same was not observed for pAPIR, which was expelled from all active sites except CfNH's (see Figs. **9-11**). This probably happened because, differently from *Ld*NH and *Lm*NH, the *Cf*NH active site entrance is narrowed by two loops and also by the side chain of the residue Arg233. pAPIR does behave differently probably because, not having substitutions on its aromatic base portion, it is unable to interact with the residues in the hydrophobic pocket, and counts only with the interactions of the ribose portion to be stabilized inside the active site. This observation corroborates the results of Milles *et* *al.* [56] showing that pAPIR is a weaker inhibitor of NH when compared to inhibitors with substitutions on the aromatic base portion. Moreover, the MD simulations results suggest that the proposed inhibitors have a good potential to become candidates for the strong inhibition of NHs.

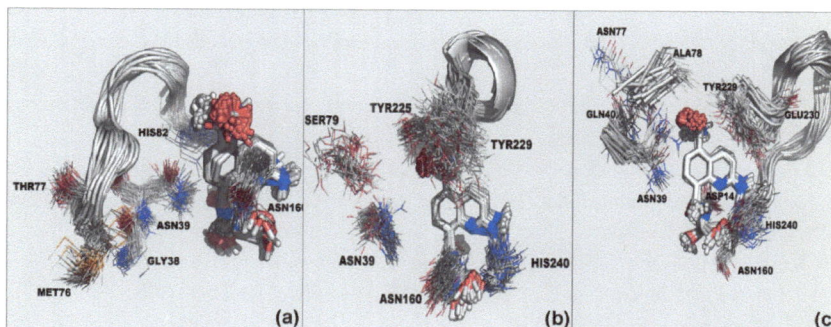

Figure 10: Dynamic Behavior of compound 1 inside the active sites of CfNH **(a)**, LmNH **(b)** and LdNH **(c)** along the 2,000 ps of MD simulations. Available in: http://jbcs.sbq.org.br/jbcs/2008/vol19_n1/10-07126AR.pdf. Copyright of the Brazilian Chemical Society.

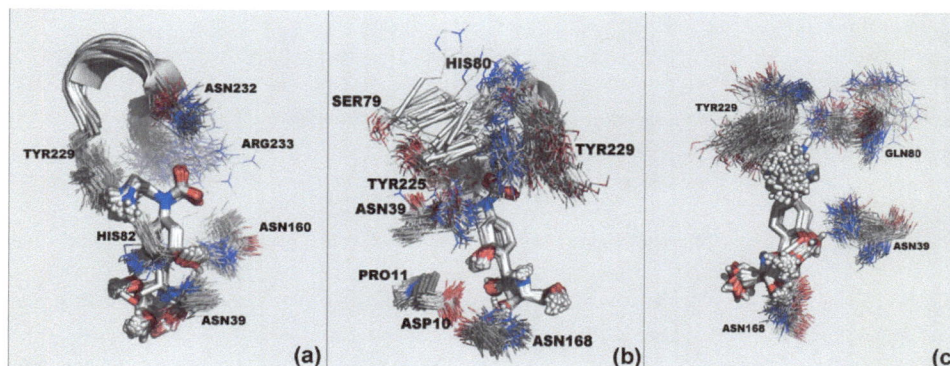

Figure 11: Dynamic behavior of compound 2 inside the active sites of CfNH **(a)**, LmNH **(b)** and LdNH **(c)** along the 2,000 ps of MD simulations. Available in: http://jbcs.sbq.org.br/jbcs/2008/vol19_n1/10-07126AR.pdf. Copyright of the Brazilian Chemical Society.

3. CONCLUSIONS

Besides the examples quickly discussed here, many additional applications of MD studies on the most diverse biological systems are available in literature today and could be useful in drug design or proposition of new targets to combat neglected diseases. Today there are specific force fields developed to simulate any biological system. Interactions between cellular membranes and proteins, for example, could be studied by the CHARMM [18] force field, while the AMBER force field [7, 8] is applicable in the studies of interactions between drugs or proteins with nucleic acids. Maybe the initial unfriendly appearance of some software and the need of professional skills not usually found in medicinal chemists are still causing an avoidance of a more intense use of these powerful tools in the study of neglected diseases. Besides, we should also keep in mind that the low interest of the large pharmaceutical companies in neglected diseases further contributes to this situation. However, we strongly believe that the constant software improvements allied to the insertion of younger scientists more accustomed to chemoinformatics into this field and, also, the government pressure for the development of more efficient chemotherapy against neglected diseases, will contribute to soon revert this situation, paving the way for a significant increase in the MD studies applied to neglected diseases available in literature.

REFERENCES

[1]   Karplus M, Petsko, GA. Molecular dynamics simulations in biology. Nature 1990; 347: 631.
[2]   Goldstein H. Classical Mechanics. 2nd ed. London: Addison-Wesley 1980.

[3] Berendsen HJC, van Gunsteren WF. Pratical algorithms for dynamic simulations. In molecular dynamics simulation of statistical mechanical systems, proceedings of the international school of physics "Enrico Fermi", Ed. by Ciccotti G. & Hoover H. G., North-Holland Phys. Publishing, Amsterdam, 1986.

[4] Allen MP, Tildesley DJ. Computer simulation of liquids, Oxford; Clarendon Press, 1987.

[5] Higgins D. Taylor W. Bioinformatics Sequence, structure and databanks, 1st ed. Oxford: Oxford University Press, 2001.

[6] Jorgeson WL. Tyrado-Reeves J. The OPLS potential functions for proteins. Energy minimizations for crystals of cyclic peptides and crambin. J Am Chem Soc 1988; 110: 1657.

[7] Cornell WD, Cieplack P, Bayly CI, Gould IRK, Merz Jr MD, Ferguson M, Spellmeyer DC, Fox TJ, Caldwell W, Kollman PA. A second generation force field for the simulation of proteins, nucleic acids and organic molecules. J Am Chem Soc 1995; 117: 5179.

[8] Weiner SJ, Kollman PA, Case DA, Singh UC, Ghio C, Alagona G, Profeta S, Weiner P. A new force field for molecular mechanical simulation of nucleic acids and proteins. J Am Chem Soc 1984; 106: 765.

[9] Alinger NL. Conformational Analysis MM2. A hydrocarbon force field utilizing V1 and V2 torsional terms. J Am Chem Soc 1977; 99: 8127.

[10] Allinger NL, Li F, Yan LJ. Molecular mechanics. The mm3 force field for alkenes. J Comp Chem 1990; 11: 848.

[11] Allinger NL, Li F, Yan L, Tai JC. Molecular Mechanics (MM3) Calculations on conjugated Hydrocarbons. J Comp Chem 1990; 11: 868.

[12] Allinger NL, Chen K, Lii JH. An improved Force Field (MM4) for saturated Hydrocarbons. J Comp Chem 1996; 17: 642.

[13] Allinger NL Chen JA, Katzenelenbogen SR, Anstead GM. Hyperconjugative effects on carbon-carbon bond lenghts in molecular mechanics (MM4). J Comp Chem 1996; 17: 747.

[14] Allinger NL, Yuh YH, LII JH. Molecular Mechanics. The MM3 Force Field for Hydrocarbons .I. J Am Chem Soc 1989; 111: 8551

[15] Nevins N, Chen K, Allinger NL. Molecular Mechanics (MM4) Calculations on Alkenes. J Comp Chem 1996; 17: 669.

[16] Nevins N, Chen K, Allinger NL. Molecular mechanics (MM4) calculations on conjugated hydrocarbons. J Comp Chem 1996; 17: 695.

[17] Nevins N, Chen K, Allinger NL. Molecular mechanics (MM4) vibrational frequency calculations for alkenes and conjugated hydrocarbons. J Comp Chem 1996; 17: 730.

[18] Brooks BR, Bruccoleri, RE, Olafson, BD, States DJ, Swaminathan S, Karplus M. CHARMM: a program for macromolecular energy, minimization and dynamics calculations. J Comp Chem 1983; 4: 187.

[19] Leach AR, Molecular Modelling: Principles and Applications 2nd ed. London: Prentice Hall 2001.

[20] Wiberg KB. A Scheme for strain energy minimization. Application to the cycloalkanes. J Am Chem Soc 1965; 87: 1070.

[21] Schlegel HB. Optimization of equilibrium geometries and transition structures. J Comp Chem 1982; 3: 214.

[22] Spiegel MR. Estatística. 3ed. São Paulo, Makron Brooks do Brasil Editora Ltda & Editora McGraw-Hill Ltda 1994.

[23] Box GEP, Hunter, WG, Hunter JS. Statistics for experiments: an introduction to design, data analysis and model building. Princeton: John Wiley & Sons, Inc. 1978.

[24] Zar JH. Biostatistical analysis. 4th ed. New Jersey: Prentice Hall, Inc. 1999.

[25] França TCC, Pascutti PG, Ramalho TC, Figueroa-Villar JD. A three-dimensional structure of Plasmodium falciparum serine hydroxymethyltransferase in complex with glycine and 5-formyl-tetrahydrofolate. Homology modeling and molecular dynamics. Biophys Chem 2005; 115: 1.

[26] Bianchini G, Bocedi A, Ascenzi P, Gavuzzo E. Mazza F, Aschi M. Molecular dynamics simulation of Leishmania major surfasse metalloprotease GP63 (leishmanolysin). Proteins: Struc Func Bioinf 2006; 64: 385.

[27] Kapoor P, Sahasrabuddhe AA, Kumar A, Mitra K, Siddiqi MI, Gupta CM. An unconventional form of actin in protozoan hemoflagellate, *Leishmania*. J Biol Chem 2008; 283: 22760.

[28] França TCC, Rocha MRM, Reboredo BM, Rennó MN, Tinoco LW, Figueroa-Villar JD. Design of inhibitors for nucleoside hydrolase from *Leishmania donovani* using molecular dynamics studies. J Braz Chem Soc 2008; 19: 64.

[29] Yao C, Donelson JE, Wilson ME. The major surface protease (MSP or GP63) of *Leishmania* sp. Biosynthesis, regulation of expression, and function. Biochem Parasitol 2003; 132: 1.

[30] Schlagenhauf E, Etges R, Metcalf P. The crystal structure of the Leishmania major surface proteinase leishmanolysin (gp63). Structure 1998; 6: 1035.

[31] Yiallouros I, Kappelhoff R, Schilling O, Wegmann F, Helms MW, Auge A, Brachtendorf G, Berkhoff EG, Beermann B, Hinz HJ, Konig S, Peter-Katalinic J, Stocker W. Activation mechanism of pro-astacin: role of the pro-

peptide, tryptic and autoproteolytic cleavage and importance of precise amino-terminal processing. J Mol Biol 2002; 324: 237.

[32]   Coombs GH and Mottram JC. In trypanosomiasis and leishmaniasis: biology and control. Hide G, Mottram JC, Coombs GH and Holmes PH, eds. Oxford: CAB International, 1997.

[33]   Zhang T, Maekawa Y, Hanba J, Dainichi T, Nashed BF, Hisaeda H, Sakai T, Asao T, Himeno K, Good RA, Katunuma N. Lysosomal cathepsin B plays an important role in antigen processing, while cathepsin D is involved in degradation of the invariant chain inovalbumin-immunized mice. Immunol 2000; 100: 13.

[34]   Zhang T, Maekawa Y, Yasutomo K, Ishikawa H, Fawzy Nashed B, Dainichi T, Hisaeda H, Sakai T, Kasai M, Mizuochi T, Asao T, Katunuma N, Himeno K. Pepstatin A-sensitive aspartic proteases in lysosome are involved in degradation of the invariant chain and antigen-processing in antigen presenting cells of mice infected with Leishmania major. Biochem Biophys Res Commun 2000; 276: 693.

[35]   da Silva-Lopez RE, Giovanni DSS. Purification and characterization of a promastigote serine protease. Exp Parasitol 2004; 107: 173.

[36]   Mottram JC, Coombs GH, Alexander J. Cysteine peptidases as virulence factors of Leishmania. Curr Opin Microbiol 2004; 7: 375.

[37]   Rafati S, Salmanian AH, Taheri T, Masina S, Schaff C, Taslimi Y, Fasel N. The Leishmania major Type I signal peptidase is a target of the immune response in humans. Mol Biochem Parasitol 2004; 135: 13.

[38]   Singh S, Sivakumar RJ. Challenges and new discoveries in the treatment of leishmaniasis. Infect Chemother 2004; 10: 307.

[39]   Alves CR, Corte-Real S, Bourguignon SC, Chaves CS, Saraiva EM. *Leishmania amazonensis*: early proteinase activities during promastigote-amastigote differentiation *in vitro*. Exp Parasitol 2005; 109: 38.

[40]   Berendsen HJC, Postma JPM, van Gunsteren WF, Di Nola A. Molecular dynamics with coupling to an external bath. J Chem Phys 1984; 81: 3684.

[41]   Amadei A, Chillemi G, Ceruso M, Grottesi A, Di Nola A. Molecular dynamics simulations with constrained roto-translational motions: Theoretical basis and statistical mechanical consistency. J Chem Phys 2000; 112: 9.

[42]   Berman HM, Westbrook J, Feng Z, Gilliland G, Bhat TN, Weissig H, Shindyalov IN, Bourne PE. The protein data bank. Nucleic Acids Res 2000; 28: 235.

[43]   Hayward S, Berendsen HJC. Systematic analysis of domain motions in proteins from conformational change; new results on citrate synthase and T4 lysozyme. Proteins 1998; 30: 144.

[44]   Amadei A, Linssen AB, Berendsen HJ. Essential dynamics of proteins. Proteins 1993; 17: 412.

[45]   McGwire BS, O'Connell WA, Chang, KP, Engman DM. Extracellular release of the glycosylphosphatidylinositol (GPI)-linked *Leishmania* surface metalloprotease, gp63, is independent of GPI phospholipolysis: implications for parasite virulence. J Biol Chem 2002; 277: 8802.

[46]   Kabsch W, Mannherz HG, Suck D, Pai EF, Holmes KC. Atomic structure of the actin: DNase I complex. Nature 1990; 347: 37.

[47]   Dean W, Goddette S, Frieden C. J. Actin polymerization. The mechanism of action of cytochalasin D. J Biol Chem 1986; 261: 5974.

[48]   Van der Spoel D, Lindahl E, Hess B, GROMACS 3.0: a package for molecular simulation and trajectory analysis. J Mol Mod 2001; 7: 306.

[49]   Berendsen HJC, Van der Spoel D, Van Drunen, R. GROMACS: a message-passing parallel molecular dynamics implementation. Comput Phys Commun 1995; 91: 43.

[50]   Parkin DW, Limberg G, Tyler PC, Furneaux, RH, Chen XY, Schramm VL. Isozyme-specific transition state inhibitors for the trypanosomal nucleoside hydrolases. Biochem 1997; 36: 3528.

[51]   Gopaul DN, Meyer SL, Degano M, Sacchettini JC, Schramm VL. Inosine—uridine nucleoside hydrolase from *Crithidia fasciculata*. Genetic characterization, crystallization, and identification of histidine 241 as a catalytic site residue. Biochem 1996; 35: 5963.

[52]   Hammon DJ, Gutteridge WE. Purine and pyrimidine metabolism in the Trypanosomatidae. Mol Biochem Parasitol 1984; 13: 243.

[53]   Mazumder D, Kahn K, Bruice TC. Computer Simulations of trypanosomal nucleoside hydrolase: Determination of the protonation state of the bound transition state analog. J Am Chem Soc 2002; 124: 8825.

[54]   Frisch MJ, Trucks GW, Schlegel HB, Scuseria, GE, Robb MA, Cheeseman JR, Zakrewski VG, Montgomery JA, Stratman RE, Vurant JC, Dapprich S, Millam JM, Daniels AD, Kudin KN, Strain MC, Farkas O, Tomasi J, Barone V, Cossi M, Cammi R, Mennuci B, Pomelli C, Adamo C, Clifford S, Ochterski J, Petersson GA, Ayala PY, Cui Q, Morokuma K, Malick DK, Rabuch AD, Raghavachari K, Foresman JB, Cioslowski J, Ortiz JV, Stefanov BB, Liu G,

Liashenko A, Piskorz P, Komaromi I, Gomperts R, Martin RL, Foz DJ, AlLaham TM, Peng CY, Nanayakkara A, Gonzalez C, Challacombe M, Gill PMW, Jonson B, Chen W, Wong MW, Andres J L, Gonzalez C, Head-Gordon M, Reploge ES, Pople JA, Gaussian 98, Revison A.6, Gaussian, Inc.: Pittsburg, PA, 1998.

[55] Schuettelkopf, AW, van Aalten DMF. PRODRG: a tool for high-throughput crystallography of protein-ligand complexes. Acta Crystallographica Section D 2004; D60: 1355.

[56] Miles RW, Tyler PC, Evans GB, Furneaux RH, Parkin DW, Schramm VL. Iminoribitol transition state analogue inhibitors of protozoan nucleoside hydrolases. Biochem 1999; 38: 13147.

CHAPTER 6

Chagas' Disease

Adriana de Oliveira Gomes, Alessandra Mendonça Teles de Souza, Alice Maria Rolim Bernardino, Helena Carla Castro and Carlos Rangel Rodrigues*

Laboratório de Modelagem Molecular e QSAR (ModMolQSAR), Faculdade de Farmácia, Universidade Federal do Rio de Janeiro, Rio de Janeiro, RJ, Brazil

Abstract: Chagas disease (American trypanosomiasis) is one of the most important parasitic diseases with serious social and economic impacts mainly in Latin America. Therefore Chagas' disease treatment is still a challenge, mainly in Brazil, where the only commercially available drug is benznidazole (Rochagan®). Since 1984, the WHO has recommended the use of crystal violet (gentian violet) in blood banks in endemic areas to prevent transmission by transfusion.

Keywords: QSAR-3D, CoMFA, Chaga's Disease.

1. INTRODUCTION

Chagas' disease is a serious health problem that affects millions of people in Central and South America. The causative agent of this disease is *Trypanosoma cruzi*, discovered by Carlos Chagas in 1909 [1]. It is characterized initially by often nonspecific symptoms [2]. In the natural history of Chagas' disease there are some defined periods that can clearly be recognized: the acute period that appears immediately after the initial infection (usually in children from endemic areas), which is clinically manifested at a low percentage (5 to 10%) with nonspecific symptoms (fever, joint pain, skin diseases, adenopathy) [3]. A cause of early death in the adult populations, trypanosomiasis has a high social-medical cost in terms of medical treatment, hospitalization, corrective surgery and pacemakers. Among all the clinical manifestations, the most important is the impact, incidence and severity of heart disease. However, the digestive forms are also frequent, and even the chronic indeterminate, due to the evolutionary character of infection [4].

Prevention of Chagas' disease should be based on a broad epidemiological vision and dynamic, in turn focused on the bio-ecological and social contexts ranging from infection by the parasite to the consequences of the latest social policies on parasitism [5].

Parasitic diseases affect hundreds of millions people around the world, mainly in underdeveloped countries. Since parasitic protozoa are eukaryotic, they share many common features with their mammalian host making the development of effective and selective drugs a difficult task. Diseases caused by *Trypanosomatidae*, which share a similar behavior regarding drug treatment, include Chagas' disease (*Trypanosoma cruzi, T. cruzi*) and leishmaniasis (*Leishmania* spp.) [5].

Chagas's disease is endemic in South American countries, where an estimated 16-18 million people are affected and 50,000 deaths occur annually. Although great progress has been made recently in the control of the vector and in the transfusional transmission of the disease, the specific treatment of infected individuals remains unsolved [6]. The causative agent of this disease is the haemoflagellate protozoan *Trypanosoma cruzi* (*T. cruzi*), which is transmitted in rural areas to humans and other mammals by reduviid bugs such as *Rhodnius prolixus* and *Triatoma infestans* [7].

Two protozoan parasites belonging to the Trypanosomatidade family (*Trypanosoma brucei gambiense* and *Trypanosososma brucei rhodesiense*) are known to cause West and East African sleeping sickness among

*Address correspondence to Carlos Rangel Rodrigues:** Laboratório de Modelagem Molecular e QSAR (ModMolQSAR), Faculdade de Farmácia, Universidade Federal do Rio de Janeiro, Rio de Janeiro, RJ, Brazil; Telefax: +55 21 2560 9897; E-mail: rangelfarmacia@gmail.com

Teodorico C. Ramalho, Matheus P. Freitas and Elaine F. F. da Cunha (Eds)

humans and a third subspecies, *Trypanosoma brucei brucei*, causes Nagana disease in cattle. These parasites are transmitted to humans through the bite of the tsetse fly, an ubiquitous insect found throughout the entire African continent. The World Health Organization estimates that about 50 million and approximately one-third of Africa's cattle are threatened by Nagana. In addition, roughly 25,000 cases of new infections or re-infections are reported annually [8].

T. cruzi presents three main morphological forms in a complex life cycle. The epimastigote form replicates within the crop and midgut of Chagas' disease vectors and is released with the insect faeces as the non-dividing highly infective metacyclic trypomastigotes that invade mammalian tissues *via* wounds provoked by blood sucking action. The parasite multiplies intracellularly as the amastigote form which is released as the non-dividing bloodstream trypomastigote form that invades others tissues [5].

Such diseases affect mainly countries under-development and almost half a billion people are presently at risk. Despite its epidemiological importance, the therapy of trypanosomatid infections remains an unmet challenge. Currently, no vaccines are available to prevent these diseases, and the recommended drugs have high toxicity and limited efficacy. In addition, rapidly-developing drug resistance has become a major problem. Although new antiprotozoan drugs are in development for these diseases, the lead identification still represents a real bottleneck, and the drug development pipeline is currently almost empty [9].

More recently, the use of modern insecticides to control the vectors, and improvements in public health and lifestyle have considerably reduced the prevalence of the disease and the acute transmission in many Latin American countries. However, chronic manifestations (10-30 years) are still a health problem priority, and involve internal organs such as heart, esophagus, colon and nervous system. Patients may be treated orally with benznidazole or nifurtimox during the acute phase. In the chronic phase, the treatment reduces the parasitaemia and stops the progression of the disease, but does not cure the internal organ manifestations which can be only by symptomatic treatment [10]. Since 1984, the WHO has recommended the use of crystal violet (gentian violet) in blood banks in endemic areas to prevent transmission by transfusion. However, crystal violet induces blood micro-agglutination and presents a potential mutagenicity profile [11].

In the last century, only a handful of drugs were in development to treat the symptoms of trypanosomiasis. Unfortunately, many of these candidates have shown toxicity or mutagenicity, have a short period of efficacy or have limited absorption. To date, no single drug has been developed that is both readily absorbed and effective for an extended period. The most promising drug developed is eflornithine, a drug candidate believed to inhibit the synthesis of polyamines (*e.g.*, spermine, spermidine, putrescine), nitrogen-containing organics essential for eukaryotic growth. This drug can effectively treat the early and late stages of *T. brucei gambiense* infections, however it is costly, difficult to administer, and ineffective against infections caused by *T. brucei rhodesiense*. Thus, alternative strategies are highly sought [8].

2. TRANSMISSION AND SYMPTOMS OF CHAGAS DISEASE

Chagas disease is caused by a protozoan parasite, *T. cruzi* transmitted to humans by triatomine insects popularly known as "barbeiro" [12].

Considering the vectorial transmission of Chagas disease, the nearly 120 species or subspecies of the vector *T. cruzi*, little more than half a dozen shows real importance. In Brazil, the main transmitter was until recently the *Triatoma infenstans*, dispersed from Rio Grande do Sul to the northeast and part of the Midwest [13]. Transmission that occurs in cities is a result of migration of rural people [14]. Other forms of transmission, such as oral (through the flesh of infected animals) and sexual, among others, shows less importance [14].

Without a safe and effective vaccine against Chagas' disease, primary prevention of the endemic disease is made mostly by the vector control and the prevention of transfusion [3].

The diagnosis of acute infection of *T. cruzi* is usually made by detection of parasites. Active Trypomastigotes can often be seen by microscopic examination of fresh non-coagulated blood.

T. cruzi can parasitize different cell types and thus presents a wide range of clinical manifestations. Trypomastigotes invades many cells *in vivo* and *in vitro*. A unique interaction occurs between evolutionary stages of the parasite and macrophages. Epimastigotes phagocytes are destroyed by macrophages, while trypomastigotes derived from the blood of the vertebrate host, vector or tissue culture are internalized [15].

The symptoms of the chronic phase may take years or decades to manifest, dominated by low parasitemia and the chances of cure are remote. Althought the clinical manifestations vary widely because the parasite is installed in various types of tissues, there is already a predominance of lesions observed in the cardiac muscle, digestive and nervous systems, which depending on the severity, are often fatal [15].

Chemotherapy of Chagas disease is an unsolved problem and the search for alternative drugs still remains. Only two nitro-heterocyclic drugs are used clinically to date, but these have severely restricted applicability to chronic patients, due to their high toxicity. Many compounds have been tested in a variety of ways, most commonly in their ability to inhibit the proliferation of epimastigotes forms [16].

Two alternatives have been proposed to eradicate Chagas' disease. The first is the prevention of transmission by eliminating the insect vector and sterilization of contaminated blood used for transfusion. The second is the chemotherapy of infected patients with absolutely effective drugs, thus eliminating the human reservoir of *T. cruzi* and at the same time, healing the patient. The proposed procedures are complementary. Campaigns for the eradication of the vector are held for several years, with important successes in Uruguay, Chile and Brazil, where the transmission has substantially reduced. Contrastingly, the parasitological cure of patients already infected, has not had the same success [17].

In the progress achieved in the biochemistry and physiology study of *T. cruzi*, in which several crucial enzymes for parasite survival, absent in the host, have been identified as potential targets for the design of new drugs, the chemotherapy to control this parasitic infection remains undeveloped. The pharmacology is based on old and quite unspecific drugs associated with long term treatments that give rise to severe side effects [7].

As the two nitro heterocyclic drugs, nifurtimox (**1**) and benznidazole (**2**) (N-benzyl-2-nitroimidazole-1-acetamide), have been used to treat this disease in therapeutic regimens, hundreds of natural and synthetic chemical compounds have been tested as antichagasic agents [17].

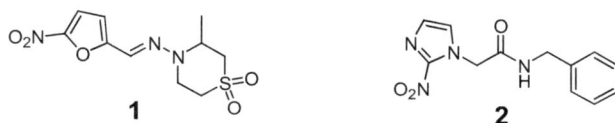

These drugs are currently in use for clinical treatment of this disease and are able to wipe out parasitaemia and to reduce serological titters, but they are not specific enough for all *T. cruzi* strains and they do not guarantee complete cure. Both drugs act *via* the reduction of the nitro group. In the case of nifurtimox, reduction generates an unstable nitro anion radical, which produces highly toxic reduced oxygen species. Benznidazole involves covalent modification of macromolecules by nitro reduction intermediates. The side effects of these drugs result from the oxidative or reductive damage in the host tissues and are thus inextricably linked to its anti-parasitic activity. Despite these limitations, some studies involving nitroimidazole derivatives have been recently described [7].

The crystal violet (**3**) is used in the treatment of contaminated blood in blood banks, but has the disadvantage of leaving a red color in blood and tissues of patients receiving donation [18].

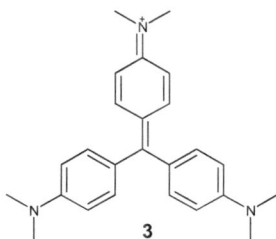

Several authors have reported that *T. cruzi* is particularly sensitive to compounds that can produce free radicals due to its lack of catalase and glutathione peroxidase, and proposed that the main mechanism by which nifurtimox acts against *T. cruzi* is through the intracellular production of superoxide anion.

A variety of nitrofurazones (5-nitro-2-furaldehyde semicarbazones) derivatives, containing different heterocyclic moieties and α,β - unsaturated amides have been synthesized due to their structural domains such as trypanocidal, bactericidal, fungicidal and squistosomicidal. The trypanocidal activity was assayed on epimastigote forms of *T. cruzi* Y strain. The most potent derivatives are compounds **4a-d**, all of them containing a 5-nitro furan [19].

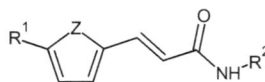

4

4a: Z=O, R^1=-NO_2, R^2=-Isobutyl

4b: Z=O, R^1= -NO_2, R^2=-Benzyl

4c: Z=O, R^1=-NO_2, R^2=-4-Chlorobenzyl

4d: Z=O, R^1=-NO_2, R^2=-Piperonyl

3. 3D-QSAR PROCEDURE

Since its introduction in 1988, Comparative Molecular Field Analysis (CoMFA) has rapidly become one of the most powerful tools for 3D-QSAR studies. The methodology has emerged as a natural extension to the classical QSAR approaches pioneered by Hansch and Free-Wilson, which exploits the three-dimensional properties of the ligands to predict their biological activities using robust chemometric techniques such as PLS, G/PLS, ANN *etc*.

The idea underlying CoMFA is that differences in a target property are often related to differences in the shapes of the non-covalent fields surrounding the tested molecules [20]. Molecules with common or related structures generally have similar physicochemical properties (the similarity principle), and thus have similar binding modes and consequently comparable biological activities. The reverse also holds true [21].

The starting point for the analysis is a set of conformations, one for each molecule in the set. Each conformation should be the presumed active structure of the molecule that must be overlaid in the suggesting binding mode. These aligned molecules are placed in the center of a lattice grid with a spacing that generally varies from 1-2Å, and a probe atom will be placed at points of this regular lattice [22].

In the context of 3D-QSAR, the biological activity may be seen as a function of the physicochemical features of the compounds of interest. The need to convert such numerical data to useful information has led to the development of methodologies that rely on statistics and applied mathematics.

Afterwards, the last step in a CoMFA study is a Partial Least Squares (PLS) analysis to determine the minimal set of grid points which is necessary to explain the biological activities of the compounds. PLS method is used on QSAR analysis where there are a great number of variables that are treated by the construction of Principal Components (PC), which are combinations of the original independent variables [23-25]. The PLS model is a two-block projection method that relates a matrix X (containing the chemical descriptors) to a matrix Y (containing the biological data) with the aim of predicting the values in Y from the information contained in X.

After building the model, a leave-one-out cross validation (LOO_{cv}) is used to test the internal predictivity of the model generated by CoMFA [26]. Each compound is systematically excluded from the training set and its biological activity is predicted by the model built. The cross validation provides the Principal Component (PC) number and q^2 (r^2_{cv}) value, which is the predictive capacity of the model generated. The q^2

has been a good indicator of the accuracy of actual predictions and a q^2 of 0.5, halfway between no model and a perfect model, is likely to be helpful in decision making.

Also a non-cross validation is performed to identify where the predictive correlation (r^2_{pred}), based on the test set molecules, is defined as the equation below:

$$r^2_{pred} = (SD\text{-}PRESS)/SD$$

Where SD is the sum of the squared deviations between the biological activities of the test set and mean activities of the training set molecules and PRESS is the sum of squared deviation between predicted and actual activity values for every molecule in the test set.

Considering the diverse CoMFA equations obtained from the variation alignments, partial atomic charges, atom types and cut-off evidence energy, the best model for qualitative analysis will be that with the highest q^2 and r^2 values, the smallest errors (SE_{cv} and ESS) and least number of outliers [27].

The QSAR produced by a CoMFA, with its hundreds or thousands of terms, is usefully represented as a three-dimensional "coefficient contour" color-coded map [28]. This is used to characterize the direction and magnitude of these differential interactions. In the steric maps green polyhedra surround positive regions where more bulk will enhance activity, whereas yellow polyhedra that surround regions indicate sterically disfavored regions. The electrostatic maps are characterized by blue and red polyhedras which corresponds to regions where an increase in positive or negative charge, respectively, is favorable for activity.

4. 3D-QSAR APPLIED TO *T. CRUZI* GROWTH CONTROL

The search for new drugs for Chagas disease treatment have evolved significantly in the last decade, mainly due to genome sequencing of *T. cruzi*. Nowadays, three main targets have been studied applying 3D-QSAR approach: Glyceraldehyde-3-phosphate Dehydrogenase (GPDH) [29], trypanothione reductase [30] and cruzain. Cruzain (EC 3.4.22), the major cysteine protease of *T. cruzi*, plays a pivotal role during the infection of host cells, replication, and metabolism, and has been considered an important target for the development of new antitrypanosomal agents. Thus, several (thio)semicarbazone and acylhydrazide derivatives have been synthesized to target this enzyme [7, 31, 32, 34].

Applying 3D-QSAR approach, Rodrigues and co-workers studied several acylhydrazides to provide useful guidelines for the design of more potent cruzain inhibitors [35]. Their QSAR-3D model was built using 112 acylhydrazides derivatives. Different types of alignments, probe atoms and lattice spacings were tested.

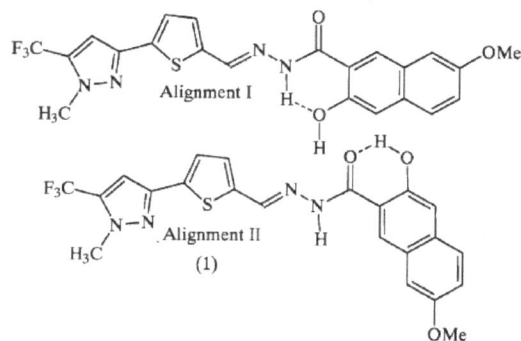

Figure 1: Alignment I and II showed for (1) derivative.

As CoMFA requires the user to identify a single relevant conformation for each ligand, some acylhydrazides in the training set were identified with o-hydroxyl groups in aromatic substituents, leading to two plausible conformations. In the first case, the hydroxyl oxygen can act as a hydrogen bond acceptor for the hydrogen on the hydrazide nitrogen (alignment I) (Fig. **1**). The other possibility is that the hydroxyl hydrogen acts as donor for hydrogen bond formation with the carbonyl oxygen present in the acylhydrazide backbone (alignment II) (Fig. **1**). In this case, alignment II, and a sp^3 probe carbon atom yielded good cross-validation (q^2=0.688) employing lattice spacing of 1Å.

To test the predictive ability of the resulting QSAR-3D model, a test set of 16 structurally diverse cruzain inhibitors excluded from the model creation work was used.

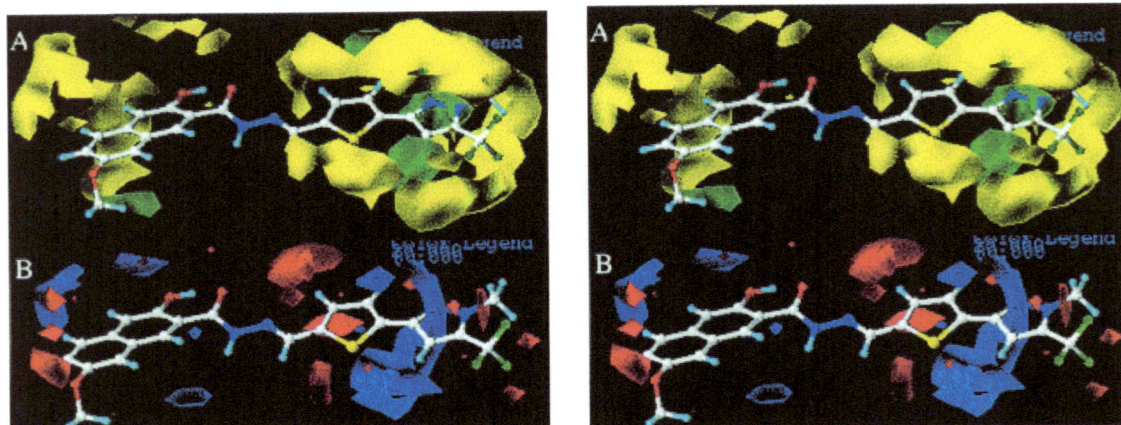

Figure 2: The CoMFA contour map built for acylhydrazides. **(A)** Steric map indicating areas where bulk is predicted to increase (green) or decrease (yellow) activity. **(B)** Electrostatic map indicating where high electron density (negative charge) (red) and low electron density (positive charge) (blue) regions are expected to increase activity.

An inspection of the steric contour plots showed that the sterically favorable regions (green polyhedra) were located next to the R1 substituent or more precisely near the pyrazole group (Fig. **2A**). It was also observed that any substitution in R2 (2-OH, 6-OMe naphthylene) will be located very close to an unfavorable region (yellow polyhedra) (Fig. **2A**).

In the electrostatic contribution contour, the red polyhedra surrounding the 4-position in the thiophene ring show that negative potentials in this position are likely to increase the inhibitory profile of the compound (Fig. **2B**). However, the strong steric hindrance at this position makes this site a chemically challenging site for substitution. The blue contours identify areas within the lattice where electropositive substituents are predicted to increase activity (Fig. **2B**). These regions include the partial positive charge associated with hydrogen atoms bound to carbon, and can be visualized in the proximity of the pyrazole substituent. We note that blue polyhedra overlap the area previously defined as being tolerant to increases in steric bulk (green polyhedra). The effect of positive steric contours and positive electrostatic contours may be interpreted as a hydrophobic effect and bulkier groups, such as hydrocarbons or phenyl rings in those regions could increase the protease inhibitory activity.

Afterward, Trossini and co-workers [35] used a data set containing 55 thiosemicarbazone and semicarbazone derivatives to propose a QSAR-3D model. In this work they employed the same probe atom as Rodrigues but a bigger lattice spacing of 2A. The alignment choice was made based on a docking study into the cruzain active site where the best conformations were used in this step.

The cross-validated correlation coefficients (q^2) of 0.78 for CoMFA and non-cross-validated correlation coefficient (r^2) of 0.95 indicates the good predictivity of the model. In addition, the relative contributions of the CoMFA steric and electrostatic fields to the biological activity were similar to the study with the acylhydrazones, with 54% and 46%, respectively, which highlights important regions in 3D space.

These contour maps were compared with the 3D protein environment in the cruzain cruzain binding site which was in agreement.

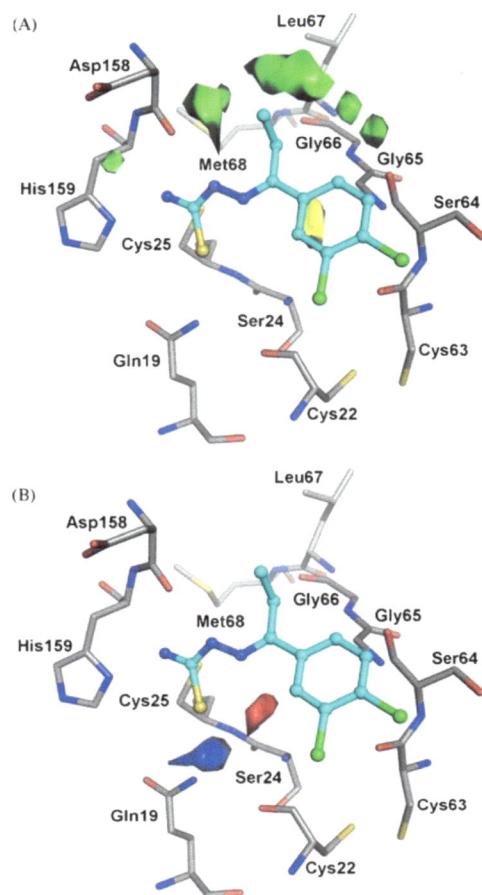

Figure 3: CoMFA contour maps. The inhibitor is represented by ball-and-stick model in the cruzain active site. **(A)** Steric contour maps are shown in yellow and green (-0.01 and 0.009 kcal/mol, respectively). **(B)** Electrostatic contour maps are shown in red and blue (-0.05 and 0.01 kcal/mol, respectively).

5. CONCLUSION

In CoMFA, a suitable sampling of the steric and electrostatic fields around a set of ligned structures provides the information required for the development of a predictive pharmacophore model. Although many CoMFA models are statistically excellent and offer good predictive performance, they are inherently limited by the need to align the database molecules correctly within 3D space. Because experimental evidence about ligand-receptor binding conformations is frequently lacking, the bioactive conformation must be postulated on the basis of information about the receptor binding site and/or the common conformational space accessible to different known ligands. This process, if performed correctly, can yield useful information about the nature of the receptor site and receptor-ligand interaction. Then, CoMFA posses statistical quality that allow the design of plausible trypanocidal drug candidates.

REFERENCES

[1]   Rodriguez JBE, Gros EG. Recents developments in the control of *Trypanosoma cruzi,* the causative agent for Chagas disease. Curr Med Chem 1995; 2: 723.

[2]   Silveira ACE, Vinhares M. Doença de Chagas: aspectos epidemiológicos e de controle. Rev Soc Bras Med Trop 1998; 31: 15.

[3]   Elizari MV. La Miocarditiopatia Chagásica – Perspectiva Histórica. Medicina 1999; 59: 25.

[4]   Dias JCP, Arq Bras. Problemas e possibilidades de participação comunitária no controle das grandes endemias no. Brasil Cardiol 1994; 63: 451.

[5]   Boiani M, Boiani L, Merlino A, Hernández P, Chidichimo A, Cazzulo JJ, Cerecetto H, González M. Second generation of 2H-benzimidazole 1,3-dioxide derivatives as anti trypanosomatid agents: Synthesis, biological evaluation and mode of action studies. Eur J Med Chem 2009; 44: 4426.

[6]   Guedes PMM, Veloso VM, Tafuri WL, Galvão LMC, Carneiro CM, Lana M, Chiari E, Soares KA, Bahia MT. The dog as model for chemotherapy of the Chagas' disease. Acta Tropica 2002; 84: 9.

[7]   Aguirre G, Boiani L, Cerecetto H, Fernández M, González M, Denicola A, Otero L, Gambino D, Rigol C, Olea-Azar C, Faundez M. *In vitro* activity and mechanism of action against the protozoan parasite *Trypanosoma cruzi* of 5-nitrofuryl containing thiosemicarbazones. Bioorg Med Chem 2004; 12: 4885.

[8]   Pandey S, Fletcher KA, Baker SH, Baker GA, DeLuca J, Fennie MF, O'Sullivan MC. Solution aggregation of anti-trypanosomal N-(2-naphthylmethyl)ated polyamines. J Photochem Photobiol A Chem 2004; 162: 387.

[9]   Bolognesi ML, Lizzi F, Perozzo R, Brun R, Cavalli A. Synthesis of a small library of 2-phenoxy-1,4-naphtoquinone and 2-phenoxy-1,4-anthraquinone derivatives bearing anti-trypanosomal and anti-leishmanial activity. Bioorg Med Chem Lett 2008; 18: 2272.

[10]   Nesslany F, Brugier S, Mouriès MA, Le Curieux F, Marzin D. *In vitro* and *in vivo* chromossomal aberations induced by megazol. Mutat Res Genet Toxicol Environ Mutagen 2004; 560: 147.

[11]   Ambrozin ARP, Vieira PC, Fernandes JB, da Silva MFGF, Albuquerque S. Trypanocidal activity of Meliaceae and Rutaceae plant extracts. Inst Oswaldo Cruz 2004; 99: 227.

[12]   Tanowitz H. Chagas' disease. Micr Reviews 1992; 4: 400.

[13]   Costa FC. Chagas' Disease Control Programme in Brazil: a study of the effectiveness of 13 years of intervention. Bull of the W. H. O. 1998; 76: 385; Willians-Blangero S. Attitudes towards chagas' disease in an endemic brazilian community. Cad Saúde Pública 1999; 15: 7.

[14]   Dias JCP. Problemas e possibilidades de participação comunitária no controle das grandes endemias no brasil. Cad Saúde Pública 1998; 14: 19; Marsden PD. The control of latin *American trypanosomiasis*. Rev Soc Bras Méd Trop 1997; 30: 521; Segura EL. Control de la transmisión de tripanosoma cruzi en la Argentina. Medicina 199; 59: 91.

[15]   Schmunis GA. Perspectivas de Eliminação – Instituto Oswaldo Cruz – 1996; 36: 969; Silveira AC, Vinhaes MC. Eliminacion de la transmision vectorial de la enfermidad de Chagas en Brasil. Medicina 1999; 59: 97.

[16]   Castro SL. The challenge of Chagas' disease chemotherapy: Na update of drugs assayed against *Trypanosoma cruzi*. Acta Tropica. 1993; 53: 83.

[17]   Tapia RA, Salas C, Morello A, Maya JD, Toro-Labbé A. Synthesis of dihydrofuroquinolinediones with tripanocidal activity and analysis of theis stereoelectronic properties. Bioorg Med Chem 2004, 12, 2451.

[18]   Morello A. Comp. The biochemistry of the model of action of drugs and the detoxication mechanisms in *Trypanosoma cruzi*. Biochem Physiol C 1988; 90: 1.

[19]   Pozas R, Carballo J, Castro C, Rubio J. Synthesis and *in vitro* antitrypanosomal activity of novel Nifurtimox analogues. Bioorg Med Chem Lett 2005; 15: 1417.

[20]   QSAR and CoMFA® Manual. Tripos 2010.

[21]   Verma J, Khedkar VM, Coutinho EC. 3D-QSAR in drug design-a review. Curr Top Med Chem 2010; 10: 95.

[22]   Leach AR, Molecular modeling: principles and applications. Universidade de Michigan. ISBN 9780582239333. Longman, 1996.

[23]   Migliavacca E. Applied introduction to multivariate methods used in drug discovery. Mini Rev Med Chem 2003; 3: 831.

[24]   Hopfinger AJ, Wang S, Tokarski JS, Jin BQ, Albuquerque MG, Madhav PJ, Duraiswami C. Construction of 3D-QSAR models using the 4D-QSAR analysis formalism. J Am Chem Soc 1997; 119: 10509.

[25]   Livingstone D. Data analysis for chemists. New York: Oxford University Press. 256 p. 1995.

[26]   Kubinyi H. QSAR and 3D QSAR in drug design. Part 1: methodology. Drug Disc Today 1997; 11: 457.

[27]   Kubinyi H. Hansch analysis and related approaches. New York: VCH. 240 p., 1993.

[28]   Cramer III RD; Patterson DE; Bunce JD. Comparative Molecular Field Analysis (CoMFA). 1. effect of shape on binding of steroids to carrier proteins. J Am Chem Soc 1988, 110: 5959.

[29]   Menezes IRA, Lopes JCD, Montanari CA, Oliva G, Pavão F, Castilho MS, Vieira PC, Pupo MT. 3D QSAR studies on binding affinities of coumarin natural products for glycosomal GAPDH of Trypanosoma cruzi. J Comp-Aided Mol Design 2003; 17: 277.

[30]   Iribarne F, Paulino M, Aguilera S, Tapia O. Assaying phenothiazine derivatives as trypanothione reductase and glutathione reductase inhibitors by theoretical docking and Molecular Dynamics studies. J Mol Graph Model 2009; 28: 371.

[31]   Du X, Guo C, Hansell E, Doyle PS, Caffrey CR, Holler TP, McKerrow JH, Cohen, FE. Synthesis and structure-activity relationship study of potent trypanocidal thio semicarbazone inhibitors of the trypanosomal cysteine protease cruzain. J Med Chem 2002; 45: 2695.

[32]   Chiyanzu I, Hansell E, Gut J, Rosenthal PJ, McKerrow JH, Chibale K. Synthesis and evaluation of isatins and thiosemicarbazone derivatives against cruzain, falcipain-2 and rhodesain. Bioorg Med Chem Lett 2003; 13: 3527.

[33]   Filho JMS, Leite ACL, Oliveira BG, Moreira DRM, Lima MS, Soares MBP, Leite LFCC. Design, synthesis and cruzain docking of 3-(4-substituted-aryl)-1,2,4-oxadiazole-N-acylhydrazones as anti-*Trypanosoma cruzi* agents. Bioorg Med Chem 2009; 17: 6682.

[34]   Rodrigues CR, Flaherty TM, Springer C, McKerrow JH, Cohen FE. CoMFA and HQSAR of acylhydrazide cruzain inhibitors. Bioorg Med Chem Lett 2002; 12: 1537.

[35]   Trossini GHG, Guido RVC, Oliva G, Ferreira EI, Andricopulo AD. Quantitative structure–activity relationships for a series of inhibitors of cruzain from *Trypanosoma cruzi*: Molecular modeling, CoMFA and CoMSIA studies. J Mol Graph Model 2009, 28: 3.

CHAPTER 7

Multivariate Image Analysis Applied to QSAR as a Tool for Mosquitoes Control: Dengue and Yellow Fever

Matheus P. Freitas[1*], Teodorico C. Ramalho[1], Elaine F. F. da Cunha[1] and Rodrigo A. Cormanich[2]

[1]*Department of Chemistry, Federal University of Lavras, P.O. Box 3037, 37200-000, Lavras, MG, Brazil and* [2]*Chemistry Institute, State University of Campinas, P.O. Box 6154, 13083-970, Campinas, SP, Brazil*

Abstract: An image-based QSAR approach, called Multivariate Image Analysis Applied to Quantitative Structure-Activity Relationships (MIA-QSAR) is described as a tool to model the toxicities of a series of organotin compounds against Aedes aegypti and Anopheles stephensi mosquito larvae, vectors of dengue and yellow fever. This methodology may help researchers to discover novel, potent mosquito controllers.

Keywords: QSAR-2D, MIA-QSAR, dengue and yellow fever.

1. INTRODUCTION

Dengue and yellow fever are tropical and subtropical diseases caused by an infection with the virus of the family of *Flaviviridae*. The virus is transmitted by the bite of mosquitoes, *e.g. Aedes aegypti*, and men and primates are the vertebrate hosts of this virus. Hemorrhagic dengue reaches 500,000 people annually, with 10% mortality for hospitalized patients, increasing to 30% mortality for non-treated patients [1]. The yellow fever is especially found in some areas of South America and Africa, and patients present cases with fever, nausea and pain; however, in some patients, a toxic phase follows, in which liver damage with jaundice can occur and lead to death. The WHO estimates that yellow fever causes 200,000 illnesses and 30,000 deaths every year in unvaccinated populations [2]; around 90% of the infections occur in Africa [3].

While a vaccine for yellow fever has been known since the middle of the 19th Century (some countries, like Brazil, require vaccinations for travelers), there are currently no drugs for the cure of dengue; patients are advised to drink water, and medicines based on acetylsalicylic acid and other non-steroidal anti-inflammatory drugs usually used when fever takes place must be avoided, because they facilitate hemorrhage. While vaccines for dengue are still commercially unavailable, preventive tasks must be done to control the vector mosquito, especially in the larval phase, such as to avoid water accumulation in possible sites where mosquitoes are supposed to lay eggs. Alternatively, the use of synthetic insecticides is a suitable strategy for mosquito control. Despite the possibility of environmental and human health concerns, *e.g.,* disruption of natural-biological control systems, resurgences in mosquito populations, widespread development of resistance, and undesirable effects on non-target organisms [4, 5], the development of novel types of insecticides that prevent insect resistance and that are environmentally friendly is of urgent worldwide interest.

Organotin compounds display strong biocidal activities and are generally very toxic, depending on the number and nature of the organic groups bound to the central Sn atom; compounds with three Sn-C bonds (R_3SnX) show the highest biological activity [6, 7]. Recently, toxicity studies demonstrated that organotins are effective against two species of mosquito larvae, *Aedes aegypti* (*Ae. aegypti*) and *Anopheles stephensi* (*An. stephensi*) [8-12]. In order to propose new active organotin compounds, Hansch and Verma [13] built Quantitative Structure-Activity Relationship (QSAR) models based on different series of organotin compounds (R_3SnX) with respect to their larvicidal activities against the second instar stage of the *Ae.*

*Address correspondence to Matheus P. Freitas; Department of Chemistry, Federal University of Lavras, P.O. Box 3037, 37200-000, Lavras, MG, Brazil. E-mail: matheus@dqi.ufla.br

Teodorico C. Ramalho, Matheus P. Freitas and Elaine F. F. da Cunha (Eds)

aegypti and *An. stephensi* mosquito larvae, using physicochemical descriptors, mainly hydrophobic (π) and Hammett electronic (σ^+) parameters of the substituents to correlate chemical structures and bioactivities.

In addition to classical physicochemical descriptors, such as those applied by Hansch and Verma [13], multidimensional (nD) QSAR, in which descriptors are generated in a grid cell for a given molecule (3D) and may account for ensemble averaging, receptor dependency and/or solvent effects (4D, 5D and 6D) [14-18], has shown to be of widespread use. However, specific programs, exhaustive conformational screening and/or difficult alignment rules are often required to obtain good models by means of these methodologies. On the other hand, an image-based QSAR approach, in which images are two-dimensional chemical structures of a pharmacophore, has given valuable and predictive QSAR models [19-22]. The so called multivariate image analysis applied to QSAR (MIA-QSAR) is easily accessible and simple to manipulate; its procedure is scrutinized here and applied to the series of organotin compounds evaluated by Hansch and Verma [13]. The regression parameters of the MIA-QSAR model may then be used to estimate the activity of novel, eventually more potent mosquito controllers.

2. MIA-QSAR PROCEDURE

In classical QSAR, physicochemical descriptors are usually calculated or measured for a set of compounds, and then correlated with the respective bioactivity through a given statistical tool. In MIA-QSAR, descriptors are images, *i.e.,* the 2D chemical structures of the series of compounds; they should be transformed in numerical values in order to allow correlation with the biological activities. Each pixel of an image is a grey value; different 2D chemical structures are shapes with pixels distributed at specific coordinates, and this variance along the series of compounds (*e.g.,* different substituent positions in an aromatic ring) explains the variance in the activities block. Although no physicochemical meaning, like group electronegativity and volume, nor conformation aspects are supposed to be considered in MIA-QSAR, it seems that structural shapes play a significant role in deriving a QSAR model. The steps required to achieve a MIA-QSAR model are depicted as follows:

1) *Building of the 2D chemical structures*: this step requires drawing chemical structures using an appropriate program for this purpose. There are numerous commercially and freely available programs used to this end, but an important task is to represent compounds systematically, for example: use the same font type and size for chemical elements of different molecules in a given series; use the same bond length (usually represented as sticks) and ring sizes for all compounds; use the same connectivity rules for similar substituent groups, in order to allow maximum similarity when aligning, *etc.* 3D information, like representation of stereogenic centers, may be schematically designated as hashed or wedge lines (bonds) in relation to the chiral center. An example of how a chemical structure can be draw is:

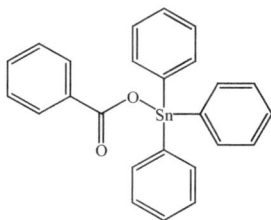

2) *Alignment of the 2D chemical structures*: each image (chemical structure) should be saved in a workspace with well defined *m×n* dimension, *e.g.,* using the Paint applicative of Windows. The result will be independent of the file extension (.bmp, .jpeg, .png, .tiff, *etc.*) [23]. Also, a common pixel among the whole series of compounds should be manually adjusted in a given coordinate of the workspace, since all images will be superposed later and the similarity moieties of all congeneric compounds must be congruent; the variable substructures correspond to variance in the activities block – the basis of a structure-based QSAR. For the series of organotin compounds used as mosquito (*Ae. aegypti* and *An. stephensi*) controllers,

the common pixel to be chosen might be one located on the Sn symbol, which is an element present in the whole series. A schematic example of a MIA-QSAR alignment is given as:

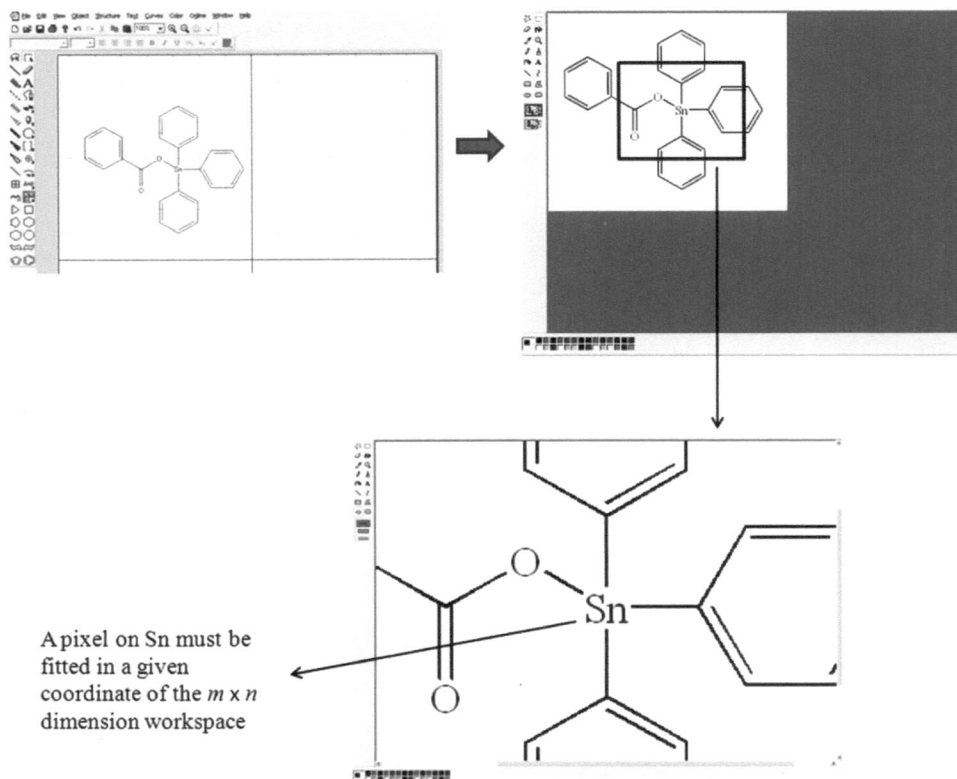

A pixel on Sn must be fitted in a given coordinate of the $m \times n$ dimension workspace

3) *Conversion of the 2D chemical structures in binaries*: in order to transform chemical structures in numbers (descriptors), each image must be read and converted into binaries. This procedure can be performed by means of the *e.g.* Matlab program using the following script lines (for images saved as *bitmap* files):

[filename,MAP] = imread ('filename.bmp','bmp');

filename =double(filename);

filename =(filename(:,:,1)+ filename(:,:,2)+ filename(:,:,3));

4) *Superposition of the 2D chemical structures (array or matrix generation)*: once images are all saved as shapes of $m \times n$ dimension and chemical structures are all aligned by a similarity center, they can be superposed in order to give data variance along the *l* vector, *i.e.*, the structural changes. Superposition of images gives a three-way array, which can be unfolded to a two-way array (a matrix). Both can be correlated with the activities column vector (*y* block) using appropriate regression tools. Array or matrix generation can be easily performed using Matlab or aother statistical platform; for example, an **X** matrix of $l \times (m \times n)$ dimension can be built by using the following script lines in Matlab:

X=[filename1(:)';

filename2(:)';

filename3(:)';

…

filenamel(:)']

A schematic example of how superposition is performed follows:

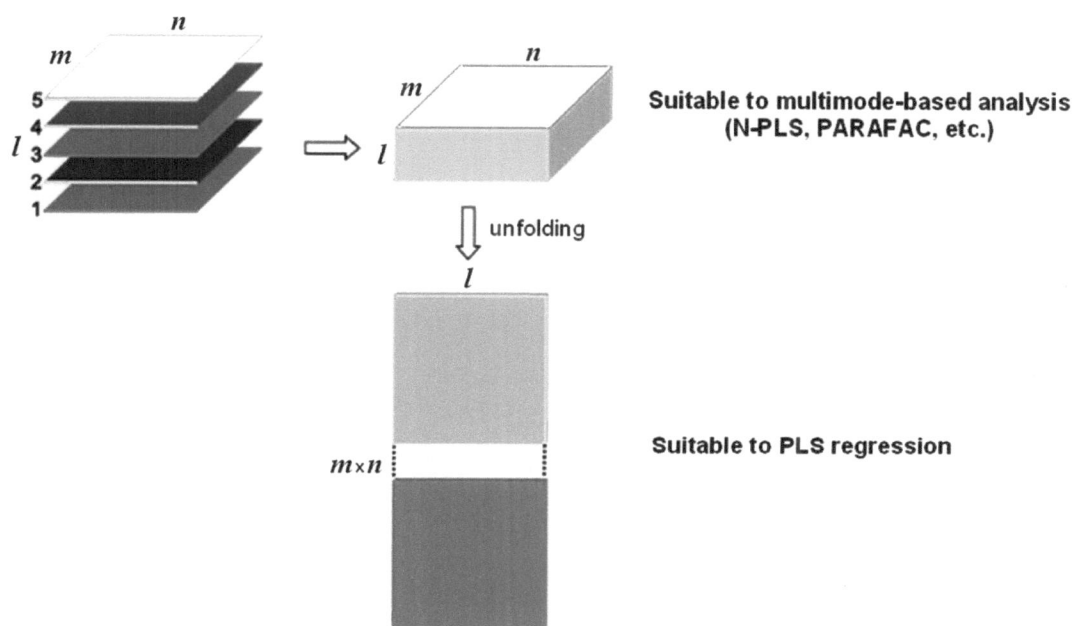

5) *Calibration of the model*: a QSAR model is achieved by correlating descriptors and bioactivities, and the regression parameters obtained are used to predict the bioactivities of novel compounds using their corresponding descriptors. However, many variables (usually thousands) are generated using MIA-QSAR, and multivariate statistics are then required in calibration. Multilinear Partial Least Squares (N-PLS) regression [24] is commonly used to correlate three-mode arrays and dependent variables, while bilinear (traditional) PLS [25] is widely used when matrices (two-way arrays) are applied. The idea of using such regression tools is to reduce the number of dimensions in a few Latent Variables (LV), linear combinations of the original variables. The optimum number of latent variables is usually chosen by analyzing the errors of estimation (RMSE) in calibration and cross-validation when successive LV are added in the model; the best number of LV is reached when errors of estimation in calibration/cross-validation do not decrease significantly after adding more LV and/or errors of prediction in cross-validation stop diminishing.

6) *Validation of the model*: in order to find out if a calibration model is reliable to predict the biological activity of novel, proposed targets, it needs to be validated. A worldwide recognized method for validation is the cross-validation, particularly the Leave-One-Out (LOO) cross-validation where, from the calibration samples, one sample is left out and the model is developed using different numbers of factors. The quality of the cross-validation is usually quantified by q^2, the leave-one-out cross-validated r^2 [26]. However, Golbraikh and Tropsha [26] state that the only way to establish a reliable QSAR model is by performing an external validation, *i.e.* to predict accurately the bioactivity of compounds not pertaining to the training set. In addition to the well known cross and external validation tests, sensitivity tests for MIA-QSAR models may be performed to show that good correlations eventually obtained using MIA descriptors do not result merely from chance. This may be assessed by scrambling the y block and succeeding with calibration. Comparison between the correlation coefficients of the calibration obtained through this procedure and the true calibration (y block not randomized), at a given number of latent variables, provides an account for the robustness of the model, and shows that the true calibration was or was not a fortuitous correlation.

7) *Prediction of the bioactivity of novel compounds using the built model*: the regression coefficients obtained may be used to predict the activities of novel compounds. Since the

compound set is a congeneric series, novel bioactive compounds may be proposed based on a miscellany of substructures of trained compounds and their biological activities reliably estimated, since variable moieties have already been calibrated. Hansch and Verma [13] have predicted the bioactivity against *Ae. aegypti* and *An. stephensi* of nine novel, proposed organotin compounds. A schematic method for proposing novel compounds using the substructure miscellany approach is:

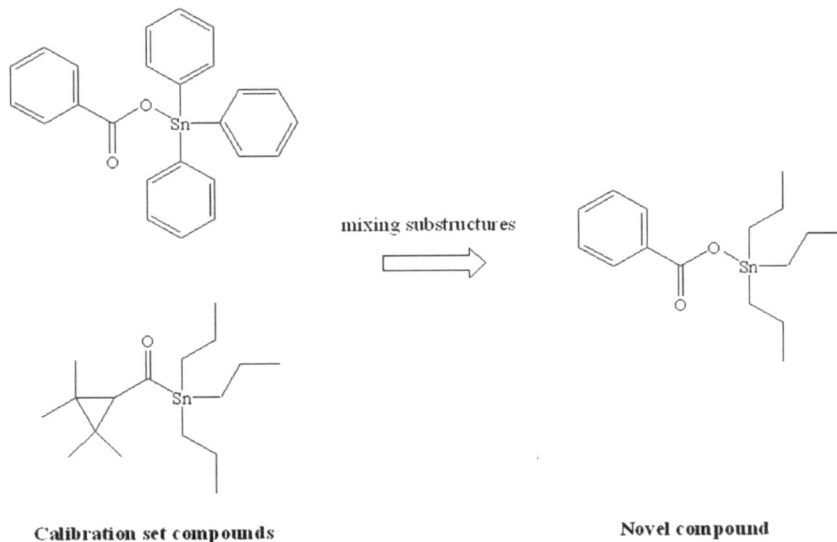

Calibration set compounds   mixing substructures   **Novel compound**

3. MIA-QSAR APPLIED TO *Ae. Aegypti* AND *An. Stephensi* CONTROL

MIA-QSAR can be applied to model the biological activities (toxicities) of compounds against the vectors of dengue and yellow fever, since there is a congeneric series of organotin compounds available [13] as mosquito (*Ae. aegypti* and *An. stephensi*) controllers (Table **1**). The procedure described above was carried out and a QSAR model was derived, which can be used to predict the biological activities of new insecticide candidates, (Hansch and Verma [13] have found that the kill mechanism is different for these two species of mosquito larvae).

Table 1: Organotin compounds and respective experimental and predicted toxicities (pIC$_{50}$, IC$_{50}$ in mol L^{-1}) against *Ae. aegypti* and *An. stephensi* mosquito larvae.

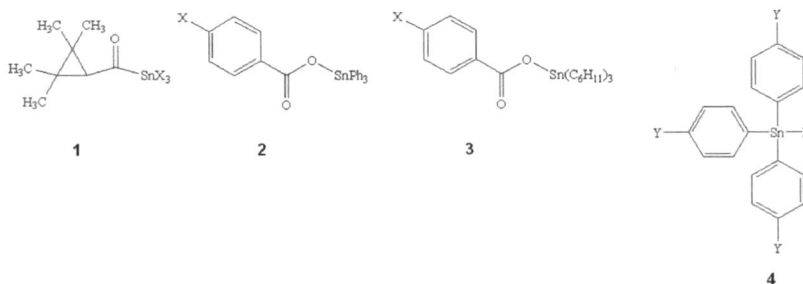

Compound	X	Y	Ae. aegypti		An. stephensi	
			Obs.	Pred.	Obs.	Pred.
1-1	CH$_3$	H	5.53	5.39	4.98	6.46
1-2	C$_2$H$_5$	H	5.45	5.32	4.79	6.04
1-3	n-C$_3$H$_7$	H	6.16	6.27	5.76	6.03
1-4	n-C$_4$H$_9$	H	6.33	6.31	5.99	5.99

1-5[a]	C_6H_5	H	6.55	6.46	5.70	5.48
1-6	C_6H_{11}	H	6.53	6.39	6.03	5.68
2-7[a]	H	H	5.75	6.04	5.19	5.19
2-8	F	H	5.74	5.80	5.09	5.80
2-9[b]	Cl	H	6.23	6.04	5.21	5.24
2-10	Br	H	5.87	6.06	5.25	5.51
2-11	I	H	5.86	6.07	5.10	5.51
2-12[a]	OCH_3	H	6.10	5.99	5.13	5.42
2-13	OH	H	5.86	5.72	5.40	5.75
2-14	NO_2	H	6.12	5.89	5.05	6.46
2-15	NH_2	H	6.12	5.96	5.26	6.04
2-16	CH_3	H	5.78	5.98	5.00	6.03
2-17[a]	$C(CH_3)_3$	H	5.95	5.68	5.12	4.94
3-18	H	H	5.73	5.80	5.87	5.99
3-19	F	H	5.55	5.62	5.55	5.68
3-20[a]	Cl	H	5.74	5.80	5.97	5.66
3-21	Br	H	5.80	5.88	5.29	5.80
3-22	I	H	5.86	5.83	5.68	5.51
3-23	OCH_3	H	5.99	5.82	6.42	5.51
3-24	OH	H	5.54	5.64	6.43	5.75
3-25	NO_2	H	5.29	5.54	5.65	6.46
3-26	NH_2	H	5.49	5.61	5.59	6.04
3-27[a]	CH_3	H	5.70	5.51	6.14	5.62
3-28	$C(CH_3)_3$	H	5.82	5.87	6.10	6.04
4-29	H	Cl	5.18	5.34	6.19	5.99
4-30	H	OH	5.39	5.31	6.42	5.68
4-31	H	OAc	5.25	5.31	7.01	5.80
4-32	CH_3	Cl	5.61	5.54	5.09	5.51
4-33	CH_3	OH	5.72	5.68	5.22	5.51
4-34	CH_3	OAc	5.58	5.50	5.01	5.75
4-35[a]	F	Cl	5.72	5.51	5.80	5.91
4-36[a]	F	OH	5.93	5.75	5.71	5.96
4-37	F	OAc	6.05	5.75	5.87	6.46
4-38	Cl	Cl	5.48	5.54	5.75	6.04
4-39	Cl	OH	5.64	5.53	5.66	6.04
4-40	Cl	OAc	5.48	5.53	5.81	5.99
4-41	SCH_3	Cl	NA	NA	4.79	5.68
4-42	SCH_3	OH	NA	NA	5.00	5.80
4-43	SCH_3	OAc	NA	NA	4.91	5.50

[a] Test set compounds.

The MIA descriptors were correlated with the toxicities column vector using PLS, and 5 latent variables were found as the optimum number of PLS components for both *Ae. aegypti* and *An. stephensi* data, according to the criterion established previously. The squared correlation coefficient of the regression line of experimental *versus* fitted pIC_{50} was significantly high, *i.e.*, r^2 of 0.805 (RMSEC = 0.14) and 0.896 (RMSEC = 0.17) for *Ae. aegypti* and *An. stephensi*, respectively, as illustrated in Fig. **1**. In order to search for the reliability of the calibration model, the *y* column vector (the activities block) was randomized and a new calibration using 5 PLS components was carried out; very poor correlations (average of 3 runs) were obtained in the Y-randomization test ($r^2_{Y\text{-randomization}}$ of 0.10±0.04 and 0.06±0.07 for *Ae. aegypti* and *An. stephensi*, respectively), indicating that the good results in the true calibrations were not chance correlations.

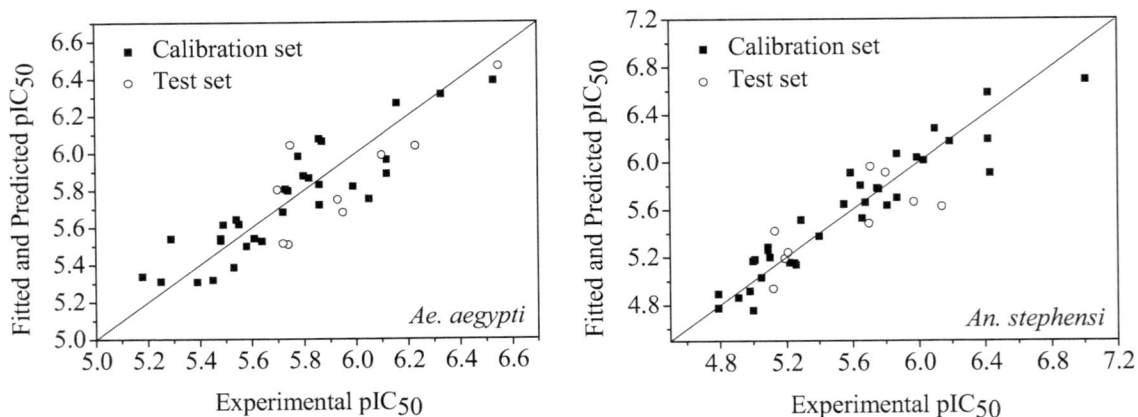

Figure 1: Plots of experimental *vs.* fitted and predicted pIC$_{50}$, using MIA-QSAR, for the *Ae. aegypti* and *An. stephensi* data.

The MIA-QSAR models were validated using the leave-one-out cross-validation approach, giving q^2 of 0.525 (RMSECV = 0.22) and 0.502 (RMSECV = 0.36) for the *Ae. aegypti* and *An. stephensi* data, respectively; q^2 values above 0.5 are acceptable [26]. Moreover, an external validation was performed to certify that the MIA-QSAR models are reliable for prediction of toxicities of test set insecticide-like compounds. The r^2_{test} values found for the *Ae. aegypti* and *An. stephensi* data were 0.656 (RMSEP = 0.38) and 0.553 (RMSEP = 0.44), respectively, confirming the suitability of the derived MIA-QSAR models as a tool to predict biological data of novel organotin insecticides. Some parallelism between MIA and physicochemical descriptors may be assumed, since the explained variance of data and predictive ability using MIA-QSAR is similar to a traditional QSAR previously established for this series of compounds [13]. Supportive analysis and comprehension of the interaction mode between the new organotin compounds and enzyme active sites may be achieved by combining structure and receptor-based approaches, namely MIA-QSAR and docking strategies, which have been successfully joined in a couple of studies [27, 28].

REFERENCES

[1]   Holmes EC, Bartley LM, Garnet GP. In Krause RM, Ed.; Academic Press: London, 1998; pp. 301.

[2]   World Health Organization – WHO/Yellow Fever, retrieved 2009-08-13.

[3]   Tolle MA. Mosquito-borne diseases. Curr Probl Pediatr Adolesc Health Care 2009; 39: 97.

[4]   Das NG, Goswami D, Rabha B. Preliminary evaluation of mosquito larvicidal efficacy of plant extracts. J Vect Borne Dis 2007; 44: 145.

[5]   Brown AWA. Insecticide resistance in mosquitoes: a pragmatic view. J Am Mosq Control Assoc 1986; 2: 123.

[6]   Blunden SJ, Chapman A. In Craig PJ, Ed.; Wiley: New York, 1986; pp. 111.

[7]   Song X, Zapata A, Hoerner J, de Dios AC, Casabianca L, Eng G. Synthesis, larvicidal, QSAR and structural studies of some triorganotin 2,2,3,3-tetramethylcyclopropanecarboxylates. Appl Organomet Chem 2007; 21: 545.

[8]   Baul TSB, Singh KS, Lycka A, Linden A, Song X, Zapata A, Eng G. Synthesis, characterization and crystal structures of triorganotin(IV) complexes of 4-[(E)-2-(3-formyl-4- hydroxyphenyl)-1-diazenyl]- and 4-{(E)-4-hydroxy-3-[(E)-4-(aryl)iminomethyl]phenyldiazenyl}-benzoic acids and toxicity studies of their tri-n-butyltin(IV) derivatives on the Aedes aegypti and Anopheles stephensi mosquito larvae (pages 788) Appl Organomet Chem 2006; 20: 788.

[9]   Duong Q, Song X, Mitrojorgji E, Gordon S, Eng G. Larvicidal and structural studies of some triphenyl- and tricyclohexyltin para-substituted benzoates. J Organomet Chem 2006; 691: 1775.

[10]   Baul TSB, Singh KS, Holcapek M, Jirasko R, Linden A, Song X, Zapata A, Eng G. Electrospray ionization mass spectrometry of tributyltin(IV) complexes and their larvicidal activity on mosquito larvae: crystal and molecular structure of polymeric $(Bu_3Sn[O_2CC_6H_4\{N=N(C_6H_3-4-OH(C(H)=NC_6H_4OCH_3-4))\}-o])_n$. Appl Organomet Chem 2005; 19: 935.

[11]   Song X, Duong Q, Mitrojorgji E, Zapata A, Nguyen N, Strickman D, Glass J, Eng G. Synthesis, structure characterization and larvicidal activity of some tris-(para-substitutedphenyl)tins. Appl Organomet Chem 2004; 18: 363.

[12]   Eng G, Song X, Duong Q, Strickman D, Glass J, May L. Synthesis, structure characterization and insecticidal activity of some triorganotin dithiocarbamates. Appl Organomet Chem 2003; 17: 218.

[13]   Hansch C, Verma RP. Larvicidal activities of some organotin compounds on mosquito larvae: A QSAR study. Eur J Med Chem 2009; 44: 260.

[14]   Cramer RD, Patterson DE, Bunce JD. Comparative Molecular Field Analysis (CoMFA). 1. Effect of shape on binding of steroids to carrier proteins. J Am Chem Soc 1988; 110: 5959.

[15]   Klebe G, Abraham U, Mietzner T. Molecular similarity indices in a comparative analysis (CoMSIA) of drug molecules to correlate and predict their biological activity. J Med Chem 1994; 37: 4130.

[16]   Hopfinger AJ, Wang S, Tokarski JS, Jin B, Albuquerque M, Madhav PJ, Duraiswami C. Construction of 3D-QSAR Models Using the 4D-QSAR Analysis Formalism. J Am Chem Soc 1997; 119: 10509.

[17]   Vedani A, Dobler M. 5D-QSAR: the key for simulating induced fit? J Med Chem 2002; 45: 2139.

[18]   Vedani A, Dobler M, Lill MA. Combining protein modeling and 6D-QSAR. simulating the binding of structurally diverse ligands to the estrogen receptor. J Med Chem 2005; 48: 3700-3703.

[19]   Freitas MP, Brown SD, Martins JA. MIA-QSAR: a simple 2D image-based approach for quantitative structure–activity relationship analysis. J Mol Struct 2005; 738: 149.

[20]   Freitas MP. MIA-QSAR modelling of anti-HIV-1 activities of some 2-amino-6-arylsulfonylbenzonitriles and their thio and sulfinyl congeners. Org Biomol Chem 2006; 4: 1154.

[21]   Freitas MP, da Cunha EFF, Ramalho TC, Goodarzi M. Multimode methods applied on MIA descriptors in QSAR. Curr Comput-Aided Drug Des 2008; 4: 273.

[22]   Freitas MP. Multivariate image analysis applied to QSAR: Evaluation to a series of potential anxiolytic agents. Chemom Intell Lab Sys 2008; 91: 173.

[23]   Goodarzi M, Freitas MP, Ferreira EB. Influence of changes in 2-D chemical structure drawings and image formats on the prediction of biological properties using MIA-QSAR. QSAR Comb Sci 2009; 28: 458.

[24]   Bro R. Multiway calibration. Multilinear PLS. J Chemom 1996; 10: 47.

[25]   Wold S. In Multivariate Analysis, Krishnaiah K. Ed.; Academic Press: New York, 1966; pp. 391.

[26]   Golbraikh A, Tropsha A. Beware of q^2! J Mol Graph Model 2002; 20: 269.

[27]   Pinheiro JR, Bitencourt M, da Cunha EFF, Ramalho TC, Freitas MP. Novel anti-HIV cyclotriazadisulfonamide derivatives as modeled by ligand- and receptor-based approaches. Bioorg Med Chem 2008; 16: 1683.

[28]   Antunes JE, Freitas MP, da Cunha EFF, Ramalho TC, Rittner R. In silico prediction of novel phosphodiesterase type-5 inhibitors derived from Sildenafil, Vardenafil and Tadalafil. Bioorg Med Chem 2008; 16: 7599.

CHAPTER 8

Molecular Modeling of the *Toxoplasma gondii* Adenosine Kinase Inhibitors

Daiana Teixeira Mancini[*], Elaine F. F. da Cunha, Teodorico C. Ramalho and Matheus P. Freitas

Department of Chemistry, Federal University of Lavras, P.O. Box 3037, 37200-000, Lavras, MG, Brazil

Abstract: *Toxoplasma gondii* (*T. gondii*) is the most common cause of secondary central nervous system infection in immunocompromised persons such as AIDS patients. Since purine salvage is essential for *T. gondii* and for other parasitic protozoa, inhibition of this salvage should block parasite growth. *T. gondii* adenosine kinase (EC 2.7.1.20) is the major route of adenosine (purine nucleoside) metabolism in this parasite. Four-Dimensional Quantitative Structure-Activity Relationship (4D-QSAR) analysis was applied to a series of 41 inhibitors of *T. gondii* adenosine kinase. Optimized 4D-QSAR models were constructed by Genetic Algorithm (GA) optimization and partial least squares (PLS) fitting, and evaluated by the leave-one-out cross-validation method. Moreover, we have used docking approaches to study the binding orientations and predict binding affinities of some benzyladenosines with adenosine kinase.

Keywords: *Toxoplasma gondii*, adenosine kinase, *docking*

1. INTRODUCTION

The parasitic protozoon *Toxoplasma gondii* (*T. gondii*) is the etiologicagent of toxoplasmosis, a parasitic disease widespread amongvarious warm-blooded animals, including humans [1]. Toxoplasmosis is known to be one of themost prevalent parasitic infections of the central nervous system and causes lethal encephalitisin immunocompromised patients such those with acquired immunodefficiency syndrome (AIDS) [2]. In spite of the tragicconsequences of toxoplasmosis, the therapy of the disease hasnot changed in the last 20 years [1]. The current treatment consists of combinations of drugs, such as Pyrimethamine (Daraprim®), trimethoprim-sulfamethoxazole (Bactrin®), Sulfadiazine (Triglobe®) or Clindamycin (Dalacin®) [4]. Newborns with congenital toxoplasmosis are treated for at least a year with a combination of antibiotics. If the woman develops toxoplasmosis during pregnancy, her doctor may prescribe medications that reduce the risk of children developing congenital toxoplasmosis. These drugs include spiramycin (Rovamicina® Periodontil®), Pyrimethamine and Sulfadiazine. To decrease the possibility of developing congenital problems (at birth) related to the drugs, the type and duration of treatment will depend on which quarter she is pregnant [3].

T. gondii is apurine auxotroph incapable of *de novo* purine biosynthesis and depends on salvage pathways for its purine requirements [4]. However, *T. gondii* can efficiently salvage purine nucleosides and nucleobases for macromolecular synthesis. The most efficiently utilized precursor is reported to be adenosine monophosphate (AMP) and adenosine kinase (AK, EC2.7.1.20) is the major enzyme in the salvage of purines in these parasites [5]. *T. gondii* adenosine kinase (TgAK) is a 363-residue (39.3 kDa) monomeric protein, that catalyzes the phosphorylation of adenosine to adenosine 5′-monophosphate (AMP), using the g-phosphate group of ATP as the phosphate donor [2]. Since purine salvage is essential for *T. gondii* and for other parasitic protozoa, inhibition of this salvage should block parasite growth.

Benzyladenosine analogues are known in the literature as subversive substrates, *i.e.*, they are preferentially metabolized to the nucleotide level and become selectively toxicfor the parasite, but not for human adenosine kinase [6-8]. Kouni *et al.* described a series of benzyl adenosine analogues that act as potent and

***Address correspondence to Daiana Teixeira Mancini;** Department of Chemistry, Federal University of Lavras, P.O. Box 3037, 37200-000, Lavras, MG, Brazil. E-mail: elaine_cunha@dqi.ufla.br

Teodorico C. Ramalho, Matheus P. Freitas and Elaine F. F. da Cunha (Eds)

selective substrates for *T. gondii* adenosine kinase [2, 9, 10]. The binding affinities of the newly synthesized analogues to *T. gondii* and human adenosine kinase were evaluated in enzyme-based assays [2, 9, 10]. Then, we used a molecular modeling technique to understand the interaction mechanism of these compounds inside TgAK active site looking to contribute to the combat against toxoplasmosis.

2. METHODS

2.1 Biological Data

The tgAK protein, including water of crystallization, Mg^{2+}, AMP-PCP and adenosine substrate were obtained from the Protein Data Bank (PDB – code LII). Table **1** contains a set of benzyladenosine analogs and their ativity (IC_{50}) values. These compounds were synthesized and then pharmacologically evaluated by Kouni *et al.* [2, 9, 10]. It is important to emphasize that the pharmacological data were obtained from the same laboratory, which eliminates potential mistaken information which might have occurred in case data had different sources. Each mouse was infected with 200 tachyzoites (0.2 mL oftoxoplasma suspension in PBS containing 0.25 parasite/field under 40x magnification of a light microscope). The compounds were administered orally (0.1 mL/10 g) every 8 h for 5 days. Enzyme assays were run under conditions where activity was linear with time and enzyme concentration. Activity was determined by following the formation of radiolabeled AMP from adenosine.

A data set of 41tgAK inhibitors (Table **1**) was taken from published results [2, 9, 10] and the QSAR model was developed using a training data set of 29 compounds, selected from the original 41 compounds. The model also was externally validated using a test data set of 12 compounds, selected from the original 41 compounds. The IC_{50} (µM) values were converted into molar units, and then expressed in negative logarithmic units (pIC_{50}). The 3D models for each compound in their neutral forms were constructed using the Hyperchem software [11] and the adenosine was used as a reference in this work. Each structure was geometry-optimized without any restriction in vacuum, and the partial atomic charges were assigned using AM1 semi-empirical method.

Table 1: Structures of the 41tgAK inhibitors and activity values(IC_{50} and K_i). Training set compound numbers are in bold and test set compound numbers are in italic.

Compd	structure	KI (µM)	IC50 (µM)	Compd	Structure	IC50 (µM)
1	6-benzylthioinosine	2.4	9.3	*2*	o-Fluoro-N6-benzyladenosine	37.8
3	o-Fluoro-6-benzylthioinosine	-	8.2	**4**	o-Chloro-N6-benzyladenosine	10.6
5	o-Chloro-6-benzylthioinosine	1.4	6.7	**6**	o-Nitro-N6-benzyladenosine	15.0
7	o-Methyl-6-benzylthioinosine	1.5	7.7	*8*	o-Methoxy-N6-benzyladenosine	8.7
9	m-Nitro-6-benzylthioinosine	3.0	6.2	**10**	m-Fluoro-N6-benzyladenosine	23.2
11	m-Methyl-6-benzylthioinosine	1.3	8.2	**12**	m-Methyl-N6-benzyladenosine	8.5
13	m-Trifluoromethyl-6-benzylthioinosine	2.9	8.7	**14**	p-Fluoro-N6-benzyladenosine	20.6
15	p-Fluoro-6-benzylthioinosine	1.2	10.3	**16**	p-Chloro-N6-benzyladenosine	25.8
17	p-Chloro-6-benzylthioinosine	2.1	8.11	*18*	p-Cyano-N6-benzyladenosine	8.3
19	p-Bromo-6-benzylthioinosine	7.5	14.3	**20**	p-Trifluoro-N6-benzyladenosine	22.9
21	p-Cyano-6-benzylthioinosine	0.9	4.3	*22*	p-Methyl-N6-benzyladenosine	10.5
23	p-Nitro-6-benzylthioinosine	1.1	12.0	**24**	p-Isopropyl-N6-benzyladenosine	22.0
25	p-Methyl-6-benzylthioinosine	3.3	7.8	**26**	p-Vinyl-N6-benzyladenosine	26.2
27	p-Methoxy-6-benzylthioinosine	2.6	3.5	**28**	p-Methoxy-N6-benzyladenosine	7.5
29	p-trifluoromethoxy-6-benzylthioinosine	141	24.8	**30**	2,4-Difluoro-N6-benzyladenosine	8.2
31	p-*tert*-Buthyl-6-benzylthioinosine	113	23.3	**32**	2,4-Dichloro-N6-benzyladenosine	9.3
33	p-acetoxy-6-benzylthioinosine	96.1	15.0	*34*	3,4-Dichloro-N6-benzyladenosine	14.3
35	2,4-Dichloro-6-benzylthioinosine	0.7	7.3	**36**	2-Chloro-4-cyano-N6-benzyladenosine	21.1
37	3,4-Dichloro-6-benzylthioinosine	2.5	10.4	**39**	2-Fluoro-4-methoxy-N6-benzyladenosine	9.5
38	2-Chloro-6-fluoro-6-benzylthioinosine	37.7	8.7	**41**	2,4-Dimethoxy-N6-benzyladenosine	8.7
40	2,4,6-trimethyl-6-benzylthioinosine	150.0	31.1			

2.2. Parameters for QSAR Studies

2.2.1. Molecular Dynamic Simulation

The AM1 optimized structures were the initial structures in each molecular dynamic simulation (MDS). The MDS resulting structures, were, then, used to construct the Conformational Ensemble Profile (CEP) of each ligand. The MDS was performed using the 4D-QSAR program [12] and the Molsim 3.0 software [13] with an extended MM2 force field [14]. The temperature for the MDS was set at 300 K, close to the temperature assays, with a simulation sampling time of 50 ps, and intervals of 0.001 ps. Appling this scheme, a total sampling of 50000 conformations of each compound was obtained. The MDS calculations were carried out applying a distance-dependent dielectric function, $\varepsilon_r = D*r_{ij}$, which was set to $3*r_{ij}$ in order to try to model the solvent effect in the absence of explicit solvent. In this expression, r_{ij} is the distance between atoms i and j, and D is the scale factor of the dielectric function. In addition, the carbon atom C-4 (Fig. 1), common to all compounds, was fixed (frozen) to prevent a large conformational change of the ligands in the absence of the protein structure.

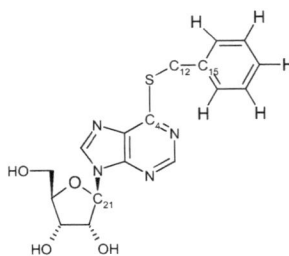

Figure 1: Structure of the compound 6-benzylthioinosine.

2.2.2. Alignment Definition

The alignment is one of the most important steps in 3D-QSAR methodologies (*e.g.*, CoMFAand 4D-QSAR). In this work, we will assume that all molecules bind to the enzyme in a similar mode, since the compounds are structural analogs. Three-ordered atom trial alignments were selected in this study. In general, the alignments are chosen to span the common framework of the molecules in the training and test sets. Alignments using atoms from the right, left, and middle of the common framework and alignments that use atoms that span the common framework should be used to ensure a complete alignment analysis [15-17]. The three ordered atom alignment definitions are (1) 21-4-12, (2) 4-12-15, and (3) 15-12-4 (Fig. 1), using compound 6-benzylthioinosine as a reference. The CEP for each compound was obtained after the MDS step was overlaid onto a cubic lattice of a selected grid cell size, according to each selected alignment. The cubic lattice serves to record the distribution of the 4D-QSAR methodology [18]. These IPEs correspond tothe interactions that may occur in the active site, and are related to the pharmacophore groups. In this work, we have selected the following trial set of interaction pharmacophore elements: i) any atom type (any); ii) nonpolar atom (np); iii) atom of polar-positive charge density (pp); iv)atom of polar-negative charge density (pn); v) hydrogen bond acceptor atom (ha); vi) hydrogen bond donor atom (hd); and vii) atoms in aromatic systems (ar). The occupancy of the grid cells by each IPE type is recorded over the conformational assembly profile, and forms the set of Grid Cell Occupancy Descriptors (GCOD), to be utilized as the pool of trial descriptors in the model building and optimization process [18]. After this, we performed two methods of analysis: Method 1 and Method 2.

2.2.3. 4 D-QSAR Model Calculations - Method 1

Partial Least-Squares (PLS) regression analysis was performed as a data (QSAR descriptors) reduction fit between the observed dependent variable measures and the corresponding set of GCOD values.The GCODs with the highest weight in each data bank from the data reduction were optimized using a combined Genetic Algorithm (GA) and Partial Least-Squares (PLS) approach [19], implemented in the 4D-QSAR program [12]. Their optimizations were initiated using 10,000 randomly generated models, each having initially four variables. Mutation probability over the crossover optimization cycle was set at 100%. The smoothing factor, the variable that specifies the number of descriptors in the QSAR models, was varied in order to determine equations with no more than six terms. Each alignment was evaluated using the procedure described above. The best models resultant from the 4D-QSAR study were based on different criteria: a) the Leave-One-Out (LOO) cross-validated correlation coefficient, or q^2, currently usedas the preferred parameter of model fitness [20]; b) number of significant and independent 4D-QSAR models; c) indices of model significance including statistical measures such as r^2, Standard Error (SE), and Lack-Of-Fit (LOF) [19], and real prediction using the test set compounds (external validation).

2.2.4. 4 D-QSAR Model Calculations - Method 2

Analysis using OPS (ordered predictor selection) algorithm was applied as the variable selection method in the construction of the PLS models [21]. First, the method obtains an informative vector with length equal to the number of response variables and each position in the vector is aligned to the best corresponding response variables for prediction of the location of the best response. After that, the original response variables (X matrix columns) are differentiated according to the corresponding absolute values of the informative vector elements obtained previously. Then, multivariate regression models are built and evaluated using a cross validation strategy. Quality parameters of the models are obtained for every evaluation and stored for future comparison. In the last step, the evaluated descriptor sets are compared using the quality parameters calculated during validations. The model with the best quality parameters should contain descriptors with the best prediction capability. Specifically, the algorithm: a) Obtaining informative vectors or their combinations from X and Y; b) Building PLS regression models; c) Calculating quality parameters by leave-N-out cross validation and d) Comparing the quality parameters for the obtained models.

2.3. Analog Docking

For the docking procedure the Molegro Virtual Docker (MVD) program was used. Through this program the most likely conformation of the enzyme ligand can be determined. The ligand conformation is identified through the candidate number evaluations (ligands conformation) and energy estimation of their interaction with the enzyme [22].

All 6-benzylthoinosine analogs (Table **1**) and the compound N^6-benzyladenosine were prepared by using the MVD [2] program. At this phase, the ligand interaction modes identification is interactive. A number of solutions (ligand conformation and orientation) are evaluated. The energy and its interactions with the protein are estimated. The best solutions are then returned for further analysis.

The docking scoring function values E_{score} are defined by equation 1:

$$E_{score} = E_{inter} + E_{intra}$$

(equation 1)

Where:

$$E_{inter} = \sum_{i \in ligante} \sum_{j \in proteina} \left[E_{PLP}\left(r_{ij}\right) + 332.0\frac{q_i q_j}{4r_{ij}^2} \right]$$

(equation 2)

Eplp stands "piecewise linear potential", which uses two different parameter sets as follows: one for approximation of the steric term (van der Waals) among atoms, and other potential for the hydrogen connection. The second term describes the electrostatic interactions among overloaded atoms. It is a Coulomb potential with a dielectric constant which depends on the distance (D(r)=4r). The numerical value of 332.0 establishes the electrostatic energy units in kilocalories per molecule [22].

For better results, the flexibility of the residues which are close to 12 Å from the connector during the docking procedure for the active enzyme has been taken into account. The conformation of each compound was selected by using its highest degree of special similarity to adenosine obtained from the PDB, which was represented by the energy interaction between the ligand position and the respective protein.

3. RESULTS AND DISCUSSION

3.1. QSAR - Method 1

The GA-PLS analysis using grid cells of 1.0 Å generated several models or equations. However, the alignment 3 (atoms 15-12-4, Fig. **1**) was the only one that gave a model with a squared coefficient of linear correlation after cross-validation (q^2) higher than 0.5. According to these results, we will only present the analysis of the best model derived from alignment 3, Model 1 (Eq. 3). According Eq. (3), the Model 1, is composed by six GCODs descriptors, with both positive and negative contributions to the activity.

Model 1

pIC50 =4.93 - 6.68(-1,1,-3any) + 17.48 (-2,0,-4 any) + 7.07(1,6,2,any)

- 3.87(-2,2,0,any) -6.77(1,0,-6,np) - 1.39 (1,0,-5,3,pn)   (equation 3)

n=29; r2=0.71; SE=0.06; q2=0.62; LOF=0.14

Model 1 represents the best 4D-QSAR equation derived from 29 compounds of the training set using the method 1. The conventional squared linear correlation coefficient, R2, of the Model 1 is equal to 0.71. This means that the analyzed results have an average fitness compared to the *in vitro* test results. Table **2** shows the calculated pIC$_{50}$values for the training set. The statistical significance ofthe relationship between the biological response and the chemical structure descriptors was further demonstrated by a cross-validation analysis. Leave-one-out cross-validation (LOO-cv) analysis of the Model 1 had a q^2 value of 0.62 with standard error of 0.36. This means that Model 1has a predictive capacity of 62%. LOO-cv correlation co efficient values over 0.5 reveals that the model is a useful tool for predicting affinities for new compounds in this set. Therefore, an external validation was performed with the test set compounds.

Table 2: Observed and predicted pIC$_{50}$ values and residual values for the training (bold) and test set compounds using Models 1 and 2.

	Observed pIC$_{50}$	Model 1		Model 2	
		Predicted pIC$_{50}$	Residual	Predicted pIC$_{50}$	Residual
1	5.032	4.927	0.105	5.095	-0.063
2	4.423	4.519	-0.096	4.625	-0.202
3	5.086	5.040	0.046	4.922	0.163
4	4.975	4.926	0.049	4.985	-0.010
5	5.174	5.234	-0.060	5.076	0.097
6	4.824	4.930	-0.106	4.772	0.051
7	5.114	5.114	0.000	5.025	0.088
8	5.060	4.918	0.141	4.842	0.217
9	5.208	5.163	0.044	5.193	0.014
10	4.634	4.841	-0.207	4.839	-0.205
11	5.086	5.007	0.078	5.150	-0.064
12	5.071	4.852	0.218	4.907	0.163
13	5.060	4.968	0.091	5.065	-0.005
14	4.686	4.880	-0.194	4.772	-0.086
15	4.987	5.241	-0.254	5.253	-0.266

16	4.588	4.879	-0.291	4.843	-0.255
17	5.091	4.936	0.154	5.060	0.030
18	5.081	4.899	0.181	4.886	0.194
19	4.845	4.831	0.013	4.892	-0.047
20	4.640	4.549	0.090	4.610	0.029
21	5.367	5.121	0.245	5.307	0.059
22	4.979	4.971	0.007	5.035	-0.056
23	4.921	4.979	-0.058	5.130	-0.209
24	4.658	4.644	0.013	4.682	-0.024
25	5.108	5.232	-0.124	5.064	0.043
26	4.582	4.552	0.029	4.545	0.036
27	5.456	5.361	0.094	5.163	0.292
28	5.143	5.184	-0.041	4.956	0.186
29	4.606	4.691	-0.085	4.5527	0.053
30	5.086	4.978	0.107	5.034	0.051
31	4.633	4.636	-0.003	4.865	-0.232
32	5.032	4.925	0.106	4.942	0.089
33	4.824	4.932	-0.108	4.797	0.026
34	4.845	4.927	-0.082	5.018	-0.173
35	5.137	5.082	0.054	5.067	0.069
36	4.983	4.935	0.047	5.078	-0.095
37	4.676	4.929	-0.253	4.947	-0.271
38	5.060	5.000	0.059	4.995	0.064
39	5.022	4.848	0.173	4.840	0.181
40	4.507	4.497	0.010	4.514	-0.007
41	5.060	5.017	0.043	4.959	0.100

The GCODs (-2,0,-4 any) and (1,6,2,any), presenting positive coefficients (Eq. 3), correspond to favorable interactions between the molecule substituent and amino acid residues in the active site of tgAK. Therefore, substituents in these positions increase the potency of the compounds. The GCODs (-1,1,-3,any), (-2,2,0,any), (1,0,-5,pn) and (1,0,-6,np), presenting negative coefficients (Eq. 3), correspond to unfavorable interactions between the molecule substituent and amino acid residues in the active site of tgAK. Therefore, substituents in these positions decrease the potency.

The occupation of these GCODs is fundamental for the activity. In order to gain a better understanding of the behavior of the data fitted to the model, the cross-correlation matrix among the different GCODs are given in Table **3**. There is no correlation (r>0.7) between the GCODs.

Table 3: Cross-correlation coefficients of the GCODs of Model 1 obtained in the 4D-QSAR analysis, using alignment 3 and cubic grid cells of 1.0.

	(-1,1,-3any)	(-2,0,-4 any)	(1,6,2,any)	(-2,2,0,any)	(1,0,-6,np)	(1,0,-5,pn)
(-1,1,-3any)	1.0					
(-2,0,-4 any)	-0.015	1.0				
(1,6,2,any)	0.306	-0.135	1.0			
(-2,2,0,any)	0.527	-0.082	0.410	1.0		
(1,0,-6,np)	-0.006	0.519	-0.170	-0.069	1.0	
(1,0,-5,pn)	-0.004	-0.024	0.070	-0.073	0.062	1.0

3.2. QSAR - Method 2

The GCODs from alignment 3 were available though the OPS-PLS algorithm developed by Ferrera *et al.* [21]. The best model, Model 2, has 8 descriptors and the statistical significance of the relationship between the biological response and the chemical structure descriptors was further demonstrated by a cross-validation analysis. LOO-cv analysis of Model 2 (Eq. 4) had a q^2 value of 0.54 for the training set. This means that Model 2 has a predictive capacity of 54%. The conventional squared linear correlation coefficient of Model 2 is equal to 0.65. Table **2** shows the calculated pIC_{50}values for the training set. The cross-correlation matrix among the different GCODs is given in Table **4**. There is correlation(r>0.7) between two pairs of GCODs (0,6,2,4)-(0,4,2,np) and (0,1,-5,pp)-(0,-1,5,pp).

Model 2

IC50 = 5.061 − 2.72 (-2,2,0,any) − 1.07 (-1,0,-6,np) + 2.95 (0,6,2,any)

-2.71 (-3,4,8,np) -1.02 (0,1,-5,3) -2.09 (-2,2,5,np) + 1,15 (0,4,2,np)

− 0.77 (0,-1,5,pp)   (equation 4)

n=29; r2=0.65; SE=0.08; q2=0.54; LOF=0.36

Table 4: Cross-correlation coefficients of the GCODs of Model 2obtained in the 4D-QSAR analysis using alignment 3 and cubic grid cells of 1.0.

	(-2,2,0,any)	(-1,0,-6,np)	(0,6,2,any)	(-3,4,8,np)	(0,1,-5,pn)	(-2,2,5,np)	(0,4,2,np)	(0,-1,5,pn)
(-2,2,0,any)	1							
(-1,0,-6,np)	-0.053	1						
(0,6,2,any)	0.206	-0.046	1					
(-3,4,8,np)	0.002	0.019	-0.306	1				
(0,1,-5,pn)	-0.079	-0.082	0.034	-0.199	1			
(-2,2,5,np)	0.551	0.006	-0.037	0.087	-0.124	1		
(0,4,2,np)	0.174	-0,064	0.934	-0.341	0.045	-0.055	1	
(0,-1,5,pn)	-0.089	-0.088	0.053	-0.188	0.992	-0.133	0.060	1

The GCODs (0,6,2,hd) and (0,4,2,np), presenting positive coefficients (Eq. 4), correspond to favorable interactions between the molecule substituent and amino acid residues in the active site of tgAK. The GCODs (-2,2,0,any), (-1,0,-6,np), (-3,4,8,np), (0,1,-5,3), (-2,2,5,np) and (0,-1,5,pp), presenting negative coefficients (Eq. 1), correspond to unfavorable interactions between the molecule substituent and amino acid residues in the active site of tgAK.

3.3. Interpretation of the Models 1 and 2

Two methods were utilized to constructed QSAR models: GA/PLA and OPS/PLS. The first method resulted in a model (Model 1) which has a more predictive capacity (q2 = 0.62) than the model generated with the second method (q2= 0.54). A graphic representation of the 3D-pharmacophore embedded in 4D-QSAR Models 1 and 2 is shown in Fig. **2** using compound **27**(more potent) as a reference. Light and dark spheres represent GCODs with negative and positive coefficients, respectively, in accordance with Eq. 3 (Model 1) and 4 (Model 2). Each GCOD is labeled as "x,y,z,IPE" which represent the cartesian coordinate positions of the selected grid cell (x, y,z) and the respective atom type (IPE).

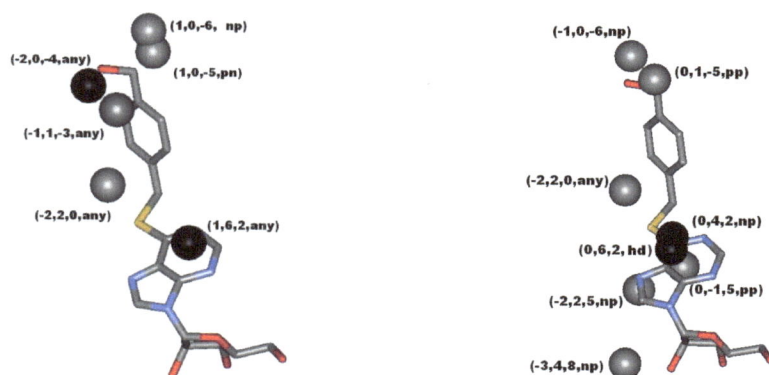

Figure 2: Models 1 (on the left) and 2 (on the right) obtained by 4D-QSAR (alignment 3 and cubic grid cells of 1.0) using compound **27** as a reference. Lightspheres indicate activity-decreasing pharmacophore sites and dark spheres indicate activity-enhancing pharmacophore sites. The hydrogen atoms are omitted for clarity. GCODs are labeled as (x,y,z,IPE) which means the cartesian coordinate positions of the selected grid cell (x,y,z) and the respective atom type (Interaction Pharmacophore Element, IPE).

GCOD (-3,4,8,np) (Fig. **2**) show the highest occupation frequency for all compounds. Since the coefficient of this GCOD is negative, non-polar substituent in this position is detrimental to the activity. GCOD (-2,2,5,np) also show the highest frequency of occupation for all compounds suggesting limitation in this region of the receptor for non-polar substituent.

GCODs (0,4,2,np), (1,6,2,any) and (0,6,2,any), which are highly interrelated (r>0.8), are located in almost the same region of space. The only difference between them is that the first GCOD represents the non-polar atom type and the second and third represents the other any atom types. These grid cells show the highest frequency of occupation for compound p-cyano-6-benzylthioinosine (IC$_{50}$ = 4.3 μM). Thus, non-polar substituents in this position increase the potency of the compounds.

GCODs (-1,0,-6,np), (0,1,-5,pp), (1,0,-6,np), (1,0,-5,pn), presents negative coefficients and represents a non-polar atom type, polar-positive charge density, non-polar, polar-negative charge density, respectively. As the potency of the compounds depends on the type of the substituent at these positions, atoms in aromatic systems are important in this site for the increase of the potency of these compounds.

3.4. Docking Study

The computer-based simulation of the molecular docking is one of the most important techniques to investigate the molecular interactions between the protein and the ligand in cases where the protein 3D structure has already been found [23]. Evaluation of the docking results was based on protein-ligand complementarity considering steric and electrostatic properties as well as calculated potential interaction energy in the complex and ligand intramolecular energy. We verified that orientation of the ribofuranosyl purine group of the ligand produced by the program was similar to the observed binding mode of adenosine in the crystal structure of tgAK. The super imposition of compounds **1** and adenosine is represented in Fig. **3**. Analyzing the hydrogen bond formed between the compounds and AK active site we observed: (i) all compounds interact with the residue Asp24, Asn73, Ser70, and Thr140; (ii) only the compounds **5, 9, 13, 33, 36** and **40** interact with the residue Gly-315; iii) the compounds **1, 5, 9, 13, 15, 21, 33, 35** and **40** interact with the residue Asp318; iv) the compounds **5, 9, 13, 15, 21, 23, 33, 35** and **36** interact with the residue Asn314; v) the compounds **1, 7, 11, 23, 25** and **33** interact with the residue Tyr-169; vi) the compounds **3, 7, 19** and **33** interact with the residue Gly69; vii) the compounds **21** and **23** interact with the residue Tyr172; viii) only the compound **1** interacts with the residue Gly69. Table 5 shows hydrogen bonding energy values between the inhibitors and AK. It was noted in this table that most of the compounds have a good interaction energy with the active site of the AK enzyme. As shown in fig. **3**, these compounds have a closeness between the oxygen of pentose and phosphate group of AMP-PCP, making them good candidates for the treatment of toxoplasmosis.

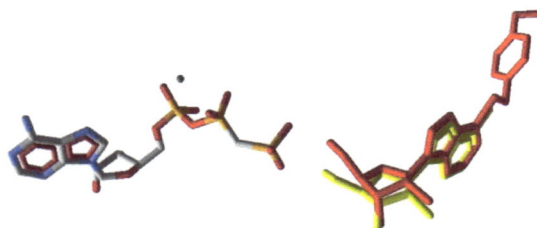

Figure 3: Structures of the Mg^{2+} (gray sphere), AMP-PCP (color by element), adenosine (yellow) and compound **1** (red) from docking study.

Table 5: pIC_{50} (pIC_{50} = -$logIC_{50}$(M)) and intermolecular interactionenergy (kcal mol^{-1}) values of compound/protein and binding of hydrogen (kcal mol^{-1}).

Compounds	pIC_{50}	Interaction Energy	Binding Hydrogen of Energy
1	5.03	-161.16	-18.05
5	5.17	-161.94	-20.00
7	5.11	-154.87	-21.13
9	5.21	-165.52	-23.72
11	5.09	-164.83	-16.93
13	5.06	-139.87	-12.40
15	4.99	-167.83	-23.25
17	5.09	-160.25	-18.26
19	4.84	-166.02	-20.36
21	5.37	-155.65	-11.76
23	4.92	-119.48	-7.53
25	5.11	-175.71	-20.13
27	5.46	-130.97	-12.67
31	4.63	-104.26	-11.12
33	4.82	-110.57	-22.49
35	5.13	-98.91	-5.90
37	4.98	-167.47	-21.77
38	5.06	-162.46	-15.57
40	4.50	-157.29	-14.37
N6-benzyladenosine	3.10	-141.26	-15.59

The compound N6-benzyladenosine was not additioned in the QSAR study because its IC_{50} value is 791 μM (out-side logarithmic scale used), but we studied the interaction of this compound into tgAK active site. It is important to mention the small structural difference (just one atom: sulfur by nitrogen) between the compound **1** and N6-benzyladenosine with a much smaller potency when compared to the other componds. N6-benzyladenosine and **1** interact with tgAK with -141.26 and -161.16 $kcalmol^{-1}$ of interaction, respectively. Fig. **4** shows the superimposition of the two structures after the docking study. It is possible to observe that there is just a conformational difference: between the substituted benzyl ring. Toward a better understanding about the activity of N6-benzyladenosine, we carried out NBO calculations on the conformation from docking studies.

Furthermore, using QM/MM methodology, we intended to evaluate the electronic contribution on interaction in the active site of the enzymes. Thus, the selected orientations for N6-benzyladenosine and compound **1** from docking calculations were then submitted to electrostatic charge calculations, the electrostatic charges were determined so as to reproduce the B3LYP/6-31G(d, p) quantum molecular Mechanical Electrostatic Potential (MEP). This means that it was necessary to produce charges that fit into the electrostatic potential at points selected according to the CHelpG scheme [24]. To get a picture of the chemical bonding in active site model structures from both compound **1** and N6-benzyladenosine, we undertook a detailed NBO analysis for each structure at its ground-state. That effect can be explained by means of second-order perturbation theory, which estimates the stabilization as the ratio between the square of the Fock matrix element and the energy difference between the interacting orbitals [25].

Figure 4: Structures of the Mg^{2+} (gray sphere), AMP-PCP (color by element), N6-benzyladenosine (yellow) and compound **1** (green) from docking study.

In fact, the participation of non-bonded electrons plays an important role in the stabilization of the complex between *T. gondii* adenosine kinase protein and inhibitor. An extra stabilization was observed due to n_{N1}/π^{*}_{C2-N3} and $\sigma_{N1-C2}/\pi^{*}_{C7-C6}$ interactions being 7.56 and 3.58 kcal mol^{-1}, respectively for N6-benzyladenosine. While for the compound **1** the values for n_{N1}/π^{*}_{C2-N3} and $\sigma_{N1-C2}/\pi^{*}_{C7-C6}$ interactions are 5.81 and 1.32 kcal mol^{-1}, respectively. Thus, the non-bonded interactions of N6-benzyladenosine are more effective in stabilizing the aromatic ring when compared with **1**. The volume of the *2p* orbitals of nitrogen is closer to the volume of the *2p* orbitals of carbon than to the *3d* orbitals of sulfur. The reason for this weaker interaction of second-row elements (carbon) with second-row elements (*e.g.*, N) as compared with third-row elements (*e.g.*, S) can be explained by means of second-order perturbation theory [26].

We believe that these regularities portray features of the electron distribution and indicate the preferred direction of the benzyl ring in N6-benzyladenosine, while in the compound **1** the benzyl ring is more flexible and gets to accommodate better in the AK active site. It should be kept in mind, then; the decrease of the potency of the N6-benzyladenosine is likely due to different benzyl ring orientation.

4. CONCLUSION

In the present study, we built and evaluated 4D-QSAR models for tgAK inhibitors, based on a series of benzyladenosione in analogues. The models generated from the best alignment (alignment 3) selected from the three trial alignments, reflect a mutual super position of the ligands along with their relative orientation towards the binding pocket. In this way, information about the geometry of the binding pocket was indirectly included in the models developed by the 4D-QSAR approach. In the best 4D-QSAR models, namely Model 1, from alignment 3 using a grid cell of 1.0 showed predictive capacity of 62%.

We used docking studies in order to understand the interaction of the series of benzyladenosine as AK. Docked structures were evaluated based on intermolecular/intramolecular energies and hydrogen bonding interactions and it was observed that the position of the hydroxyl group of the ribofuranosyl purine ring, next to the cofactor, is important for activity of the compounds well as the position of the benzyl ring. Studies employing NBO calculations presented that the compound N6-benzyladenosine is less potent than compound **1** because the non-bonded interactions of N6-benzyladenosine are more effective in stabilizing the ring aromatic when compared with **1**. Thus, the benzyl ring in compound **1** is more flexible and gets to accommodate better in the AK active site. Considering that the emergence of resistant strains of *T. gondii* poses a serious threat to the control of this disease, benzyladenosine analogs are possible new drug targets to combat toxoplasmosis.

ACKNOWLEDGEMENTS

We thank the Brazilian agencies Coordenação de Aperfeiçoamento de Pessoal de Nível Superior (CAPES), Conselho Nacional de Desenvolvimento Científico e Tecnológico (CNPq) and Fundação de Amparo à Pesquisa do Estado de Minas Gerais (FAPEMIG) for their support. We thank to Prof. A.J. Hopfinger who kindly supplied the 4D-QSAR program for academic use.

REFERENCES

[1]   El Kouni MH, Guarcello V, Al Safarjalani ON, Naguib FNM. Metabolism and selective toxicity of 6-Nitrobenzylthioinosine in *Toxoplasma gondii.* Antimicro Agents Chem 1999; 43: 2437.

[2]   Kim YA, Sharon A, Chu CK, Rais RH, Al Safarjalania ON, Naguib FNM, El Kouni MH. Synthesis, biological evaluation and molecular modeling studies of N^6-benzyladenosine analogues as potential anti-toxoplasma agents. Biochem Pharmacol 2007; 73: 1558.

[3]   Petersen E. Prevention and treatment of congenital toxoplasmosis. Expert Rev of Anti infect Ther 2007; 5: 285.

[4]   Ngo HM, Ngo EO, Bzik DJ, Joiner KA. *Toxoplasma gondii:* are host cell adenosine nucleotides a direct source for purine salvage? Exp Parasit 2000; 95: 148.

[5]   Reddy MCM, Palaninathan SK, Shetty ND, Owen JL, Watson MD, Sacchettini JC. High resolution crystal structures of *Mycobacterium tuberculosis* adenosine kinase: insights into the mechanism and specificity of this novel prokaryotic enzyme. J Biol Chem 2007; 282: 27334.

[6]   El Kouni MH. Potential chemotherapeutic targets in the purine metabolism of parasites. Pharmacol Therap 2003; 99: 283.

[7]   El Kouni MH, Guarcello V, Al Safarjalani ON, Naguib FNM. Metabolism and selective toxicity of 6-nitrobenzylthioinosine in *Toxoplasma gondii.* Antimicro Agents Chem 1999; 43: 2437.

[8]   Al Safarjalani ON, Naguib FNM, EL Kouni MH. Uptake of nitrobenzylthioinosine and purine beta-L-nucleosides by intracellular *Toxoplasma gondii.* Antimicro Agents Chem 2003; 47: 3247.

[9]   Yadav V, Chu CK, Rais RH, Al Safarjalani ON, Guarcello VG, Naguib FNM, El Kouni MH. Synthesis, biological activity and molecular modeling of 6-benzylthioinosine analogues as subversive substrates of *Toxoplasma gondii* adenosine kinase. J Med Chem 2004; 47: 1987.

[10]   Rais HR, Al Safarjalani ON, Yadav V, Guarcello V, Kirk M, Chu CK, Naguib FNM, El Kouni MH. 6-Benzylthionosine analogues as subversive substrate of *Toxoplasma gondii* adenosine kinase: activities and selective toxicities. Bio chem Pharmacol 2005; 69: 1409.

[11]   Designed by MakoLab (C) 1985-2007, Hypercube, Inc. All Rights Reserved.

[12]   4D-QSAR User's Manual v.1.00. The Chem21 Group Inc., 1780 Wilson Dr., Lake forest, IL 60045, 1997.

[13]   MOLSIM User'sGuide v.3.0. Doherty DC & The Chem 21 Group Inc., 1780 Wilson Dr., Lake Forest, IL 60045, 1997.

[14]   Weiner SJ, Kollman PA, Nguyen DT. An all atom force field for simulations of proteins and nucleic acids. J Comp Chem 1986; 7: 230.

[15]   da Cunha EFF, Albuquerque MG, Antunes OAC, Alencastro RB. 4D-QSAR models of HOE/BAY-793 analogues as HIV-1 protease inhibitors. QSAR Comb Sci 2005; 24: 240.

[16]   da Cunha EFF, Martins RCA, Albuquerque MG, Alencastro RB. LIV-3D-QSAR model for estrogen receptor ligands. J Mol Model 2004; 10: 297.

[17]   da Cunha EFF, Sippl W, Ramalho CT, Antunes OAC, Alencastro RB, Albuquerque MG. 3D-QSAR CoMFA/CoMSIA models based on theoretical active conformersof HOE/BAY-793 analogs derived from HIV-1 protease inhibitor complexes. Eur J Med Chem 2009; 44: 4344.

[18]   Hopfinger AJ, Wang S, Tokarski JS, Jin B, Albuquerque M, Madhav PJ, Duraiswami C. Construction of 3D-QSAR models using the 4D-QSAR analysis formalism. J Am Chem Soc 1997; 119: 10509.

[19]   Rogers D, Hopfinger AJ. Application of genetic function approximation (GFA) to quantitativestructure-activity relationships. J Chem Inf Comput Sci 1994; 34: 854.

[20]   Senese CL, Hopfinger AJ. A simple clustering technique to improve qsar model selection andpredictivity: application to a receptor independent 4D-QSAR analysis of cyclic urea derived inhibitors of HIV-1 protease. J Chem Inf Comput Sci 2003; 43: 2180.

[21]   Martins JPA, Barbosa EG, Pasqualoto KFM, Ferreira MMC. LQTA-QSAR: A New 4D-QSAR Methodology. J Chem Inf Model 2009; 49: 1428.

[22]   Thomsen R, Christensen MH. MolDock: A new technique for high-accuracy molecular docking. J Med Chem 2006; 49: 3315.

[23]   Silveira NJF, 117f. (Tese de Doutorado) Universidade Estadual Paulista, UNESP. Brazil, 2005.

[24]   Singh UC, Kollman PA. An approach to computing electrostatic charges for molecules. J Comp Chem 1984; 5: 129.

[25]   Wang P, Zhang YL, Glaser R, Reed AE, Schleyer PVR, Streitwieser A. The effects of the first- and second-row substituents on the structures and energies of PH4X phosphoranes. An ab initio study. J Am Chem Soc 1991; 113: 55.

[26]   Josa D, da Cunha EFF, Ramalho TC, Souza TCS, Caetano MS. Hypothesis paper: homology modeling of wild-type, d516v, and h526l *Mycobacterium tuberculosis* RNA polymerase and their molecular docking study with inhibitors. J Biomol Strct Dyn 2008; 25: 373.

CHAPTER 9

An Overview of Tropical Parasitic Diseases: Causative Agents, Targets and Drugs

Carlton Anthony Taft[1*] and Carlos Henrique Tomich de Paula da Silva[2]

[1]Centro Brasileiro de Pesquisas Físicas, Rua Dr. Xavier Sigaud, 150, Urca, 22290-180, Rio de Janeiro, Brazil and [2]School of Pharmaceutical Sciences of Ribeirão Preto, University of São Paulo, 14040-903, Ribeirão Preto, Brazil

Abstract: There is a need for improved treatments for diseases including protozoan parasitic diseases, such as chagas, malaria, trypanosomiasis, leishmaniasis, onchocerciasis, schistosomiasis, filariasis, toxoplasmosis, cryptosporidiosis, filardiasis and giardiasis. The existing chemotherapy for these diseases is not sufficiently effective due to factors such as toxicity, drug resistance and different strain sensitivities. New chemotherapies are needed to help in the control and prevention of these parasitic diseases.

Keywords: Tropical diseases, causative agents, drug targets.

1. INTRODUCTION

Despite the difference in visibility and epidemiology, most of the diseases (Chagas, malaria, trypanosomiasis, leishmaniasis, onchocerciasis, schistosomiasis, filariasis, toxoplasmosis, cryptosporidiosis, filardiasis and giardiasis) discussed in this work share a similar history of strategies for their treatment and control [1-66].

Although vaccine development has been imperative for decades, it is more difficult today due to the large degree of antigenic variations exhibited by these parasites. Chemotherapy remains the treatment option for controlling the infection.

We now proceed to make a brief introduction of these parasitic diseases.

1.1. Chagas' Disease

Chagas disease is an insect-born parasitic disease threatening millions of lives. Due to migration (insect vectors and mammalian hosts), blood transfusion, HIV co-infection and organ transplantation, the disease is spreading worldwide. Chagas is an infectious disease caused by a parasite, which was named by Chagas after his mentor, parasitologist Oswaldo Cruz. Endemic in several Latin American countries, cases are found from Mexico to Argentina, including parts of the Caribbean. In 2007 in the United States one in 4,700 blood donors tested positive for Chagas disease. During the early stages of the disease, the victims do not experience specific symptoms. Depending on their immune status, the disease can either be deadly or pass unnoticed. The disease is commonly fatal in the chronic stage. After 10-20 years, about 40% of infected individuals suffer serious intestinal and/or cardiac symptoms. For immunocompromised patients, the probability increases up to 70%. Transmission can also occur *via* placenta or by blood transfusion [2, 3, 6, 8, 11, 15, 17, 19, 23, 24, 38].

1.2. Leishmaniasis

The leishmaniasis has been considered a tropical affliction that constitutes one of the six entities on the World Health Organization (WHO) tropical disease research list of the most important diseases occurring in 88 countries of temperate and tropical regions. An estimated 350 million population is at risk and 10 million people are affected from this disease worldwide. The annual global burden of Visceral Leishmaniasis (VL) is about 500,000. The risk of VL among AIDS patients increases by 100 - 1000 times in endemic areas, while VL accelerates the onset of AIDS in HIV infected people [8-10, 16, 25, 26, 40].

*Address correspondence to Carlton Anthony Taft; Centro Brasileiro de Pesquisas Físicas, Rua Dr. Xavier Sigaud, 150, Urca, 22290-180, Rio de Janeiro, Brazil. E-mail: catff@terra.com.br

Teodorico C. Ramalho, Matheus P. Freitas and Elaine F. F. da Cunha (Eds)

1.3. Malaria

Malaria has affected humans since their evolutionary emergence. The earliest reports on malaria refers to plenomegaly with fever from China in the Nei Ching Canon of Medicine in 1700 B.C and from Ancient Egypt in the Ebers Plapyrus in 1570 B.C [32]. The tropical disease malaria, with an annual death toll of more than one million people, is considered one of the most significant infectious diseases worldwide. It is one of the main global causes of death from infectious diseases. About 40% of the world population is at risk from malaria, causing around 1 million deaths each year, predominantly in infants. Malaria has a broad distribution in both subtropical and tropical regions. The countries of sub-Saharan Africa carry the highest burden [12-14, 27, 29, 32, 34, 36, 39, 41, 45].

1.4. Trypanosomiasis

Human African trypanosomiasis or sleeping sickness (HAT) is caused by two kinetoplastid flagellates. African trypanosomes are among the most ancient eukaryotic organism known. The number of new cases yearly is around 500,000 while 60 million individuals are at risk. Sleeping sickness is fatal if untreated and has re-emerged in recent years. Most of the available drugs are toxic, with lack of efficacy and parasite resistance [11, 18, 23].

1.5. Cryptosporidiosis

Cryptosporidiosis is a parasitic disease which affects the intestines of mammals and is typically an acute short-term infection. It is spread through the fecal-oral route, often through contaminated water. The main symptom is self-limiting diarrhea in people with intact immune systems. In immunocompromised individuals, such as AIDS patients, the symptoms are particularly severe and often fatal. It is one of the most common waterborne diseases [21, 28].

1.6. Filariasis

Filariasis (Philariasis) is a parasitic and infectious tropical disease that is caused by worms. Lymphatic Filariasis puts at risk more than a billion people in more than 80 countries. Over 120 million have already been affected and over 40 million of them are seriously incapacitated and disfigured by the disease [47].

1.7. Toxoplasmosis

Toxoplasmosis is a parasitic disease which infects genera of warm-blooded animals, including humans, but the primary host is the feline (cat) family. Animals are infected by eating infected meat, by ingestion of feces of a cat that has itself recently been infected, or by transmission from mother to fetus. Although cats are often blamed for spreading toxoplasmosis, contact with raw meat is a more significant source of human infections in many countries.

Up to one third of the world's human population is estimated to carry a Toxoplasma infection. During the first few weeks, the infection typically causes a mild flu-like illness or no illness. After the first few weeks of infection have passed, the parasite rarely causes any symptoms in otherwise healthy adults. However, people with a weakened immune system, such as those infected with advanced HIV disease or those who are pregnant, may become seriously ill, and it can occasionally be fatal. The parasite can cause encephalitis (inflammation of the brain) and neurologic diseases and can affect the heart, liver, and eyes (chorioretinitis) [48].

1.8. Schistosomiasis

Schistosomiasis is a parasitic disease and although with a low mortality rate, can damage internal organs and in children, impair growth and cognitive development. The urinary form of schistosomiasis is associated with increased risks for bladder cancer in adults. It is the second most socioeconomically devastating parasitic disease after malaria [47].

1.9. Onchocerciasis

Onchocerciasis, also known as river blindness and Robles' Disease, is the world's second leading infectious cause of blindness [47].

1.10. Giardiasis

Giardiasis infects over 2.5 million people annually. In humans, it is caused by the infection of the small intestine by a single-celled organism and occurs worldwide with a prevalence of 20-30% in developing countries. It has a wide range of human and other mammalian hosts making it very difficult to eliminate [46].

1.11. Amebiasis

Amebiasis is an intestinal illness that is typically transmitted when someone eats or drinks something contaminated with a microscopic parasite. In many cases, the parasite lives in a person's large intestine without causing any symptoms. But sometimes, it invades the lining of the large intestine, causing bloody diarrhea, stomachaches, cramping, nausea, loss of appetite, or fever. In rare cases, it can spread into other organs such as the liver, lungs, and brain [47].

2. CAUSATIVE AGENTS

2.1. *Trypanosoma cruzi*

The hemoflagellate protozoan *T. cruzi* is the causative agent of the Chagas's disease, which is transmitted to humans either by blood-sucking reduviid bugs or directly by the transfusion of infected blood. After 10-20 years, a chronic form of the disease often develops causing irreversible damage to colon, heart and esophagus, which can lead to death. In rural areas, the disease is transmitted by bugs such as *Rhodnius prolixus* and *Triatoma infestans*.

T. cruzi multiplies within the insect gut as an epimastigote form and is spread as a non-dividing metacycle trypomastigote from the insect excrements by contamination of intact mucosa or wounds produced by the blood-sucking activity of the vector. In the mammalian host, it proliferates intracellularly in the amastigote form and releases into the blood stream as a non-dividing infective trypomastigote [11, 47].

2.2. *Trypanosoma brucei*

These parasites are the cause of human *African trypanosomiasis*, *i.e.*, sleeping sickness, which is fatal if untreated. The causative agents are phenotypically and morphologically similar. *T. brucei* is characterized by the presence of a DNA-containing organelle in the mitochondrion. It also contains a cell membrane (complex grid of microfilaments and microtubules). The outer surface is covered with a single glycoprotein. The parasite has the capability to synthesize aminoacids, and polyamines, compounds essential for proliferation and differentiation of the bloodstream stages. The main thiol is trypanothione, whose metabolism plays an important role in trypanosomal survival.

T. brucei is transmitted by species of tsetse flies. The insect vector takes a blood meal from mammals injecting metacyclic trypomastigotes, which transform into bloodstream trypomastigotes, transported to other sites and profilerating by binary fission to different body fluids such as blood, lymph and spinal fluid. The procyclic cell acts as a prototype for the basic architecture of *T. brucei*. Regulatory pathways controlling the eukaryotic cell cycle involve regulatory proteins such as cyclin dependent kinases and inhibitors possessing distinct roles in the different life cycle stages [4, 11, 47].

2.3. *Protozoan leishmania*

Leishmaniasis is caused by 20 species of Leishmania (order of Kinetoplastida, family Trypanosomatidae) and transmitted by 30 species of sand fly. Leishmania species are divided into species by geographic location of endemic species. The parasite is carried by the female phlebotamine sandfly of the genus *phlebotamus* or *Lutzomyia*. Only two Leishmania species can maintain anthroponotic human-human cycle, *i.e.*, *L. donovani*, and *L. tropica*.

In the mammalian host, the parasite proliferates as amastigotes intracelluarly within a phagolysosome compartment of macrophages. The extracellular flagellated promastigotes proliferates in the gut of their sand fly vectors, where they multiply and differentiate before emerging as an infectious species. The spectrum of disease can be divided in cutaneous (CL), mucocutaneous (MCL) and visceral leishmaniasis (VL). *L chagasi* is the most important agent for visceral leishmaniasis [11, 25, 47].

2.4. Plasmodium

Malaria is caused by Apicomplexa parasites of the genus Plasmodium which is transmitted by female Anopheles mosquito. They alternate during their life cycle, between the insect vector and the human host, requiring rapid adaptation to the new environment in order to coexist with the respective host. Infection is initiated with the bite of a mosquito and the injection of sporozoites from the salivary glands into the host's bloodstream. The sporozoites invade hepatocytes and undergo an asexual replication that results in the production of schizonts. The released merozoites invade red blood cells and initiate another asexual replication phase in the erythrocyte. Three different classes of enzymes present in the parasite food vacuole are responsible for hemoglobin degradation. Among these plasmepsins are the key enzymes to initiate the proteolylsis of hemoglobin. Among the numerous Plasmodium species that infect reptiles, birds and mammals, four of them are human-specific: *P. falciparum, P. vivax, P. malariae and P. ovale.* The most virulent agent is *P. falciparum* [29, 47].

2.5. *Cryptosporidium parvium*

C. parvium is the causative agent of cryptosporidiosis. Knowledge obtained from studying parasites closely related to *C. parvum*, such as other coccidian (notably *Toxoplasma* and *Eimeria*) or, more generally, other apicomplexa (especially the extensively studied malaria parasite *P. falciparum*) has been used to guide investigations of *Cryptosporidium* itself. The feeding organelle appears to play an important part in the parasite's well-being in a mammal. The infective sporozoite stage contains the organelles typically associated with the apical complex. These structures seem to enable the infective stage of the parasite to enter a host cell and modify the intracellular environment to make it suitable for the parasite's proliferation. There is some evidence for the involvement of both proteinases and sialidases in cell entry and penetration of the mucous layer [21, 47].

Cryptosporidium is the organism most commonly isolated in HIV positive patients presenting diarrhea. The parasite is transmitted by environmentally hardy microbial cysts (oocysts) that, once ingested, exist in the small intestine and result in an infection of intestinal epithelial tissue [28].

2.6. *Toxoplasma gondii*

Toxoplasmosis is a parasitic disease caused by the protozoan *T. gondii* (a species of parasitic protozoa in the genus Toxoplasma). They are parasitic sporozoans that can cause disease in mammals [46-48].

2.7. *Onchocerca volvulus*

Onchocerciasis is caused by *O. volvulus*, a nematode that can live for up to fifteen years in the human body. It is transmitted to humans through the bite of a black fly. The worms spread throughout the body, and when they die, they cause intense itching and a strong immune system response that can destroy nearby tissue, such as the eye [47].

2.8. Schistosoma

Schistosomiasis (also known as bilharzia, bilharziosis or snail fever) is a parasitic disease caused by several species of fluke of the genus *Schistosoma* [47].

2.9. Filariae

Filariasis is caused by thread-like filarial nematode worms in the superfamily Filarioidea, also known as "*filariae*". There are 9 known filarial nematodes which use humans as the definitive host. These are divided into 3 groups according to the niche within the body that they occupy: Lymphatic Filariasis, Subcutaneous Filariasis, and Serous Cavity Filariasis. Lymphatic Filariasis is caused by the worms *Wuchereria bancrofti, Brugia malayi, and Brugia timori.* These worms occupy the lymphatic system, including the lymph nodes, and in chronic cases these worms lead to the disease *Elephantiasis.* Subcutaneous Filariasis is caused by Loa loa (the African eye worm), *Mansonella streptocerca, O. volvulus, and Dracunculus medinensis* (the guinea worm). These worms occupy the subcutaneous layer of the skin, the fat layer. Serous Cavity Filariasis is caused by the worms which occupy the serous cavity of the abdomen. In all cases, the transmitting vectors are either blood sucking insects (fly or mosquito) or Copepod crustaceans in the case of *Dracunculus medinensis.* The worms live almost exclusively in humans and lodge in the lymphatic system, the network of nodes and vessels that maintain the delicate fluid balance between the tissues and blood and are an essential component for the body's immune system defense. They live for 4-6 years, producing millions of immature microfilariae (minute larvae) that circulate in the blood [47].

2.10. *Entamoeba histolytica*

Amebiasis is transmitted by a microscopic parasite called *E. histolytica.* The parasite is an amoeba, a single-celled organism [46-47].

2.11. *Giardia lamblia*

Giardiasis in humans is transmitted by a single-celled organism called *G. lamblia* [46-47].

3. DRUG TARGETS

Trypanosomatids are relatively early branching eukaryotic organisms and their cell organization differs considerably from that of mammals. Biochemical pathways present in trypanosomatids and absent from their hosts should thus provide excellent targets for rational drug design. However, there is a very different biology of the parasites inside their hosts. South American trypanosomes live in the cytosol of a variety of cell types. African trypanosomes live in the bloodstream and cerebrospinal fluid. Leishmania species live within phagolysomes of macrophages.

The route to drug target identification has usually been made through comparative genetic and biochemical studies. Genes and proteins identified in parasites and known to be absent from, or strikingly different in, the mammalian host were considered ideal targets. Target validation is a key step of any rational drug design program. The validation of a given target involves the gene knockout approach of investigating whether a specific protein is or is not essential for the parasite's survival and therefore a suitable drug target [8].

A major breakthrough in identifying new drug targets came with recent publication of the 'Tritryp' (*T. brucei, T. cruzi* and *L. major*) genome, which revealed that each genome contains 8300-12000 protein-coding genes, of which approximately 6500 are common to all three genomes. It has lead to the postgenomic era for trypanosomatid drug discovery. In addition, the sequence of the genomes of two other species of Leishmania (*L. infantum and L. braziliensis*) was reported in 2007 [50-52].

In the following sections we briefly discuss sterol biosynthesis, squaline synthase, protein prenyhlation, trypanotione pathway, trypanosome carboxpeptidases, superoxide dismutases, superoxide dismutases, glutathione biosynthesis, spermidene biosynthesis, trypanothione biosynthesis, glutationylspermidine and trhypanothione synthase, dihydrofolate reductase, lactate dehydrogenase, proteases, transporters, dpentose phosphate pathway, glycolytic enzymes, phosphoribosyltransferases, tubulin, folate biosynthesis, serine oligopeptidase, pyrimidine biosynthesis, gametocyte, pteridine reductase, merzoite surface protein, apicoplast targets, mitochondrial targets, membrane targets, hematin, shikimate pathway, redox systems, polyamine biosynthesis, nucleic acid metabolism, purine salvage, cyclin-dependent protein kinases, farnesyl transferase, helicases, erythrocyte G protein, isoprenoid biosynthesis, fatty acid biosynthesis, electron transfer chain and alternative oxidase, kinetoplast DNA replication machinery, histone deacetylases, cell cycle, glycogen synthase kinase, carbohydrate biosynthesis, proteasomes, trans-sialidase, mannitol cycle, lysosomes, epigenetic codes, really big targets: the parasitic worms, whole-organism screening, RNA interference and using proteomics and transcritomics.

3.1. Sterol Biosynthesis

The final product in mammalian hosts is ergosterol rather than cholesterol. Depletion of endogenous sterols such as ergosterol produces growth inhibition of the parasite whereas the metabolites cannot be replaced by cholesterol. Selective inhibition of a crucial enzyme involved in the sterol biosynthesis of the parasite will impair *T. cruzi* (TC) proliferation.

Recent attention has focused on sterol 14α-demethylase (cyp51). In TC, this cytochrome P450 enzyme catalyzes a three-step reaction of oxidative removal of the 14α-methyl group from 24-methylene dihydrolanosterol. Inhibition of sterol biosynthesis in TC is becoming one of the promising directions in the development of anti-Chagasic drugs [17].

3.2. Squaline Synthase

Another interesting target to control diseases associated with infections caused by pathogenic parasites is squalene synthase (SQS). This enzyme catalyzes the first commited step in sterol biosynthesis by coupling two molecules of farnesyl phosphate to form squalene. The accumulated isoprenoids, such as farnesyls pyrophosphate and precursors, are readily metabolized with no toxic effects. An effective inhibition of mammalian SQS can impair both *T. cruzi* (epimastigotes) and *L. mexicana* (promastigotes) growth [11].

3.3. Protein Prenylation

Protein prenylation is an attractive target for drug design in many tropical parasitic diseases. Protein prenylation is responsible for the attachment of farnesyl and geranylgeranyl units to the C-terminal cisteine residues of several proteins giving rise to proteins that are important signaling molecules involved in key cell processes for osteoclasts function. The attached prenyl groups play a significant role in anchoring proteins to membranes and also take action in protein-protein interactions [11].

3.4. Trypanothione Pathway

The uniqueness of trypanothione metabolite and its biosynthetic pathway in parasites of the order Kinetoplastida confers a great usefulness as a molecular target to the involved enzymes of its biosynthesis. There is an opportunity to design highly selective antiparasitic drugs against leishmaniasis and trypanosomiasis without toxic effects due to the absence of this metabolite in the host mammalian cells [18,11].

Trypanothione has functions in parasite metabolism such as cellular redox equilibrium, the mechanism of defense against xenobiotics and the antioxidant defense. Aerobic organisms need to regulate their cellular redox equilibrium whereas some low molecular weight thiols switch between oxidized and reduced forms of these sulfur-containing species. The most important function in glutathione is to avoid disulfide formation in intracellular proteins [11].

3.5. Trypanosoma Carboxypeptidases

Trypanosoma carboxypeptidases have been proposed as potential drug targets for treatment of the Chagas disease, since eukaryote homologues of carboxypeptidases were discovered in the protozoan *T. cruzi* [44].

3.6. Superoxide Dismutases

Many aerobic organisms, such as trypanosomatids, are exposed to highly reactive oxygen species such as O_2^-, H_2O_2 and HO. The defense against radicals is to trap them by low molecular weight molecules such as vitamins or thiols, in particular glutathione or trypanothione. Superoxide dismutases (SOD) have been identified in *T. cruzi* and other parasites. Overexpression of Fe-SOD in *T. cruzi* yields parasites more susceptible to redox drugs due to their inability to detoxify H_2O_2 generated from O_2 [11].

3.7. Glutathione Biosynthesis

Glutathione reductase activity is also not found in trypanosomes. It has been suggested that trypanothione plays the same role in parasites as glutathione does in mammals. Glutathione biosynthesis is catalyzed by enzymes that require Mg^{2+} as cofactor. There are differences between *T. brucei* and the mammalian enzyme in the binding pocket [11].

3.8. Spermidine Biosynthesis

Spermidine biosynthesis is also an interesting target. Polyamine spermidine is essential for normal proliferation of various parasites [11].

3.9. Trypanothione Biosynthesis

The last two steps of trypanothione biosynthesis are the most relevant in terms of molecular targets, because they involve enzymes that have no counterpart in mammals. Trypanothione Reductase (TR) is a crucial and distinctive enzyme in trypanosomatids. However, TR has a counterpart in mammals: Glutathione Reductase (GR). Regarding active site and substrate specificity for TR and GR, it has been suggested that the most important causes of substrate specificity are attributed to the width and charge distribution at the substrate binding site [11].

3.10. Glutathionylspermidine Synthase

Glutathionylspermidine is also an interesting target. Different catalytic domains are present in the enzyme [11].

3.11. Dihydrofolate Reductase

Dihydrofolate Reductase (DHFR) catalyses the reduction of dihydrofolate to tetrahydrofolate, which is methylated to form methylene tetrahydrofolate, *i.e.* a vital cofactor to convert deoxyuridine monosphosphate into thymidine monosphosphate. DHFR constitutes an interesting target for drug design, because DHFR inhibition prevents biosynthesis of thymidine, leading to cell death. The leishmanial and trypanosomal DHFR and the corresponding human enzyme show structural differences. Consequently, a selective inhibition of leishmanial and trypanosomal DHFR would lead to growth impairment of the parasite, because these microorganisms do not have a transport mechanism of this cofactor from the host [25, 28, 41, 49].

3.12. Lactate Dehydrogenase

Parasites such as malaria rely upon anaerobic glucose metabolism for their energy needs during their erythrocytic stages, consuming glucose at higher rates than mammalian cells. Oxidation of accumulated reduced nicotinamide adenine dinucleotide (NADH) is necessary for the process to continue. Plasmodium lactate dehydrogenase (pLDH) converts NADH to NAD^+ through the reduction of pyruvate to *L*-lactic acid. Lactate dehydrogenase thus appears to be an attractive drug target for parasites such as malaria [18].

3.13. Proteases

Proteases are considered attractive targets for drug discovery due to their roles in critical biological processes, such as hormone production, blood coagulation, complement activation and cell invasion. Proteases also play important roles in metabolism, survival, replication and pathology of parasites and, therefore, may be excellent targets for development of antiparasitic agents. Cysteine proteases have been detected in most parasitic protozoans. Early work indicated that trypanosomes were susceptible to protease inhibitors. The cloning and overexpression of cruzain, the major cysteine protease activity in *T. cruzi*, paved the way for structure-based drug discovery effort based on this target. The cellular effects of cysteine protease inhibitors on *T. cruzi* suggest that inhibition of cruzain has a role in parasite killing. Cysteine protease inhibitors are also effective against African trypanosomes. Proteases appear to be critical to the survival and growth of *Plasmodium* and have thus been the target of intense antimalarial drug discovery efforts. Targeting proteases in Leishmania is also of interest.

Malaria parasites degrade between 60-80% of the hemoglobin during the intraerythrocytic cycle. The amino acids derived from hemoglobin digestion are incorporated into parasite proteins and may also be important for parasite energy metabolism. Hemoglobin degradation begins within the parasite digestive vacuole and appears to be mediated by the combined action of cysteine and aspartic proteases. Inhibition of these targets is of interest to drug discovery [4, 8, 12, 14, 37, 46].

3.14. Transporters

Pathogenic protozoan parasites often scavenge host molecules and are sometimes totally dependent for their metabolic needs with high affinity transporters for the procurement of essential host-derived molecules. If the transporters possess higher catalytic efficiencies than their mammalian counterparts, the transporters could potentially be exploited to ferry toxic molecules in the parasite. Characterizing the structural requirements for uptake of drug candidates by parasites is of interest for new targets [11, 12, 14, 18].

3.15. Pentose Phosphate Pathway

The pentose phosphate pathway synthesis generates NADPH, a required cofactor for trypanothione reductase. Inhibition of this pathway may be a reasonable drug discovery strategy [18].

3.16. Glycolytic Enzymes

The bloodstream-form African trypanosome relies on glycolysis process for its energy requirements. Glycolytic enzymes are thus attractive targets for chemotherapeutic intervention against these parasites. *T. brucei* glyceraldehydes-3-phosphate dehydrogenase (GAPDH), an enzyme in the glycolytic pathway, should also be a good antitrypanosomal drug target [18, 25].

3.17. Phosphoribosyltransferases

In general, parasitic protozoa need to salvage purines from the host. In Leishmania and trypanosomes, purines are converted to nucleoside monophosphates by reactions including phosphoribosyltransferases. Since purine salvage is essential for parasites and apparently not for the human host, targeting phosphoribosyltransferases in the kinetoplastids appears to be a good strategy for drug discovery [18].

3.18. Tubulin

Tubulin is a protein that assembles into tube like polymers under proper conditions playing a role in various cellular processes including the segregation of chromosomes at mitosis and the maintenance of cellular morphology. Tubulin is a proposed cellular target of several clinically useful agents including parasitic drugs. However, there are differences between the drug susceptibility of mammalian tubulin and parasite tubulin. Determination of antimicrotubule agents against the parasite protein is of interest [18, 25].

3.19 Folate Biosynthesis

Some potential targets include biosynthesis of reduced folate cofactors, *T. brucei* glycolytic enzyme, phosphoglycerate kinase, topoisomerase-mediated lesions in parasite, glycosylphosphatidylinositol anchor of the African trypanosome`s surface coat and myristate-metabolizing enzymes in *T. brucei and falcipain* [18, 28].

3.20. Serine Oligopeptidase

A cytosolic serine oligopeptidase from *T. brucei* (OP-Tb) was isolated. Since high intracellular pentamidine concentrations have been measured in *T. brucei*, inhibition of OP-Tb could be important to the action mechanism of drugs [18].

3.21. Pyrimidine Biosynthesis

Malaria parasites are unable to acquire pyrimidines through salvage and thus depend entirely on *de novo* pyrimidine synthesis. Differences also exist between pyrimidine biosynthesis in plasmodium and mammalian cells. Studies of pyrimidine biosynthesis in plasmodium, in particular the expression and characterization of enzymes in this pathway, could be productive approaches toward discovering new antimalarials [4, 18, 28].

3.22. Gametocyte

Within the last two decades, the increase of drug resistance in malaria parasites has forced researchers to broaden their tactics to combat the disease including transmission blocking strategies aimed at the sexual stages.

These strategies on the level of either drugs or vaccines are designed to disrupt parasite reproduction and further development in the mosquito midgut, thus breaking the life cycle of the parasite. In recent years, knowledge on the malaria sexual phase has benefited from a dramatic resurgence provided by proteomic, microarray and annotation projects that arose out of the genome sequence projects for the multiple malaria species. As a result, a number of new sexual stage antigens have been identified and progress has been made in the identification and functional characterization of enzymes and regulatory proteins that are involved in gametocyte differentiation and fertilization. The uptake by the blood-feeding mosquito triggers important molecular and cellular changes in the gametocytes, thus allowing the rapid adjustment of the parasite from the warm-blooded host to the insect host and subsequently initiating reproduction [36].

3.23. Pteridine Reductase

Pteridine Reductase (PTR1) is a target for drug development against *Trypanosoma* and *Leishmania* species. In mammals, biopterin and reduced derivatives are cofactors for aromatic amino acid hydroxylations, the biosynthesis of neurotransmitters and nitric oxide signaling, and oxidation of glycerol ethers. Biopterins are essential for metacyclogenesis and implicated in resistance to reactive oxygen and nitrogen species in *Leishmania*.

A mechanism that contributes to trypanosomatid resistance to typical antifolates is amplification of the gene encoding the NADPH-dependent pteridine reductase 1 (PTR1), an enzyme unique to these parasites. It is suggested that dual DHFR-PTR1 inhibition may provide a successful treatment for trypanosomatid infections. Three scaffolds were identified to support the design of new PTR1 inhibitors [9].

3.24. Merozoite Surface Protein

Merozoite Surface Protein 1 (MSP1) of the malaria parasite *P. falciparum* is an important vaccine candidate antigen. Antibodies specific for the C-terminal maturation product, $MSP1_{19}$ has been shown to inhibit erythrocyte invasion and parasite growth [22].

3.25. Apicoplast Targets

The apicoplast is a chloroplast-like organelle of apicomplexan parasites. The *P. falciparum* apicoplast genome is much smaller than its plastid ancestor. Most of the proteins of this organelle are encoded in the nuclear genome and the proteins are subsequently transported to the apicoplast with pathways such as fatty acid, isoprenoid and heme synthesis apicomplexan parasites, but are absent in the human host. The parasite-specific metabolic pathways thus make an ideal source of drug targets and provide obvious opportunities for chemotherapy [14].

3.26. Mitochondrial Targets

Protein synthesis and electron transport are two main functions of mitochondria which appear to be essential for survival and constitute potential targets for antimalarial chemotherapy. Plasmodium mitochondria use a different homolog of parasite respiratory chain from their mammalian host and can be used as targets of antimalarial drugs [8, 14, 18].

3.27. Membrane Targets

Malaria parasites undergo extensive phospholipid synthesis to produce the membranes necessary to enclose the parasitophorous vacuole, cytosol and multiple subcellular compartments. Growing and dividing malaria parasites require large amounts of phospholipids, which are synthesized from plasma fatty acids. The most abundant lipid in plasmodial membranes is phosphatidylcholine (PC). Molecules can be designed to target the parasite's supply of PC [14,18].

3.28. Hematin

Heme, hematin or hemozoin are other possible drug targets for antimalarials. Heme has been implicated in the mode of action of endoperoxide antimalarials, such as artemesinin, which have been proposed to form radical adducts with heme that act against the parasite. Hematin is believed to be the target of various antimalarials and there is evidence suggesting that these drugs act by preventing detoxification of hematin [14].

3.29. Shikimate Pathway

The shikimate pathway in *P. falciparum* provides several targets for the development of new antiparasitic agents, since this pathway is not present in mammals and consists of various conserved enzymatic steps. *Plasmodium falciparium* requires a functioning biosynthetic pathway for growth [14].

3.30. Redox Systems

Oxidative stress is an important mechanism for destruction of malaria and other intracellular parasites. In intra-erythrocytic-stage malaria, parasites encounter reactive oxygen species. To prevent oxidative stress, the parasite has its own battery of defense tactics and produces its own antioxidant enzymes. The malaria parasite contains three antioxidant enzymes: superoxide dismutase, glutathione peroxidase and catalase. The parasite's antioxidant defense could thus be a potential target for antimalarial chemotherapeutics [11, 12, 18].

3.31. Polyamine Biosynthesis

Based on the critical role of polyamines in key processes such as cell growth, differentiation and macromolecular synthesis, the enzymes of the polyamine biosynthesis pathway were suggested as targets in the treatment of parasitic diseases [14,18].

3.32. Nucleic Acid Metabolism

Nucleotides are the precursors of DNA and RNA biosynthesis. Nucleic acid metabolism pathways differ between *P. falciparum* and the human host. The two pathways involve a range of essential enzymes that can be targeted for therapeutic intervention [14,18].

3.33. Purine Salvage

Purine salvage and pyrimidine synthetic pathways also offer potential drug targets. Malaria parasites cannot synthetize purines and rely on salvage of host purines for nucleic acid synthesis. The principal source of purines in *P. falciparum* appears to be hypoxanthine and hypoxanthine-guanine phosphoribosyltransferase (HGPRT) considered as potential drug targets [4, 8, 18].

3.34. Cyclin-Dependent Protein Kinases

Cyclin-Dependent Protein Kinases (CDKs) are essential regulators of cell growth and differentiation, and attractive targets for drug discovery in infectious diseases. More recently, they have become the focus of rational drug design for development of new antimalarial agents [12].

3.35. Farnesyl Transferase

Farnesyl transferase is a heterodimeric zinc protein that catalyzes the transfer of a farnesyl residue from farnesyl pyrophosphate to a cysteine side chain. Some proteins involved in intracellular signal transduction are only active when anchored to the membrane by a farnesyl residue. They have been recently shown to be antimalarial targets [4].

3.36. Helicases

Helicases contain multiple functional domains and a variety of enzymatic activities, and have essential roles in the metabolism of DNA and RNA. Helicase inhibitors may offer a feasible route for novel, more efficient anti-resistance malarial drugs [14].

3.37. Erythocyte G protein

It has been suggested that inhibiting erythrocyte guanine nucleotide regulatory protein Gs signaling blocked invasion by the human malaria parasite. In addition to invasion, the Gs signaling is needed for intracellular parasite proliferation and thus may present a novel antimalarial target. Gs antagonism offers a novel strategy and has the potential to develop combination therapies with existing antimalarials [14].

3.38. Isoprenoid Biosynthesis

Plasmodium protein farnesyltransferase (PfPFT) inhibition is a potential drug target. They act by inhibiting post-translational modification pathways and constitute a promising new class of antimalarials [12, 14].

3.39. Fatty Acid Biosynthesis

It has been reported that *T. brucei* has adapted microsomal elongases for bulk fatty acid synthesis. *L. major* and *T. cruzi* apparently use the same enzyme family for fatty acid synthesis. Nearly all of the fatty acids synthesized in *T. brucei* are produced by the elongase pathway which could represent an innovative pathway to search for novel drug targets [8,18].

3.40. Electron Transfer Chain and Alternative Oxidase

Most parasites do not use the oxygen available within the host, but employ their own system's anaerobic metabolic pathways. The enzymes in these parasite-specific pathways are potential targets for chemotherapy. The respiratory systems of parasitic protozoa typically show great diversity in electron pathways when compared to their host. These aspects of electron transport (ETC) complexes and their related enzymes represent promising targets for chemotherapy.

A cytochrome-independent *Trypanosoma* alternative oxidase (TAO- not existent in the host), of *T. brucei* is the terminal oxidase of the respiratory chain of bloodstream forms of the African trypanosomes and is a leading drug target [8, 18].

3.41. Kinetoplast DNA Replication Machinery

Kineplastids show a DNA known as kinetoplast DNA (kDNA) which comprises a giant network of interlocked DNA rings with a rather unique topology. The discovery of new proteins involved in the kDNA replication process is a valuable source for novel anti-trypanosomatid drug targets [8].

3.42. Histone Deacetylases

Class I and II histone deacetylases (HDAC) are regulatory components of corepressor complexes involved in cell cycle progression and differentiation. In *T. brucei* there are four genes encoding histone deacetylase orthologues. HDAC1 and HDAC3 appear to have the critical essential residues for deacetylase activity and have been indicated as potential chemotherapy targets for *T. brucei* [8].

3.43. Cell Cycle

Trypanosomatid cell cycle progression might also be investigated as a possible source for new drug targets. The TriTryp genome sequences have greatly impacted trypanosomatic cell cycle research, leading to faster functional analyses, particularly in *T. brucei*, where RNA interference (RNAi) is possible [8].

3.44. Glycogen Synthase Kinase

It has been reported from genetic and chemical data that glycogen synthase kinase-3 (GSK-3) is a potential drug target for *T. Brucei* parasite [8, 11, 18, 49].

3.45. Carbohydrate Biosynthesis

Carbohydrate metabolism enzymes are compartmentalized in the glycosome. Enzymes such as *T. brucei* enolase and hexokinase have been envisaged as possible proteins to address anti-trypanosomiasis lead identification. In this pathway, other targets include pyruvate kinase, phosphofructokinase, phosphoglycerate kinase [4, 11, 8, 28].

3.46. Proteasomes

Proteasomes are one of the cellular complexes controlling protein degradation from archaebacteria to mammalian cells. *T. brucei* proteasome has been indicated as a candidate target for *T. brucei* [8].

3.47. Trans-Sialidase

One of the targets discussed as a possible entry to novel therapeutics against Chagas' disease is transialidase (TcTS). This enzyme plays a key role in the ability of the parasite to evade the host immune response, mainly by transferring sialic acid residues from host glycoconjugates to the parasite surface [15].

3.48. Mannitol Cycle

The energy metabolism of *C. parvum* and other coccidians involves the mannitol cycle. There is net mannitol synthesis in the parasite's sexual stages, whereas mannitol comnsumption, presumably primarily as an energy substrate, takes place during sporulation of the oocyst after excretion by the host. A key feature of the cycle is that the enzyme that initiates mannitol biosynthesis is regulated by the binding of a specific inhibitor [28].

3.49. Lysosomes

Parasitic protozoa also possess complex intracellular lysosomes/endosomes/vesicles involved in digestion, transport and recycling of molecules similar to those of mammalian cells. Unique characteristics are ascribed to lysosomes of different parasites and may even differ between parasite stages. Transport of hydrolases and proteins to parasite lysosomes is directed either from the Golgi complex *via* endosomal vesicles or from endocytic vesicles originating on the cell surface. Inhibition of lysosomal proteases demonstrated that different proteolytic machineries catabolize distinct classes of proteins and this selectivity may be exploited for the development of effective anti-parasitic drugs. Lysosomal molecules are either validated or potential drug targets for Chagas' disease, sleeping sickness, leishmaniasis, toxoplasmosis, malaria, amebiasis and giardiasis [46].

3.50. Epigenetic Codes

Ubiquitin-like with PHD and ring finger domains 1 (UHRF1), also known as 1CBP90 or Np95, plays a central role in transferring methylation status from mother cells to daughter cells. *T. gondii*, which causes toxoplasmosis, utilizes UHRF1 to control the cell cycle phase and enhance its proliferation. Thus, knockdown of UHRF1 can be effective at stopping the proliferation of the parasites in infected cells [5].

3.51 Really Big Targets: The Parasitic Worms

Nematode and trematode worms are some of the most abundant life forms on earth. Intestinal nematodes, which make up the bulk of human infections are readily treated with relatively safe orally administered drugs. These organisms exist within the lumen of the intestine and only need to be induced to release their hold to be rapidly excreted.

3.52. Using Whole-Organism Screening

In addition to the target-based approach to antiparasitic drug discovery and development, another effort is centered on developing parasite-based assays suitable for high throughout screening. These allow diversity screening of compound libraries containing drugs already approved for humans for potential antiparasitic efficacy and should require less development before entering clinical trials as antiparasites. This route opens new ways for lead discovery and optimization, including natural-product libraries [4].

3.53. Using RNA Interference

In another approach, the introduction of double-stranded RNA (dsRNA) into some cells or organisms results in degradation of its homologous mRNA, a process called RNA interference (RNAi). The dsRNAs are processed into short interfering RNAs (siRNAs) that subsequently bind to the RNA-Induced Silencing

Complex (RISC) causing degradation of target mRNAs. Because of this sequence-specific ability to silence target genes, RNAi has been used to study gene functions and has the potential to control disease pathogens or vectors [43].

3.54. Using Proteomics and Transcriptomics

Most notably, proteomics and transcriptomics have become important tools in gaining increased understanding of the biology of the systems involved, thus accelerating the pace of discovery of drugs/vaccines. These approaches can provide information regarding genes and proteins that are expressed and under which conditions [16].

4. AVAILABLE DRUGS

4.1. Anti-Chagas' Chemotherapy

The Chagas' disease chemotherapy is based on nifurtimox and benznidazole belonging to the class of nitroaromatic compounds with limited efficacy to the acute phase of the disease and only to some pathogen strains. For chronic patients there are side effects such as anorexia, vomit and diarrhea. The efficacy and tolerance of benznidazole is related to the age of the patients. It is believed that these nitroheterocyclic derivatives function as prodrugs activated within the parasite. Two-electron reductions of the nitro group are mediated by a NADH-dependent nitroreductase leading to DNA damage and resistance emergence [47].

Commercially available antifungal agents ketocona and terbinafine have also been indicated as anti-Chagas drug acting as sterol-membrane biosynthesis inhibitors. Megazol and guanylhydrazone hybrids, which have demonstrated potent trypanocidal activity, were also reported as antichagasic compounds. The design concept for these compounds primarily explored the introduction of the arylhydrazone moiety onto the heterocyclic framework aimed at conferring a radical scavenger activity to the new molecule, which could reduce the oxidative stress induced by formation of toxic nitro radical species [8].

4.2. Anti-Trypanosomiasis Chemotherapy

The drugs to treat sleeping sickness are scarce, toxic and encounter parasite resistance. Among the four licensed drugs, suramin, which does not cross the blood-brain barrier and pentaminidine are used against stage I disease. Tetarsoprol and eflornithine are used against the neurological stage 2 disease. Pentamidine uptake into trypanosomes involves high and low affinity transporters [8,53].

Melarsoprol was introduced to replace earlier arsenical compounds due to resistance. Eflornithine acts as an inhibitor of trypanosomal ornithine decarboxilase, interfering with the polyamine's metabolic pathway. This expensive drug is effective against *T. b. gambiense* [8].

4.3. Anti-Leishmaniasis Chemotherapy

In the beginning of the 20th Century, generic pentavalent antimonial compounds (Sb^V) have been the main drug to treat infected people. However, there is resistance of *Leishmania* parasites against these compounds. Pentostam (sodium stibogluconate) and glucantime (meglumine antimoniate) have been the mainstay therapy for VL. These are first-line treatments. There are also organic complexes of Sb^V with improved solubility and uptake properties. Due to high cost, a generic sodium antimony gluconate (SAG) is presently being used, which also suffers with the emergence of resistance [8,10,65].

Pentamidine and amphotericine are the second-line drugs which also suffer from toxicity problems. Different expensive colloidal and lipid formulations have been prepared. Thus, amphotericine B liposomal is a milestone in leishmaniasis treatment. However, the cost is prohibitive.

More effective formulation of amphotericin B, in which deoxycholate has been replaced by other lipids yields liposomal amphotericin B (Ambiosome), amphotericin B lipid complex (Abelcet) and amphotericin B colloidal dispersion (Amphocil). They are well taken up by the reticuloendothelial system and poorly taken up by the kidney, the major target of toxicity.

Recently, miltefosine, an alkylphosphocholine derivative, with side effects and contraindications, has been registered in India for oral treatment of VL. The combination therapy of liposomal amphotericin B and oral miltefosine has shown good results for the visceral disease form.

Paramomycin, an aminoglycoside antibiotic was registered in India in 2006 for leishmaniasis indicating good efficacy and safety. Sitamaqine, a 8-aminoquinole is an orally administrable compound. However, its development has been slow [8, 10, 65].

4.4. Anti-Malaria Chemotherapy

Extracts from the sweet wormwood shrub, *Artemesia annua,* had been used in Chinese herbal medicine hundreds years B.C. and more recently was used by the Chinese army as an effective anti-malarial. In the 1970s the active component, artemisinin was isolated from Artemesia. Artemisinin and its sesquiterpene lactone congeners are thought to exert their toxic effects by increasing the oxidative stress of the parasite [46]. Artemisinin synthesis is costly and difficult. The mevalonate pathway in *Saccaromyces cerevisiae* has been engineered to secrete high levels of artemisinic acid, the immediate precursor that is relatively easily converted to artemisinin. This approach along with additional peroxides should help deliver affordable antimalarials.

The use of cinchona bark to combat malaria has been long known, although the active quinoline alkaloid, quinine was not isolated until the early part of the 18th century and full chemical synthesis was not realized until the mid 1990s. Quinine is the therapy of choice to fight malaria, although the emergence of break through resistant strains has occurred. It can be also useful as a suppressant and can be used in combination therapies, although other amino-quinolones, such as chloroquine, mefloquine and amodiaquine are the drugs of choice. Since its discovery in the early 1930s, chloroquine is the most commonly used antimalarial and its efficacy is undoubtedly due to its many thousand fold rapid accumulation in the parasite food vacuole [46].

The antimalarial drugs used as first-line treatment by medical practitioners are AS/Q, quinine tablets, amodiaquine, artemether-lumefantrine, artesunate-mefloquine, artesunate suppositories, dihydroartemisinin, halofantrine, sulphadoxine/pyrimenthamine, pyrimenthamine alone, injectable quinine, injectable artemether. The doses vary for adults, pregnant women, children and medical practitioners in urban zones [12, 14, 18, 22, 27, 29, 32-36, 39, 41, 45, 66].

However, the most effective way to reduce disease and death from infectious diseases is to vaccinate susceptible populations. However, development of a malaria vaccine is confounded by the complexity of the parasite lifecycle, the broad range of possible antigens and the mechanisms of immune evasion by the parasite. The lifecycle is characterized by a sporozoite/pre-erythrocytic stage, an asexual multiplication stage and a sexual/sporogonic stage.

In particular for malaria targets in *P. falciparum*, we have for parasite membrane biosynthesis, the phospholipid synthesis (choline transporter) with inhibitor G-25. For parasite proteases we have plasmepsin, falcipains with Leupeptin and pepstatin inhibitors. For the shikimate pathway, we have 5-enolpyruvyl shikimate 3-phosphate synthase with the glyphosate inhibitor. For the apicoplast we have Fab H and Fab I with Thiolactomycin inhibitor. For the redox system we have thioredoxin reductase and Gamma-GCS with 5,8-dihydroxy-1,4-napthoquinone and buthionine sulfoximine inhibitors. For the mitochondrial system we have cytochrome c oxidoreductase with atovaquone inhibitor. For purine metabolism, we have HGPRT with immucillin-H inhibitor. For pyrimidine metabolism we have thymidylate synthase with 5-fluoroorotate inhibitor. For cyclin-dependent protein kinase with Pfmrk we have oxindole derivatives, thiophene sulfonamide inhibitors. For the isoprenoid biosynthesis we have the DOXP reductoisomerase with fosmidomycin inhibitor [12].

4.5. Anti-Schistosomiasis Chemotherapy

A drug for schistosomiasis is called praziquantel (Biltricide). It is orally bioavailable, safe and requires only a short therapy. Oxamniquine and praziquantel are also used [47].

4.6. Anti-Amebiaisis Chemotherapy

Metronidazole (oral or intravenously) has been the drug of choice with success rates of 50-90%. It is not effective against cysts. A third of patients continues shedding cysts after completion of treatment and require a luminal agent. Newer nitroimidazole derivatives, such as tinidazole, secnadizole and ornidazole, have also been effective [46].

4.7. Anti-Giardiasis Chemotherapy

Treatment with tinidazole, metronidazole or paromomycin is recommended in the US, while nitroimidazoles, such as secnidazole and albendazole, are available in other countries. Drug resistance is an increasing concern [46].

4.8. Anti-Toxoplasmosis Chemotherapy

Reactivation toxoplasmosis in patients with AIDS or immune-compromised by transplantation medication are usually treated with a combination of pyrimethamine, sulfadiazine and folinic acid for 4-6 weeks. Alternatives in sulfa allergic patients include Clindamycin, Trimethoprim or Atovaquone A [46].

4.9. Anti-Filariasis Chemotherapy

Both albendazole and diethylcarbamazine (DEC) are effective in killing the adult-stage filarial parasites [47].

4.10. Anti-Cryptosporidiosis Chemotherapy

For the energy metabolism, considering the glycolytic enzymes and the mannitol cycle, the phosphonic analogue and nitrophenide drugs are respectively available. For gene expression, folate metabolism, electron transport chain, tubulin 1 transporters of the ATP-binding cassette superfamily, we have the drugs apicidin, pyrimethamine/sulphonamide, atovaquone, dinitroanilines and cyclosporine analogues, respectively. For plastid-associated activities, *i.e.* protein synthesis, DNA gyrase and shikimate pathways, we have the drugs clindamycin, ciprofloxacin and glyphosate, respectively.

Compounds such as spiramycin, clarithromycin, paromomycin and nitazoxanide display modest activity in model systems, but limited efficacy in clinical trials with immunocompetent patients and poor efficacy in immunocompromised patients. The current gold standard for anticryptosporidial acitivity is paromomycin [21].

4.11. Anti-Onchorciasis Chemotherapy

Suramin and Irvermectin are the available drugs [47].

5. EFFORTS TO DISCOVER MORE EFFECTIVE INHIBITORS

5.1. Selected Research: New Drug Leads for Chagas' Disease

The authors reported molecular modeling and QSAR tools to study 18 dithiocarbamates suppressors of the growth of *T. cruzi epimastigotes,* which have been reported in the literature as Superoxide Dismutase (SOD) inhibitors. The Principal Component Analysis (PCA) showed that the descriptors superficial area, heat of formation, logarithm of the partition coefficient, charge of the nitrogen atom from the dithiocarbamate group, and charges of the two carbon atoms adjacent to that nitrogen are responsible for the classification between the higher and lower trypanomicid activity. Using Multiple Linear Regression (MLR) and docking methods, it was possible to identify the probable bioactive isomers that suppress the growth of *T. cruzi* epimastigotes. Our best Partial Least Square (PLS) model obtained with these six descriptors yields a good correlation between experimental and predicted biological activities and compares two different SODs as possible targets for interaction with the dithiocarbamates [3].

Gentian violet is used as an *in vitro* trypanocidal agent to prevent transfusion-associated Chagas disease. The commercially available anti-fungal agents ketoconazole and terbinafine, both acting as sterol-membrane biosynthesis inhibitors, have been studied as anti-Chagas drugs [8].

The conjugation of megazol and guanylhydrazone hybrids yields antichagastic compounds. Megazol is a 5-nitroimidazole compound endowed with a potent trypanocidal activity. Hydrazones have been demonstrated to have trypanocidal activity. The design concept for these compounds primarily explored the introduction of the arylhydrazone moiety to the heterocyclic framework aimed at conferring a radical scavenger activity to the new molecules which could reduce the oxidative stress induced by formation of toxic nitro radical species. There was also interest in the interaction with cruzipain (CP) and *T. cruzi* Thiol Reductase (TR) [55].

Combining the active thiosemicarbazone and *N*-amidohydrazone derivatives towards CP and TR with a benzofuraxan pharmacophore yield interesting hybrids. Two series of compounds were generated and tested against two strains of *T. cruzi* and their inhibitory activity against CP and TR. Thiosemicarbazone was the most promising [56,57].

The multitarget approach yields the discovery of dual ligands from hybrids, including terbinalfine, a promising anti *T. cruzi* agent acting by reducing the parasite's endogeneous membrane sterol levels. The ergosterol biosynthesis pathway is a validated target for drug therapy against *T. cruzi* [58].

A design strategy was aimed at combining, in a single molecule, the nitrofuran moiety (potential for the generation of free radicals and redox cycling) with a heteroallyl group which could inhibit the parasite's squalene epoxidase resulting in a heteroallyl-containing 5-nitrofuranes active against *T. cruzi* proliferative stages [8,58].

Another example of Multiple-Target-Directed Ligands (MTDL) against *T. cruzi,* based on the combination of two molecular scaffolds takes advantage of the independent biological activities displayed by [1,2,3] trizoles and naphthoquinones [59]. In folk medicine, plants containing naphthoquinones are widely used to treat many diseases, including parasitic infections. Lapachol and lapachone extracted from Brazilian trees of the genus *Tabebuia* have been exploited as protypes in medicinal chemistry programs aimed at developing anti-Chagas lead candidates. The cyclic reduction-oxidation of the quinine moiety is at the basis of their biological mechanism of action, which includes involvement in the electron transport and oxidative-phosphorylation processes [60].

The development of metal complexes of clotrimazole and ketoconazole showed positive synergistic effects in anti-trypanosome therapy [61].

Work was also reported on series of platinum-based complexes and of 3-(5-nitrofuryl) acroleine thiosemicarbazones, which seem to have considerable promise in the treatment of Chagas disease [62]. In this MTDL 'metal complexation approach', the metal complexes act through dual or even multiple action mechanisms by combining the pharmacological properties of the ligand and the metal.

Different series of metal derivatives of aromatic amine *N*-oxides, as well as 5-nitrofuryl containing thiosemicarbazones, were developed and indicated trypanocidal activity [63].

Recent work has reported novel coordination compounds of the ligand pyridine-2-thiol *N*-oxide and gold(I) triphenylphosphine, as well as palladium and platinum complexes, for their potential against Chagas disease indicating that the trypanocidal action of the complexes could mainly rely on the inhibition of the enzyme NADH-fumarate reductase, an enzyme absent in the mammalian host [63].

A synthetic route to prepare isoquinoline analogs of the clinical candidate cancer drug tipifarnib was recently reported. It was shown that these compounds kill *T. cruzi* amastigotes grown in mammalian host cells at concentrations in the low nanomolar range. These isoquinolines represent new leads for the development of drugs to treat the Chagas disease [23].

5.2. Selected Research: New Drug Leads for Trypanosomiasis

Pafuramidine is a serious new oral drug candidate. It is an *O*-methyl amidoxime prodrug which is converted *in vivo* into the diamidine furamidine. It is also being developed for malaria [8,54]. Combinations of eflornithine and melarsoprol or nifurtimox have been the focus of recent clinical studies. The eflornithine-nifurtimox combination is investigated for the treatment of the neurological phase of HAT.

Some work aimed at designing adenosine analogues to simultaneously bind the targets of *T. brucei* phosphoglycerate kinase, glyceraldehyde-3-phosphate dehydrogenase and glycerol-3-phosphate dehydrogenase in complex with their adenosyl-bearing substrates. Natural quinine lapachol inspired other series of potential MTDLs. In terms of meeting a criterion for generating a library of anti-trypanosomatid compounds, the quinine unit was selected as a core structure for combinatorial derivation. Naphthoquinones and related quinine compounds constitute one of the natural product classes that have significant activity against Leishmania and Trypanosoma. Various anthra- and naphthoquinones, which incorporated in the 2-position a selection of aromatic groups mimicking the structural elements of the general biocide triclosan, were synthesized and investigated, some of which indicated promising anti-trypanosomatid activity [8].

Some recent trends point towards a metal-drug synergism paradigm whereas in addition to synergism of action another favorable issue is the possible stabilization of the drug. This feature might lead to a longer residence time of the drug in the body, allowing it to reach the biological targets more efficiently and may also result in a decrease in toxicity. Thus, the new platinum-sterol hydrazone complex might exert a synergistic mechanism of action by combining inhibition of the sterol biosynthesis pathway and dual interaction with the DNA of the parasite.

In a very recent work, a series of compounds, most newly synthesized, were identified as inhibitors with PTR1-species specific properties explained by structural differences between the *T. brucei* and *L. major* enzymes. The most potent inhibitors target *T. brucei* PTR1, and two compounds displayed antiparasite activity against the bloodstream form of the parasite. PTR1 contributes to antifolate drug resistance by providing a molecular bypass of dihydrofolate reductase (DHFR) inhibition. Therefore, combining PTR1 and DHFR inhibitors might improve therapeutic efficacy. By testing two new compounds with known DHFR inhibitors, a synergistic effect was observed for one particular combination highlighting the potential of such an approach for treatment of African sleeping sickness [9].

Several antifungal azoles have recently entered clinical trials for antitrypanosomal chemotherapy. Development of resistance has been observed in azole derivatives and floconazole. Although strong inhibition of TCCYP51 with azoles can decrease the probability of resistance, alternative set of inhibitors has also being recommended. Indomethacin amide derivatives may also find new application as potential anti-trypanosomal agents. Additional modifications targeted to decrease COX-2 inhibitory potency may be attractive for the acute stage because drugs traditionally aimed to ease nonspecific symptoms of inflammation and fever can actually provide treatment for the disease [6].

Trypanothione reductase is an enzyme involved in the protection of *Trypanosoma* and *Leishmania* species against oxidative stress and is considered to be a validated drug target. The endophytic fungus *Alternaria* sp. (UFMGCB55) was isolated from the plant *Trixis vauthieri* DC (*Asteraceae*), known to contain trypanocidal compounds. The organic extract of the culture of *Alternaria* sp. was able to inhibit TR by 99%. Biphenyl altenusin is the first in its class to have shown inhibitory activity, opening new perspectives for the design of more effective derivatives that could serve as drug leads for new chemotherapeutic agents to treat trypanosomiasis and leishmaniasis [42].

5.3. Selected Research: New Drug Leads for Leishmaniasis

All available drugs, excluding SAG, are not true anti-leishmanials. Most are expensive with side effects. There is a paramount interest for the search of more effective antileishmanial drugs. Natural product literature provides growing research on plant derived antileishmanial alkaloids. Many alkaloids are reported to have excellent antileishmanial activity. However, they have yet to be evaluated in clinical studies [10].

Harmaline is reported to possess significant antileishmanial activity (amastigote-specific) based on the ability to interfere with amino acid metabolisms.

An iboga-type indole alkaloid, coronaridine, isolated from Apocynaceae, shows activity against promastigote and amastigote forms of the parasite.

Dihydrocorynantheine, corynantheine and corynantheidine, isolated from the bark of Rubiacae, exhibits excellent *in vitro* antileishmanial activity against *L. major*.

The indole analogues harmane, pilocarpine and buchtienine from stem bark and leaf of Apocynaceae shows significant *in vitro* antileishmanial activity promastigotes of *L. donovani*.

Monterpenoid indole alkalids ramiflorine isolated from stem bark of Apocynacae show significant antileishmanial acitivity against promastigotes of *L. amazonensis*.

The bis-indole alkaloid gabuine, the metabolite conodurine, isoquinoline alkaloids (*O*-methylmoschatoline and liriodenine), several alkaloids containing isoquinoline skeleton (berberine, metabolite anonaine, isodomesticine, norisodomesticine, nantenine, neolitsine, lirioferine, mehylaurotetanine, norlirioferine, isoboldine), a dimeric aporpine alkaloid (unonposine), napthylisoquinoline alkaloids (ancistrogriffine, ancistroealaine, anticistrocladidine), bisbenzylisoquinoline alkaloids (metabolite daphanandrine, isotetradrin, puertogaline), quinoline alkaloids (chimanine-D and chimanine-B), steroidal alkaloids (sarachine, holamine, holacurtine, benzoquinolizidine alkaloids, diterpene alkaloids and C_{20}-diterpene), pyrimidine-β-carboline alkaloid (*N*-hydroxyannomontine, annomontine) and piperine as well as benzoxazol piperine alkaloids all show varying anti-leishmanial activity [10].

The potentials of plant products as a source of anti-leishmanial drugs are needed to be discussed with respect to biochemical differences between protozoa and host. More research is needed to have a better understanding of the leishmanial targets for plant products.

Peganine hydrochloride dehydrate, besides being safe, was found to induce apoptosis in both the stages of *L. donovani*. Molecular docking studies suggest that a binding interaction with DNA topoisomerase I of *L. donovani* forms a stable complex, indicating a possible role in apoptosis [26].

Bisphosphonates, risedronate and pamidronates, imiquimod have been reported to have activity [25].

5.4. Selected Research: New Drug Leads for Malaria

A series of thiazole-derived N-Boc aminoacids were synthesized and evaluated as targeted potential antimalarial against *plasmepsins II* enzyme of malaria parasite *P. falciparum* [27]

There is a robust effort to identify new aminoquilinoline derivatives. The logic is that although resistance to chloroquine has arisen in *P. falciparum*, much of the chemical space around the aminoquinoline scaffold has not been systematically explored. Quinoline analogs have been known to overcome chloroquine resistance in malaria parasites [4].

The sesquiterpene artemisinin is the active antipyretic component of an herbal remedy that has been used in China for nearly 2000 years. Its semisynthetic derivatives artemether and sodium artesunate are the most effective drugs currently available for treating multi-drug resistant *P. falciparum malaria*. Artemether and artesunate are C10 ether and ether derivatives, respectively, and both are rapidly converted *in vivo* to the active metabolite dihydroartemisinin (DHA). There is a search for a new generation of artemisinin-based therapeutics. Chemical strategies include modification of the C10 position, C10 carba analogs, artemisinin dimers, aza-artemisinins, C9-modified analogs, C10 ammino substituents, C10 piperazine, morpholine and thiomorpholine heterocycles [4].

The sequencing of the genomes yields 'structural antigens', which represent small protein domains that can be chemically synthesized and isolated from the context of the whole protein, can fold in the same native structure. In particular, based on the known *P. falciparum* genome, putative α-helical coil regions, 30-40 amino acids long, in proteins presented in asexual malaria blood stages. Peptides of such regions frequently fold into the 'native' structure. A hundred such peptides were synthesized and all of them were recognized at various degrees (5-80%). The results obtained demonstrate that a bioinformatics/chemical synthesis strategy can rapidly lead to the identification of new proteins that can be targets of potential vaccines and/or drug against malaria and other infectious organisms [13].

Artemisinin-based combination therapy (ACT) has been advocated as the therapy of choice for handling widespread drug resistance in *P. falciparum* [8].

The synergy between fast-acting artemisinin and slow-acting quinine yields an artemisin-quinine hybrid which was more effective than either drug alone or both drugs together in two different drug-sensitive and drug-resistant strains of *P. falciparum* parasite [64].

5.5. Selected Research: New Drug Leads for Cryptosporidiosis

The drug discovery for *C. parvum* is challenging because the parasite cannot be maintained continuously in cell culture. Mining the sequence of the *C. parvum* genome has revealed that the only route to guanine nucleotides is *via* inosine-5'-monophosphate dehydrogenase (IMPDH). Moreover, phylogeneic analysis suggests that the IMPDH gene was obtained from bacteria by lateral gene transfer. The unexpected evolutionary divergence of parasites and host enzymes were exploited by designing a high-throughput screen to target the most diverged portion of the IMPDH active site. Four parasite-selective IMPDH inhibitors were identified that display antiparasitic activity with greater potency than paromomycin, the current gold standard for anticryptosporidial activity [21].

6. PERSPECTIVES

Drug resistance is a serious problem. A less competent drug-resistant parasite could evolve as a competent isolate. More studies are necessary to establish a validated protocol to measure cell proficiency both in the presence and absence of drugs. Such standardization is essential. However, the continual evolution of resistant strains might lead to compensatory mutations that eventually result in return to normal cell potency. In the search for these markers, the phenotype of individual species, subjected to drug pressure must be monitored. In particular for patients, it is necessary to evaluate characteristics such as metabolite usage and enzymatic pathway fluxes, in parallel with resistance selection and infectivity. Their prognostic value for treatment outcome should be useful.

The gap between target identification and validation, discovery and optimization should be reduced with biochemical studies of the targets and inhibitors. Knowledge about the transport processes operating in an infected cell may also be useful for rational drug design. Understanding the mechanism underlying drug resistance should also help us to circumvent the emergence of resistance to new generations [40].

The mechanism of immunity is likely constituted by a combination of cellular responses to several developmental stages of the parasite. The fine specificity and duration of protective immunity are clearly subjects for future research. The most effective way to reduce disease and death from infectious diseases is to vaccinate susceptible populations.

Natural products are useful source of chemical diversity for programs aiming at drug discovery for neglected diseases.

Computational strategies such as virtual screening or structure-based drug design may be productively used without having complete information on the drug/enzyme binding interaction for identifying new and selective inhibitors.

Usually, a combination regimen (drug cocktail) is composed of two or three different drugs that combine different therapeutic mechanisms. An alternative strategy might be the use of single-pill combinations (fixed-dose combinations), which incorporate different drugs into the same formulation. A third approach is based on the assumption that a single compound may be able to hit multiple targets.

Combination therapy has been shown to be a possible strategy for both preventing and overcoming chemotherapy-induced resistance. However, combination therapies could suffer from major drawbacks related to pharmacokinetics and pharmacodynamics.

Multifunctional compounds represent an innovative approach to addressing chemotherapy-induced drug resistance. A single compound endowed with a multifunctional profile is thus able to simultaneously modulate the activity of two or more biological counterparts. The probability of simultaneous mutations in both genes encoding for both target proteins is greatly reduced. The use of a single drug that has multiple biological properties would obviate the challenge of administering multiple single-drug entities, which could have different bioavailability, pharmacokinetics and metabolism. In terms of ADMET or pharmacokinetic optimization, it should not in principle be different from the development of another single lead molecule [8].

REFERENCES

[1]   O'Connell D. Neglected diseases. Nature 2007; 449: 157.

[2]   http://www.who.int/neglected_diseases/en/.

[3]   da Silva CHTP, Sanches SM, Taft CA. A molecular modeling and QSAR study of supressors of the growth of *Tripanosoma cruzi* epimastigotes J Mol Graph Model 2004; 23: 89.

[4]   Renslo AR, McKerrow JH. Drug discovery and development for neglected parasitic diseases. Nature Chem Biol 2006; 2: 701.

[5]   Unoki M, Brunet J, Mousli M. Drug discovery targeting epigenetic codes: the great potential of UHRF1, which links DNA methylation and histone modifications, as a drug target in cancers and toxoplasmosis. Biochem Pharmacol 2009; 78: 1279.

[6]   Konkle ME, Hargrove TY, Kleschenko YY, von Dries JP, Ridenour W, Uddin MJ, Caprioli RM, Marnett LJ, Nes WD, Villata F, Waterman MR, Lepesheva GI. Indomethacin amides as a novel molecular scaffold for targeting *T. cruzi* sterol 14-alpha-demethylase. J Med Chem 2009; 52: 2846.

[7]   Sivaprakasam P, Tosso PN, Doerksen RJ. Structure–activity relationship and comparative docking studies for cycloguanil analogs as PfDHFR-TS inhibitors. J Chem Inf Model 2009; 49: 1787.

[8]   Cavalli A, Bolognesi ML. Neglected tropical diseases: multi-target-directed ligands in the search for novel lead candidates against Trypanosoma and Leishmania. J Med Chem 2009; 52: 7339.

[9]   Tulloch LB, Martini VP, Iulek J, Huggan JK, Lee JH, Gibson CL, Smith TK, Suckling CJ, Hunter WN. Structure-based design of Pteridine Redutase inhibitors targeting African sleeping sickness and the leishmaniases. J Med Chem 2010; 53: 221.

[10]   Mishra BB, Singh RK, Srivastava A, Tripathi VJ, Tiwari VK. Fighting against leishmaniasis: Search of alkaloids as future true potential anti-leishmanial agents. Mini Rev Med Chem 2009; 9: 107.

[11]   Linares GEC, Ravaschino EL, Rodriguez JB. Progresses in the field of drug design to combat tropical protozoan parasitic diseases. Curr Med Chem 2006; 13: 335.

[12]   Jana S, Paliwal J. Novel molecular targets for antimalarial chemotherapy. Int J Antimicrob Agents 2007; 30: 4.

[13]   Corradin G, Villard V, Kajava AV. Protein structure based strategies for antigen discovery and vaccine development against malaria and other pathogens. Endocr Metab Immune Disord Drug Targets 2007; 7: 259.

[14]   Sahu NK, Sahu S, Kohli DV. Novel molecular targets for antimalarial drug development. Chem Biol Drug Des 2008; 71: 287.

[15]   Neres J, Brewer ML, Ratier L, Botti H, Buschiazzo A, Edwards PN, Mortenson PN, Charlton MH, Alzari PM, Frasch AC, Bryce RA, Douglas KT. Discovery of novel inhibitors of *T. cruzi* trans sialidase from in silico screening. Bioorg Med Chem Lett 2009; 18: 589.

[16]   Kumari S, Kumar A, Samant M, Singh N, Dube a. Discovery of novel vaccine candidates and drug targets against visceral leishmaniasis using proteomics and transcriptomics. Curr Drug Targ 2008; 9: 938.

[17]   Chen CK, Doyle PS, Yermalitskaya LV, Mackey ZB, Ang KKH, McKerrow JH, Podust LM. *T. cruzi* CYP51 inhibitor derived from a Mycobacterium tuberculosis screen hit. PLoS Negl Trop Dis 2009; 3: e372.

[18]   Werbovetz KA. Target-based drug discovery for malaria, leishmaniasis, and trypanosomiasis. Curr Med Chem 2000; 7: 836.

[19]   Bettiol E, Samanovic M, Murkin AS, Raper J, Buckner F, Rodriguez A. Identification of three classes of heteroaromatic compounds with activity against intracellular *T. cruzi* by chemical library screening. PLoS Negl Trop Dis 2009; 2: e.384.

[20] Martins LPA, Marcili A, Castanho REP, Therezo ALS, Oliveira JCP, Suzuki RB, Teixeira MMG, Rosa JÁ, Speranca MA. Rural Triatoma rubrovaria from Southern Brazil Harbors *T. cruzi* of Lineage IIc. Am J Trop Med Hyg 2008; 79: 427.

[21] Umoejiego NN, Gollapalli D, Sharling L, Volftsun A, Lu J, Benjamin NN, Stroupe AH, Riera TV, Striepen B, Hedstrom L. Targeting a prokaryotic protein in a eukaryotic pathogen: identification of lead compounds against cryptosporidiosis. Chem Biol 2008; 15: 70.

[22] Autore F, Melchiorre S, Kleinjung J, Morgan WD, Fraternali F. Interaction of malaria parasite-inhibitory antibodies with the merozoite surface protein $MSP1_{19}$ by computational docking. Proteins 2007; 66: 513.

[23] Chennamaneni NK, Arif J, Buckner FS, Gelb MH. Isoquinoline-based analogs of the cancer drug clinical candidate tipifarnib as anti-*T. cruzi* agents. Bioorg Med Chem Lett 2009; 19: 6582.

[24] Pita SSR, Cirino JJV, Alencastro RB, Castro HC, Rodrigues CR, Albuquerque MG. Molecular docking of a series of peptidomimetics in the trypanothione binding site of T. cruzi Trypanothione Reductase. J Mol Graph Model 2009; 28: 330.

[25] Croft S, Coombs GH. Leishmaniasis - current chemotherapy and recent advances in the search for novel drugs Trends Parasitol 2003; 19: 502.

[26] Misra P, Khaliq T, Dixit A, SenGupta S, Samat M, Kumari S, Kumar A, Kushawaha PK, Majumder HK, Saxena ak, Narender T, Dube A. Antileishmanial activity mediated by apoptosis and structure-based target study of peganine hydrochloride dihydrate: an approach for rational drug design. J Antimicrob Chemother 2008; 62: 998.

[27] Karade HN, Acharya BN, Sathe M, Kaushik MP. Design, synthesis, and antimalarial evaluation of thiazole-derived amino acids. Med Chem Res 2008; 17: 19.

[28] Coombs GH. Biochemical peculiarities and drug targets in *Cryptosporidium parvum*: lessons from other coccidian parasites. Parasitol Today 1999; 15: 333.

[29] Ollomo B, Durand P, Prugnolle F, Douzery E, Arnathau C, Nkoghe D, Leroy E, Renaud F. A new malaria agent in African hominids. PLOS Pathog 2009; 5: e1000446.

[30] Dacks JB, Doolittle WF. Reconstructing/Deconstructing the earliest Eukaryotes: How comparative genomics can help. Cell 2001; 107: 419.

[31] Montono LGP, Santana L and Gonzalez-Diaz H. Scoring function for DNA-drug docking of anticancer and antiparasitic compounds based on spectral moments of 2D lattice graphs for molecular dynamics trajectories. Eur J Med Chem 2009; 44: 4461.

[32] Faik I, de Carvalho EG, Kum JFJ. Parasite-host interaction in malaria: genetic clues and copy number variation. Genome Medicine 2009; 1: 82.

[33] Deen JL, von Seidlein L, Dondorp A. Therapy of uncomplicated malaria in children: a review of treatment principles, essential drugs and current recommendations. Trop Med Int Health 2008; 9: 1111.

[34] Sauerwein RW. Clinical malaria vaccine development. Immunol Lett 2009; 12: 115.

[35] Sayang C, Gausseres M, Vernazza-Licht N, Malvy D, Bley D, Millet P. Treatment of malaria from monotherapy to artemisinin-based combination therapy by health professionals in urban health facilities in Yaoundé, central province, Cameroon. Malar J 2009; 8: 176.

[36] Kuehn A, Pradel G. J. The coming-out of malaria gametocytes. J Biomed Biotechnol 2010; 2010: 1.

[37] Liu K, Shi H, Xiao H, Chong AGL, Bi X, Chang YT, Tan KSW, Yada RY, Yao SQ. Functional profiling, identification, and inhibition of plasmepsins in intraerythrocytic malaria parasites angew. Chem Int Ed 2009; 48: 8293.

[38] Brak K, Derr ID, Barrett KT, Fuchi N, Debnath M, Ang K, Engel JC, McKerrow. Doyle PD, Brinen LS, Ellman JA. Nonpeptidic tetrafluorophenoxymethyl ketone cruzain inhibitors as promising new leads for chagas disease chemotherapy. J Med Chem 2010; 53: 1763.

[39] Kasam V, Salzemann J, Botha M, Da Costa A, Deliesposti G, Isea R, Kim D, Maass A, Kenyon C, Rastelli G, Hofmann-Apitius M, Breton V. WISDOM-II: Screening against multiple targets implicated in malaria using computational grid infrastructures. Malar J 2009; 8: 88.

[40] Natera S, Machuca c, Padron-Nieves M, Romero A, Diaz E, Ponte-Sucre AG. Leishmania spp.: proficiency of drug-resistant parasites. Int J Antimicrob Agents 2007; 29: 637.

[41] Bag S, Tawari NR, Sharma R, Goswami K, Reddy MVR, Degani MS. *in vitro* biological evaluation of biguanides and dihydrotriazines against Brugia malayi and folate reversal studies. Acta Tropica 2010; 113: 48.

[42] Cota BB, Rosa LH, Caligiorne RB, Rabello ALT, Alves TMA, Rosa CA, Zani CL. Altenusin, a biphenyl isolated from the endophytic fungus Alternaria sp., inhibits trypanothione reductase from *T. cruzi* FEMS Microbial Lett 2008; 285: 177.

[43] Kang S, Hong YS. RNA interference in infectious tropical diseases. Kor J Parasitol 2008; 46: 1.

[44] Rawlings ND. Unusual phyletic distribution of peptidases as a tool for identifying potential drug targets Biochem J 2007; 401: e5.

[45] Birkholtz LM, Bastien O, Wells G, Grando D, Joubert F, Kasam V, Zimmermann M, Ortet P, Jacq N, Saidani N, Roy S, Hofmann-Apitius M, Breton V, Louw AI, Marechal E. Integration and mining of malaria molecular, functional and pharmacological data: how far are we from a chemogenomic knowledge space? Malar J 2006; 5: 110.

[46] Doyle PS, Sajid M, O'Brien TO, DuBois K, Engel JC, Mackey ZB, Reed S. Drugs targeting parasite lysosomes. Curr Pharm Des 2008; 14: 889.

[47] Nwaka S and Hudson A. Innovative lead discovery strategies for tropical diseases. Nat Rev Drug Discov 2006; 5: 941.

[48] Unoki M, Brunet J, Mouli M. Drug discovery targeting epigenetic codes: The great potential of UHRF1, which links DNA methylation and histone modifications, as a drug target in cancers and toxoplasmosis. Biochem Pharmacol 2009; 78: 1279.

[49] Xingi E, Smirlis D, Myrianthopoulos V, Magiatis P, Grant KM, Meijer L, Mikros E, Skaltsounis AL, Stoeriadou K. 6-Br-5methylindirubin-3′oxime (5-Me-6-BIO) targeting the leishmanial glycogen synthase kinase-3 (GSK-3) short form affects cell-cycle progression and induces apoptosis-like death: Exploitation of GSK-3 for treating leishmaniasis. Int J Parasitol 2009; 39: 1289.

[50] Berriman M, Ghedin E, Hertz-Fowler C, Bilandin G, Renauld H, Bartholomeu DC, Lennard NJ, Caler E, Hamlin NE, Haas B, Bohme U, Hannick L, Aslett MA, Shallom JC, Atkin RJ, Barron AJ, Bringaud F, Brooks K, Carrington M, Cherevach I, Chillinworth TJ, Churcher C, Clark LN, Corton CH, Cronin A, Davies RM, Doggett J, Djikeng A, Feldblyun T, Field MC, Fraser A, Goodhead I, Hance Z, Harper D, Harris BR, Hauser H, Hostetler J, Ivens A, Jagels K, Johnnson D, Johnson J, Jones K, Kerhornou AX, Koo H, Larke N, Landfear S, Larkin C, Leech V, Line A, Lord A Macleod A, Mooney PJ, Moule S, Martin DM, Morgan GW, Mungall K, Norbertczak H, Ormond D, Pai G, Peacock CS, Peterson J, Quail MA, Rabbinowitsch E, Rajandream MA, Reitter C Sal berg SL, Sanders M, Schobel S, Sharp S, Simmonds M, Simpson AJ, Tallon L, Turner CM, Taigt A, Tivey AR, Van Aken S, Walker D, Wanless D, Wang S, White B, Whie O, Whitehead S, Woodward J, Wortman J, Adams MD, Embley TM, Gull K, Ullu E, Barry JD, Fairlamb AH, Oppedoes F, Barrell BG, Donelson JE, Hall N, Fraser CM, Melville SE, EL-Sayed NM. The Genome of the African Trypanosome *T. brucei*. Science. 2005; 309: 416-422.

[51] El-Sayed NM, Myler PJ, Bartholomeu DC, Nilsson D, Aggarwal G, Tran AN, Ghedin F, Worthey EA, Delchet AL, Blandin G, Westenberger SJ, Caler E, Cerqeira GC, Branche C, Haas B, Anupama A, Arner E, Aslund L, Attipoe P, Bontempi E, Bringaud F, Burton P, Cadag E, Campbell DA, Carrington M, Crabtree J, Darban H, da Silveira JF, de Jong P, Edwards K, Englund PT, Fazelina G, Feldblyum T, Ferella M, Frasch AC, Gull K, Hord D, Hou L, Huang Y, Kindlund E, Klingbeil M, Kluge S, Koo H, Lacerda D, Levin MJ, Lorenzi H, Louie T, Machado CR, McCulloch R, McKenna A, Mizuno Y, Mottram JC, Nelson S, Ochaya S, Osoegawa K, Pai G, Parsons M, Pentony M, Pettersson U, Pop M, Ramirez JL, Rinta J, Robertson L, Salzberg SL, Sanchez DO, Seyler A, Sharma R, Shetty J,Simpson AJ, Sisk E, Tammi MT, Tarleton R, Teixeira S, Van Aken S, Vogt C, Ward PN, Wickstead B, Wotman J, White O, Fraser CM, Stuart KD, Anderssson B. The genome sequence of *T. cruzi*, etiologic agent of chagas disease. Science 2005; 309: 409.

[52] Ivens AC, Peacock CS, Worthey EA, Murphy L, Aggarwal G, Berriman M, Sisk E, Rajandream MA, Adlem E, Acrt R, Anupama A, Apostolou Z, Attipoe P, Bason N, Bauser C, Beck A, Beverley SM, Bianchettin G, Borzym K, Bothe G, Bruschi CV, Collins M, Cadag E, Ciarloni L, Clayton C. Coulson RM, Cronin A, Cruz AK, Davies RM, De Gaudenzi J, Dobson DE, Duesterhoeft A, Fazelina G, Fosker N, Frasch AC, Fraser A, Fuchs M, Gab el C, Goble A, Goffeau A, Harris D, Hertz-Fowler C, Hilbert H, Horn D, Huan Y, Klages S, Knights A, Kube M, Larke N, Litvin L, Lord A, Louie T, Marra M, Masay D, Matthews K, Machaeli S, Motram, JC, Muller-Auer S, Munden H, Nelson S, Norbertczak H, Oliver K, O'Neil S, Pentony M, Pohl TM, Price C, Purnelle B, Quail MA, Rabbinowitsch E, Reinhardt R, Rieger M, Rinta J, Robben J, Robertson L, Ruiz JC, Rutter S, Sauders D, Schafer M, Schein J, Schwartz DC, Seeger K, Setyler A, Sharp S, Shin H, Sivam D, Squares R, Squares S, Tosato V, Vogt C, Volckaert G, Wambutt R, Warren T, Wedler H, Woodward J, Zhou s, Zimmermann W, Smith DF, Blackwall JM, Stuart KD, Barrell B, Myler PJ. The genome of the kinetoplastid parasite, *Leishmania major*. Science 2005; 309: 436.

[53] Delespaux V, de Loning HP. Drugs and drug resistance in *African trypanosomiasis*. Drug Resist Updates 2007; 10: 30.

[54] Ansede JH, Anbazhagan M, Brun R, Easterbrook JD, Hall JE, Boykin DW. O-Alkoxyamidine prodrugs of furamidine: *in vitro* transport and microsomal metabolism as indicators of in vivo efficacy in a mouse model of *T. brucei* rhodesiense infection. J Med Chem 2004; 47: 4335.

[55]   Carvalho SA, da Silva EF, Santa-Rita RM, de Castro SL, Fraga CA. Synthesis and antitrypanosomal profile of new functionalized 1,3,4-thiadiazole-2-arylhydrazone derivatives, designed as non-mutagenic megazol analogues. Bioorg Med Chem Lett 2004; 14: 5967.

[56]   Carvalho SA, Lopes FA, Salamao K, Romero NC, Wardell SM, de Castro SL, DA Silva EF, Fraga CA. Studies toward the structural optimization of new brazilizone-related trypanocidal 1,3,4-thiadiazole-2-arylhydrazone derivatives. Bioorg Med Chem 2008; 16: 413.

[57]   Porcal W, Hernandez P, Boiani L, Boiani M, Ferreira A, Chidichimo A, Cazzulo JJ, Olea-Azar C, Gonzalez M, Cereceito H. New trypanocidal hybrid compounds from the association of hydrazone moieties and benzofuroxan heterocycle. Bioorg Med Chem 2008; 17: 6995.

[58]   Gerpe A, Odreman-Nunez I, Draper P, Boiani L, Urbina JA, Gonzalez M, Cerecetto H. Heteroallyl-containing 5-nitrofuranes as new anti-*T. cruzi* agents with a dual mechanism of action. Bioorg Med Chem 2008;16: 569.

[59]   Da Silva EN, Menna-Barreto RF, Pinto MC, Silva RS, Teixeira DV, de Souza MC, de Simone CA, de Castro SL, Ferreira VF, Pinto AV. Naphthoquinoidal [1,2,3]-triazole, a new structural moiety active against *T. cruzi*. Eur J Med Chem 2008; 43: 1774.

[60]   Pinto AV, Menna-Barreto RFS, De Castro SL, Naphthoquinones isolated from Tabebuia: a review about the synthesis of heterocyclic derivatives, screeening against *T. cruzi* and correlation structure-trypanocidal activity. Phytomedicines. In: Recent progress in medicinal plants. In. Govl JV, Ed.; Studium Press: Houston, TX, 2007. pp 112.

[61]   Sanchez-Delgado RA, Anzellotti A. Metal Complexes as chemotherapeutic agents against tropical diseases: trypanosomiasis, malaria and leishmaniasis. Mini-Ver Med Chem 2004; 4: 23.

[62]   Vieites M, Otero L, Santo D, Toloza J, Figueroa R, Norambuena E, Olea-Azar c, Aguirre G, Cerecetto H, Gonzalez M, Morello A, Maya JD, Garat B, Gambino D. Platinum-based complexes of bioactive 3-(5-nitrofuryl)acroleine thiosemicarbazones showing anti-*T. cruzi* activity. J Inorg Biochem 2009; 103: 411.

[63]   Vieites M, Smircich P, Parajon-Costa B, Rodriguez J, Galaz V, Olea-Azar C, Otero L, Aguirre G, Cerecetto H, Gonzalez M, Gomez-Barrio A, Garat B, Gambino D. Potent *in vitro* anti-*T. cruzi* activity of pyridine-2-thiol N-oxide metal complexes having an inhibitory effect on parasite-specific fumarate reductase. J Biol Inorg Chem 2008; 13: 723.

[64]   Walsh JJ, Coughlan D, Heneghan N, Gaynor C, Bell A. A novel artemisinin-quinine hybrid with potent antimalarial activity. Bioorg Med Chem Lett 2007; 17: 3599.

[65]   Le Pape P. Development of new antileishmanial drugs - current knowledge and future prospects. J. Enzime Inhib. Med Chem 2008; 23: 708.

[66]   Ibezim E, Odo U. Current trends in malarial chemotherapy. Afr J Biotechnol 2008; 7: 349.

INDEX

www.ingramcontent.com/pod-product-compliance
Lightning Source LLC
Chambersburg PA
CBHW080020240326
41598CB00075B/471